Forms of Knowledge in Early Modern Asia

Forms of Knowledge in Early Modern Asia

Explorations in the Intellectual History of India and Tibet, 1500–1800

Edited by
Sheldon Pollock

Duke University Press · *Durham and London* · 2011

Printed in the United States of America
on acid-free paper ♾
Typeset in Arno by Tseng Information Systems, Inc.
Library of Congress Cataloging-in-Publication Data
appear on the last printed page of this book.

In memory of
our dear friend
and colleague,
Aditya Behl
(1966–2009)

Contents

Acknowledgments

A number of the papers in this collection were first presented at the seminar Forms of Knowledge in Early Modern South Asia, organized at the University of Chicago in the academic year 2002–3. I gratefully acknowledge the support of the Committee on Southern Asian Studies, without which the seminar could not have taken place.

Valerie Millholland, our editor at Duke University Press, has been expert in her guidance and gracious in her patience. Tim Elfenbein did a remarkable job managing the complex editing and production of the volume, and I am truly grateful to him. The index was prepared with great care and efficiency by Katherine Ulrich. Three anonymous reviewers generously offered criticisms and suggestions for improvement.

I wish to thank Arthur Dudney, Elaine Fisher, Andrew Ollett, and Audrey Truschke, my research assistants at Columbia, for their help in preparing the volume for the press.

Earlier versions of the essays of Muzaffar Alam and Sanjay Subrahmanyam, Imre Bangha, Allison Busch, Sumit Guha, Janet Gyatso, and Sunil Sharma were published in *Comparative Studies of South Asia, Africa, and the Middle East* 24, no. 2 (2004). A longer version of Aditya Behl's essay was published in *Notes from a Mandala: Essays in Honor of Wendy Doniger*, edited by David Haberman and Laurie Patton (Newark: University of Delaware Press, 2010). An earlier version of Sheldon Pollock's essay was published in *Contributions to Indian and Cross-Cultural Studies: Volume in Commemoration of Wilhelm Halbfass*, edited by Karin Preisendanz (Vienna: Akademie der Wissenschaften, 2005). An abbreviated version of the essay by V. Narayana Rao, David Shulman, and Sanjay Subrahmanyam was published in *South-Indian Horizons: Felicitation Volume for François Gros*, edited by Jean-Luc Chevillard (Pondicherry: Institut français de Pondichéry, École française d'extrême-orient, 2004). Individual authors' preferences in diacritics and styles of transliteration have been respected in the book.

Introduction

SHELDON POLLOCK

The impact of British colonialism on culture and power has been the dominant arena of inquiry in the past three decades in South Asian studies. A large body of scholarship has been produced in the colonialism-and-*X* mode: colonialism and economy, colonialism and caste, colonialism and religious categories, art, empiricism, gender, historicality, law, literature, the nation, numeracy, science, sexuality, and so on down the alphabet. A good deal of this scholarship has been both substantively and theoretically exciting and provocative and has changed the way we understand the transformative interactions between India and the West, starting from the consolidation of British power in the subcontinent around 1800.[1] But as many of its practitioners would be ready to admit, colonial studies has long been skating on the thinnest ice, given how far it presupposes knowledge of the precolonial realities that colonialism encountered and how little such knowledge we actually possess.

As I have argued in various forums for some fifteen years—though it will seem breathtakingly banal to frame the issue in the only way it can be framed—we cannot know how colonialism changed South Asia if we do not know what was there to be changed.[2] In the domain of culture viewed broadly, and more specifically with respect to systematic forms of thought, understanding how Western knowledge and imagination won the day presupposes a comprehension more deeply grounded in epistemological and social facts than we now possess of how South Asian knowledge and imagination lost, which in turn requires a better understanding of what exactly these forms of thought were, how they worked, and who produced them. To date, hypotheses on the demise of Indian science and scholarship with the advent of colonialism seem largely dependent on interpretations dominant since the time of Max Weber, which take for granted the presumed uniqueness of Western rationality, technology, rights-bearing citizenship, or capacity for capitalism—

in short, Western modernity—and the inevitability of its eventual global conquest. These interpretations, however, were derived more from assumptions than from actual assessments of data, as Weber, who was quick to emphasize the provisional nature of his ideas, would likely have been the first to acknowledge. Worse, they were based on now discredited notions about the character and history of precolonial Indian economy and society.[3]

What is perhaps worst, these contrastive assessments of non-Western intellectual and cultural history assume a scholarly consensus about the nature of Western modernity itself. As recent work shows all too clearly, however, this consensus has epistemic and empirical lacunae of its own, if there can be said to be any consensus still left. Thinkers, especially sociological thinkers (for whom, as one wry observer has put it, "history tends to be the mildly annoying stuff which happens between one sociological model and another"), are far less readily inclined to bother with the boring task of excavating premodernity than to sit back and simply imagine it—and indeed to imagine it purely as a counterpositive to their preconceptions about modernity. This criticism applies almost without exception to every major social theorist of Western modernity, including Ernst Gellner, Anthony Giddens, Jürgen Habermas, Niklas Luhmann, and even Bruno Latour, whose dazzling account of why we have never been modern is based on a sense of nonmodernity—what it is or was, when, and where—that is completely unspecified and speculative.[4]

For all these reasons, attempting to understand the "forms of knowledge" in South Asia prior to the coming of colonial modernity is a self-evidently valuable enterprise. Why, however, in the face of all the confusion about modernity, we aim here to investigate forms of knowledge in *early modern* South Asia may be less self-evident.

For the past decade or so the very idea of early modernity has been a much disputed topic of conversation among scholars, both regionalists and generalists. Many object to the apparent teleology of the idea, committing us as it is supposed to do to some inevitable developmental goal.[5] Of course, our inquiry is perforce teleological in the sense that it aims to understand what occurred in the past that enabled us to get us to the *telos*—if that is still the right word here—at which we have arrived. There is no way to forget the end of this story just because we concentrate on the beginning; indeed we would not even know where to begin the story if we did not know how it has ended, since we would not know what the story was. Others object that many so-called early modernities never became full modernities except when mediated through Western modernization. But what if Western modernization short-circuited other processes of dynamic transformation? No given present was bound to come out of any given past, but our present has come out, and we want to know how and why it has.

Few deny that over the three centuries up to 1800 the world as a whole witnessed unprecedented developments: the opening of sea passages that were global for the first time in history and of networks of trade and commodity production for newly globalizing markets; spectacular demographic growth (the world's population doubled); the rise of large stable states; and the diffusion of new technologies (including gunpowder and printing) and crops from the Americas. If this is a list of material transformations (borrowed from the late John Richards)[6] of what is supposed to make life "modern" rather than just new or different from the past, what part of the world failed to experience early modernity? On the other hand, if we descend from that broad definition of the early modern to the narrow—the presence of fossil fuel technology, constitutional governance, and religious freedom and secularization[7]—there will be no case of early modernity aside from Britain. We may instead want to insist that modernity is additionally, or exclusively, a condition of consciousness. But what kind of consciousness? If we stipulate this a priori, in light of European experience—a new sense of the individual, a new skepticism, a new historical sensibility, to name three master categories—and go forth to find them in South Asia, we are likely to succeed, since one usually finds what one is looking for. Conversely, if we set out to find some highly specific characters—an Indian Montaigne, a Chinese Descartes, an Arab Vico—and somehow do not, well, too bad then, there will be no pre-European South Asian modernity at all.

It is probably the case that much of the current discussion of early modernity is irrelevant for our purposes here, or even an obstruction. As Frederick Cooper has argued with great intelligence, the notion of modernity may have had an important historical role in making claims, but it is virtually useless as an analytic concept (as our sociological speculators show).[8] We are therefore perfectly justified in seeking to understand how variegated the world was at the moment before what would become the dominant form of modernity—colonial, capitalist, Western—achieved global ascendancy, even if that question can be posed only in the moment after. We can call the era "early modern" simply in the sense of a threshold, where potentially different futures may have been arrested or retained only as *masala* for that dominant form. But we may be able to go further. Since the material world changed dramatically during the few centuries prior to this threshold moment, and changed universally, there is good reason to ask how the systems devised for knowing the world responded—or indeed why they failed to respond if they failed—to the world that was changing objectively between these dates. At the same time there is good reason to resist the teleology (here indeed an infelicity) in the term *early modern* and so refuse to assign the period between 1500 and 1800 any shared structure or content a priori, let alone to insist on finding in it Western modernity in embryonic form (such as the Chinese Descartes). Definitional consistency is precisely the trap we must avoid. What we require is historical synchronicity; we do not require and have no reason to expect conceptual symmetry.

In short the era constitutes an entirely reasonable periodization for intellectual history without leading us to posit any necessary uniformity in the history of intellection that transpired. Everyone began to participate in a world economy, to live in vaster and more complex states, to confront a demographic explosion, a diffusion of unprecedented technology, and larger movements of people in a newly unified or at least unifying world. How did people experience these transformations in the realm of thought? That is what we need to uncover. I believe there may be remarkable parallels awaiting discovery, aside from the shocking fact that the period—an empty vestibule, it has been thought, between premodern high tradition and modern Westernization—has been all but unstudied across much of Asia. But we should not worry if such parallels are not found. A "negative" outcome—resistance, say, or stability in the face of dynamic change elsewhere—producing a global version of what Ernst Bloch famously characterized as modernity's constitutive "Gleichzeitigkeit des Ungleichzeitigen" (a multiplex simultaneity of things that are nonsimultaneous), would be as important as a "positive" one, since we are interested in knowing why people may wish to preserve forms of knowledge in the face of changing objects of knowledge no less than in knowing why they may be prepared to transform them. A negation of Western modernity is, obviously, not necessarily a failure.

It is indeed astonishing, then, that while colonial criticism depends on precolonial knowledge, so little of that knowledge has been produced for early modern South Asia, the period prior to 1800, just before British colonial power changed the rules of the knowledge game. It is not as if we do not have the materials to do so. In the sphere of imagination and its written expression South Asia boasts a literary record far denser, in terms of sheer number of texts and centuries of unbroken multilingual literacy, than all of Greek, Latin, and medieval European culture combined. In recognition of this richness an international collaborative research project completed in 2003 undertook a remapping of the literary field across southern Asia, especially for the late precolonial period and in relationship to larger cultural and political processes.[9]

With respect to science and scholarship, however, especially during the critical early modern period, in-depth research on most disciplines is only just commencing. Again the requisite materials have long been available in abundance. In fact it can be argued that with the coming of the Pax Mughalana from the second half of the sixteenth century, a new and dynamic era of intellectual inquiry was inaugurated in many parts of the subcontinent. Whole libraries of the manuscripts produced over the following three centuries exist today—and lie unedited, even unread. The factors contributing to this indifference are worth weighing with care. One is certainly the vastly diminished capacity of scholars today—one of the most disturbing if little remarked legacies of colonialism and modernization—to actually read the languages and scripts in which the materials are preserved. But other factors have contributed to the apathy. These include the old Orientalist-Romanticist credo that the importance of any Indian artifact or text or form of thought is directly proportional to its antiquity: the older it was, or such was the belief, the closer it would bring us to Indo-Germanic *Urzeit* and the cradle of European life. Even more important is the colonial-era narrative of Indian decline and fall before 1800, so central to the ideology of British imperialism and its supposed modernizing mission, which of necessity devalued the late precolonial period as an unworthy, because historically defeated, object of study. One highly instructive example, noted by Allison Busch in her essay in this volume, concerns the achievements of Hindi literary science of the period of neoclassicism, the so-called *rītikāl*, or Era of High Style, circa 1650–1850. The astonishing and completely new cultural formation of this epoch was disdained and dismissed by colonized Indian intellectuals no less than by their colonial masters, who viewed it as decadent, depraved, even emasculating. As a result many of the most important works of the period lie unedited to this day, and most of the fundamental questions, whether internal to the intellectual history of India or external and comparative, remain unasked. (What did it mean, for example, for a vernacular language to travel far beyond its place of origin, to become a cosmopolitan idiom available for courtly usage from Bengal to Maharashtra? Why did both northern India and France see the rise of powerful neoclassical movements of astonishing similarity at precisely the same period?) And this neglect and the ignorance it entails is true across the board. Our intellectual and cultural histories of the period remain grotesquely stunted.[10]

To gain some understanding of the style and substance of Indian thought during these centuries, a research project called Sanskrit Knowledge Systems on the Eve of Colonialism was initiated in 2001, with support from the National Endowment for the Humanities and the National Science Foundation. The group's ongoing work aims to examine seven disciplines in their bibliographical, prosopographical, and substantive dimensions in order to better understand how scholars in the fields of language analysis, logic and epistemology, hermeneutics, poetics, moral-political thought, life science, and astral science understood their objects of study, what knowledge they produced, and in what specific social contexts.[11] Restricting this research program to Sanskrit materials had at once pragmatic and historical justification. If the project was to remain historically responsible as well as manageable, it was as necessary to narrow the scope to a core language as it was to narrow it to core disciplines.

To be sure, Sanskrit was not the only language of science and scholarship in early modern South Asia, and those who communicated in Sanskrit did not constitute the only community that generated systematic knowledge. We are just beginning to understand how the division of language labor functioned and to clarify who used which languages for which purposes. Persian and vernacular intellectuals produced no less sophisticated work, sometimes in conversation with their Sanskrit-using colleagues—a conversation that seems to have taken place principally in the fields of astronomy and mathematics—but more often, it seems, segregated from them. (Precisely how and to what extent interaction occurred between these different communities now designated by their linguistic or religious preferences are problems in need of serious investigation.) Yet again, despite the quality and quantity and cultural-historical significance of these materials, very little scholarly attention has been devoted to them. While the comparative religion industry, in the United States at least, continues to claim an ever larger market share in the academy, it is almost impossible to find scholars who understand the importance of research on any aspect of precolonial science and scholarship in Persian, Arabic, or the regional languages. In Indo-Persian studies, for example, only in the past several years has research been undertaken on early modern aesthetics, historiography, philology, philosophy, or political thought.[12] The same must be said of most regional language traditions, with the notable exception of Telugu and Tamil, thanks to the pioneering collaborative efforts of Velcheru Narayana Rao, David Shulman, and Sanjay Subrahmanyam.[13] In Tibetan studies, the early modern era may be more richly cultivated, but this has not necessarily been the case of its nonreligious traditions of science and scholarship.

Forms of Knowledge in Early Modern Asia thus enters into a strikingly underdeveloped scholarly field, with subfields that have their sometimes radically different scholarly histories and needs. The contributors have accordingly understood their brief variously. Some have aimed for a large-scale assessment of a whole problematic, whereas others have sought to achieve larger generalizations through the study of representative texts or persons. A review of the main concerns of these quite varied offerings will be of use in orienting the reader.

Part I, "Communication, Knowledge, and Power," begins with a general review of the problem of science and language choice. One of the key factors in the modernization of knowledge production in seventeenth-century Europe was the transformation of the vernaculars into languages of science, as for example in the work of Bacon, Descartes, and Galileo. As I argue in the first essay, although South Asia shared a comparable history of vernacularization in the area of literary production, Sanskrit persisted into the early modern period as the exclusive code for most areas of science, and scholarship more generally, outside the Persianate cultural sphere. It seeks first to delineate the boundaries of this relationship in terms of disciplines and regions, and then to lay out the presuppositions in Sanskrit-language philosophy that militated against the vernacularization of scientific discourse. A useful orientation to the latter problem, which summarizes the dominant position of Sanskrit intellectuals on the eve of colonialism, is the work of the great scholar Khaṇḍadeva on scriptural hermeneutics from mid-seventeenth-century Varanasi.

Sumit Guha continues this theme by opening a historical inquiry into the very complex history of language awareness and language use in the Marathi-speaking regions. In its competition for prestige and status, Marathi had to contend not only with Sanskrit and the newly ascendant Persian of the southern sultanates, but also with Arabic and the increasingly widespread Dakhani, the southern form of what in the north would come to be known as Urdu. Innovative texts like the (Sanskrit) prosimetrical work of Jayarama Pindye, which shows the author's familiarity with a dozen languages, are exemplary of an astonishing linguistic efflorescence. In the midst of this linguistic sea and the changes it was working across all the competing idioms—where, as Guha puts it, a tension existed "between hybridization tending toward assimilation and distinction tending toward establishing identity"—Marathi literati began tentatively to use their language, long a code for religious poetry, as a vehicle for political theory and history, and eventually to seek to provide greater lexicographical discipline, especially in Tanjavur (Tanjore), the easternmost region of the empire. The world of political discourse, for its part, shows a tendency by Marathas—in their continuation by other means of their war with the Mughals—toward the resuscitation of Sanskrit, swimming here against the stream that was elsewhere consigning that language to historical irrelevance. In the end the evidence suggests a serious self-awareness of language distinctions and a growing linkage of language and social identification.

The role of the polity in early modern South Asia, and its modes of governance, which are of implicit importance in Guha's essay, come in for explicit assessment in the essay by Velcheru Narayana Rao, David Shulman, and Sanjay Subrahmanyam. One of the most remarkable polities in early modern South Asia, the Vijayanagara Empire entered a fresh and interesting phase of its existence in the early sixteenth century. New challenges of a fiscal, military, and diplomatic order presented themselves, not least because of the arrival of the Portuguese on the western shores of India. In this context it is useful to know how the problem of imperial management was addressed by Vijayanagara rulers such as Krishnadevaraya (r. 1509–29). Rao, Shulman, and Subrahmanyam look to the masterwork in Telugu by this emperor, *Āmuktamālyada.* They focus above all on the section concerning *nīti,* or statecraft, and show how received wisdom on the subject was transformed in light of the new challenges and circumstances of the period.

Part II of the book, "Literary Consciousness, Practices, and Institutions in North India," addresses the intellectual and social history of the literary system of classical Hindi. Allison Busch offers a study of the Hindi *rītigranth* (book of method), a major vehicle of precolonial north Indian intellectual life. Expanding patronage networks during the Mughal period fostered the conditions for the development of a new vernacular embodiment of the Sanskrit discipline of literary science (*alaṅkāraśāstra*). The crystallization of this trend was the *rītigranth,* which was produced in astonishing quantities by a wide range of poets and emerged as the most significant genre of Hindi courtly literature. Busch explores the epistemological world of early modern vernacular thinkers in an attempt to understand both the literary vision they were trying to actualize and the larger intellectual community in which they functioned. New vernacular intellectual practices posed a challenge to traditional language hierarchies, in which Sanskrit had long held unquestioned dominance. She traces the development of vernacular scholarly writing as it both built upon and marked differences from earlier Sanskrit texts. Forging a scholarly discipline in a language medium not sanctioned by tradition took courage, and it also initially engendered feelings of insecurity, an "anxiety of innovation" that is reflected in both the style and the substance of the works.

We can know nothing about early modern knowledge without knowing the texts in which that knowledge is stored. The history of early modern textuality, however, or the history of manuscript culture, as I would prefer to call it, is among the more critically underdeveloped domains of the South

Asian humanities.[14] Imre Bangha's work on the *Kavitāvalī* indicates some of the paths toward progress in this area. The *Kavitāvalī* is a series of some 350 loosely connected quatrains, in strict meters and with the timbre of a personal voice, by the renowned late sixteenth-century poet Tulsidas. The collection gained its form toward the end of the poet's life. An exhaustive inventory of the manuscript material shows the existence of a shorter and a longer recension, both different from the modern published version, and which reveals the process of editing they have undergone. The shorter recension reflects a purist tendency, and the longer what Bangha terms a collector tendency. In a good half of the manuscripts poems were suppressed on the grounds of aesthetic shortcomings rather than because of deviation from religious doctrine. Clearly, in some vernacular traditions faithfulness to the received text was often far less pressing a concern to early modern scribal culture than poetic excellence.

It is rare for scholars in South Asian studies to be able to reconstruct the actual institutions of cultural production. One of the exceptions is the Bhuj Brajbhāṣā Pāṭhśālā (Braj Language School in Bhuj, Gujarat), whose history is carefully reconstructed by Françoise Mallison. From 1749 until 1948 the Bhuj Brajbhāṣā Pāṭhśālā (also called Kāvyaśālā, Poetry School) each year turned out ten or more court poets belonging to bardic castes hailing from the provinces surrounding Kutch, while training other poets belonging to many other castes or creeds. The initiative to educate professional writers and poets for official duty was taken by a local prince, Lakhpati Simha (r. 1741–61), who sought to achieve fame for the Rajput culture of his small kingdom, isolated from the rest of India and yet open to and knowledgeable about the world beyond the Indian Ocean. What the school produced in technical and pedagogical literature as well as works of poetry has been dispersed among many institutions, so that the sources of its history have not yet been fully examined in any way. Even so the materials that are accessible provide us with an almost unique glimpse into the transmission of knowledge and literary activity in early modern western India.

Cultural training is also central to the Persianate cultural order, as explored in Part III, "Inside the World of Indo-Persian Thought." The central concern of the essay by Muzaffar Alam and Sanjay Subrahmanyam lies in the making of a particular form of knowledge in early modern India, namely that carried out by the so-called *munshī*, or scribe. While studies abound on the chancellery literati in Ming and Qing China, few such treatments can be found to explain how the service class of the Mughals was trained, the men who kept accounts, managed estates, and were the real backbone of the state and of subimperial households. To answer this question the authors turn to the little-known autobiographical account of a seventeenth-century *munshī* named Nek Rai, who lived in the early years of the reign of Aurangzeb. A close reading of his account together with more normative materials allows a rare glimpse into the mind of the *munshī*, and also the tensions that characterized the process of acculturation that produced him.

One of the most remarkable texts to emerge from the cosmopolitan and polyglot world of Mughal India is the *Dabistān-i Maẕāhib*, "The School of Religions," studied by Aditya Behl. Produced from within a Zoroastrian sect that had been persecuted in Safavid Iran and found refuge in India, the *Dabistān* presents the cosmology, angelology, and religious system of an esoteric Zoroastrianism said to have preceded Zoroaster. The text's author, Mūbad Shāh, whose identity remained hidden for several centuries, traveled throughout Mughal India, sometimes in disguise, and mingled with a wide variety of religious virtuosi and holy men. He uses his system to present a wide-ranging survey, arranged typologically, of seventeenth-century Indian religious beliefs and narratives about religious identity and difference. Not least interesting is the quasi-ethnographic side of the work, for some presumably privileged information about other religions seems to have been gathered by Zoroastrians masquerading as members of these other orders. The *Dabistān* presents an extraordinary quest to define the boundaries of religious truth and adjudicate the truth claims of an entire period. The variety of strategies used describes an arc from similarity to incommensurable difference, and the complexity of the responses to other religions confounds any simple notion of tolerance as the leitmotif of the Mughal era.

Ethnography and encyclopaedism also mark the genre of poetry that Sunil Sharma studies in his essay. He is concerned with the rhetorical connections between the Indo-Persian love lyric, commerce, and the city as a medium for the transmission of knowledge about various forms of cultural and social interaction in urban cores. In the sixteenth and seventeenth centuries Muslim centers of power in India were described in a special poetic language that was embedded in tradition but at the same time representative of a new historical mode of thought. This period corresponds to an epoch of Indo-Persian literary innovation and experimentation that later came to be known as the *sabk-i Hindī* (the Indian style) and takes into account the works of selected Iranian émigré poets writing for Indian patrons as well as Indian poets who wrote in Persian and Urdu. Following the political shifts in the eighteenth century this mode of writing developed into two strands, one in the so-called Urdu poetry of decline and another as a full-fledged ethnography of empire, as in the late

Mughal chronicle *Khulāṣat al-tavārīkh*, and under colonial patronage, as in the first Urdu chronicle *Ārāish-e Maḥfil* by Afsos, produced at Fort William College. Poetry in early modern South Asia could often be a form of knowledge too, no less than the dull "prose of the world" demanded in the early modern West.

The very modernity of the West, as I noted earlier, is a matter of scholarly disagreement: What actually are the elements that add up to modernity? Most scholars have not considered very deeply whether any of those elements might in fact be derived from the nonmodern world. Mohamad Tavakoli-Targhi takes on this fascinating question by looking closely at the transregional (South and West Asian) formation of Persianate modernity in the seventeenth and eighteenth centuries, before the rise of Orientalism and nationalism rendered such a concept utterly impossible. Part of this story concerns the very intense exchanges between Persianate and European scholars of the period that led to the translation of Descartes and Gassendi into Persian in the 1650s and the importation of European astronomical models in the early eighteenth century and resultant engagements with the theories of Copernicus and Newton. But another part concerns entirely independent discoveries. We now know that it was Mirza I'tisam al-Din, an Indian scholar, who translated for Jones the work that formed the basis of Jones's bestseller, *A Grammar of the Persian Language* (1771), and more significantly that it was Siraj al-Din Khan Arzu's *Muṣmir* (before 1756) that first established the "affinity" between Sanskrit and Persian, which Jones later used to win renown as "the creator of the comparative grammar of Sanskrit and Zend." How many other such texts and persons have been lost in the willful amnesia of Orientalism?

In Part IV, "Early Modernities of Tibetan Knowledge," we turn to some remarkable developments in the intellectual history of Tibet, which for many intellectual and cultural purposes is as much or more within the South Asian sphere as it is the Inner Asian or East Asian, as indeed several of the essays here demonstrate. Kurtis Schaeffer looks at the division of knowledge among Tibetan intellectual historians. Such historians, dating from the thirteenth century to the seventeenth, divided cultural practice into five major arts and sciences (a systematic organization of knowledge and practice comparable to the seven liberal arts of medieval Europe), comprising language, logic, material arts, medicine, and the "esoteric art" of Buddhism itself. In their systematic treatises Tibetan intellectuals reveal their conception of the complex relationships among disciplinary practices, from astronomy to logic, medicine, and finally meditation. It was their understanding of the nature of the bodhisattva and his moral perfection that linked these practices, which were thought of as elements of a person's ethical training. By the late seventeenth century scholars in turn linked the bodhisattva ideal, and thus the arts and sciences, to the ideology of the Tibetan central government. Thus if culture can be thought of as a complex interplay between tradition, institutional power, and human practice, then in articulating the relationship between the arts and sciences scholars of Tibetan knowledge systems were engaged in nothing less than the construction of culture.

Debates in Tibetan medical writings from the sixteenth century to the eighteenth form the core of Janet Gyatso's contribution. She analyzes the role of empirical evidence, its growing importance in Tibetan medicine, and how it came into conflict with traditional assumptions about scriptural authority. At the same time Gyatso explores various senses of a notion of "experience" in medical theory and practice that came to the fore in the same period and offers an assessment of the influence of Buddhist literary and educational practices on medicine, as well as indications from medical practitioners of their sense of distance from traditional Buddhist authority. Her essay ends with an exploration of the relationship of the new professionalization of medicine to contingencies associated with the emerging Tibetan Buddhist state under the Fifth Dalai Lama and thereafter during the Qing dynasty.

During the eighteenth century Tibetan encounters with the growing Manchu, British Indian, and Russian Empires brought about a sudden awareness in Tibetan learned circles of peoples and places that had literally no place in received knowledge of the world. Though aspects of this new knowledge have been explored in earlier Tibetological scholarship, the epistemological questions raised by this assimilation of new material into established conceptual schemes have not yet been considered. Part of the interest here lies in the analogy we find with the European problem of assimilating the post-Columbian world into Ptolemaic schemes, something that has been much discussed in work on the history of science and exploration during the past several decades. In his essay Matthew Kapstein puts the Tibetan geographical literature of the eighteenth century into dialogue with recent work on the history of geography and cartography in the West. His focal point is the so far unstudied *General Geography* (*'dzam gling spyi bshad*) of Sum-pa mkhan-po (1704–87), a work that introduced Tibetan readers to such marvels as the frozen wastes of Siberia and the polar bears of the Arctic Sea.

Forms of Knowledge in Early Modern Asia thus covers a wide range of intellectual and historical concerns, from the development of new forms of language and ethnic self-identification, the scientificization of vernacular cultural sensibilities, the actual practices of editing and textual circulation in the era before printing, to the pedagogy and the production of bureaucrats in the Persianate sphere, the intersection of imagination and information, what amounts to a kind of protoethnography but in poetry, and the development of science in relationship to empiricism outside of the usual European framework, where experience and religion seem to have come newly, even "modernly," into tension with each other. Most readers are unlikely to have heard of any of the fascinating characters who appear in the following pages, including Jayarama Pindye, the multilingual poet at the Maratha court; Cintamani Tripathi, the poetician of *rīti*; Nek Rai, the clerk and autobiographer; the Iranian émigré poet Nuruddin Muhammad Zuhuri, ethnographer of Indian city life; and Dar-mo sMan-rams-pa, one of an inner group of physicians close to the Fifth Dalai Lama. Yet these figures will now enter the historical record for what they tell us about the creative reinvention of the world of South Asian thought in the late precolonial period—what I believe we will one day come to understand was an iceberg of creativity, of which the voices we hear in the following pages represent the merest tip.

Our book is the first such collection of its kind, and like all firsts it is tentative and experimental. The contributors all share the aim that, whatever the merits of their particular arguments, they will have succeeded in demonstrating something of the allure and excitement of the general problematic itself. And they hope thereby to stimulate deeper research into one of the most complex eras and areas in global intellectual history, when a whole world of knowledge, of centuries-long standing and singular prestige and importance, began to make its own adjustments to the early modern world before undergoing a transformation more profound and disruptive than any it had previously known.

Notes

Portions of this introduction appeared previously in *Comparative Studies of South Asia, Africa, and the Middle East* 24, no. 2 (2004) and *International Association of Asian Studies Newsletter* (Leiden) 43 (2007).

1. For a recent review, see Washbrook, "Orients and Occidents."

2. See, for example, my "Deep Orientalism?"

3. Weber's notorious generalizations include the following: "Only in the West does science exist at a stage of development which we recognize to-day as valid"; "Ratio-

nal chemistry has been absent from all areas of culture except the West"; "All Indian political thought was lacking in a systematic method comparable to that of Aristotle" ("Vorbemerkung"). Notable attempts at revision in economic history, to take only that dimension, include Bayly, *Indian Society and the Making of the British Empire*; Bose, *South Asia and World Capitalism*; Subrahmanyam, *Merchants, Markets and the State in Early Modern India*; Washbrook, "From Comparative Sociology to Global History." For a deeper critique of the very concept of modernity, see Kaviraj, "Outline of a Revisionist Theory of Modernity."

4. See Latour, *We Have Never Been Modern.*

5. Starn, "The Early Modern Muddle."

6. Richards, "Early Modern India and World History."

7. Goldstone, "The Problem of the 'Early Modern' World."

8. Cooper, "Modernity."

9. Pollock, *Literary Cultures in History*. For a new account specifically of political formations in early modern India, though concentrating on the eighteenth century and using much colonial archival material, see Subrahmanyam, *Penumbral Visions.*

10. Despite a growing interest in the early modern as a conceptual problematic, long-standing positivist and supposedly materialist tendencies in Indian historiography have drastically narrowed the scope of inquiry. Edited volumes assessing the state of the field of research for the last century of our period, like Alavi's *The Eighteenth Century in India* and Marshall's *The Eighteenth Century in Indian History*, resolutely exclude all questions of late precolonial scientific, literary, or intellectual culture. Even where the transformation of the latter is directly thematized, precolonial history is ignored; see Chatterjee, *Texts of Power*; Prakash, *Another Reason*; Arnold, *Science, Technology and Medicine in Colonial India.*

11. For further particulars, see Pollock, this volume.

12. Exemplary work includes Alam, "The Culture and Politics of Persian in Precolonial Hindustan" and *The Languages of Political Islam*; Tavakoli-Targhi, *Refashioning Iran.*

13. See, for instance, Rao et al., *Symbols of Substance* and *Textures of Time.*

14. For a general account, see Pollock, "Literary Culture and Manuscript Culture."

References

Alam, Muzaffar. "The Culture and Politics of Persian in Precolonial Hindustan." *Literary Cultures in History: Reconstructions from South Asia*, ed. Sheldon Pollock. Berkeley: University of California Press, 2003.

———. *The Languages of Political Islam: India 1200–1800*. London: Hurst, 2004.

Alavi, Seema. *The Eighteenth Century in India*. New Delhi: Oxford University Press, 2002.

Arnold, David. *Science, Technology and Medicine in Colonial India*. Cambridge: Cambridge University Press, 2000.

Bayly, C. A. *Indian Society and the Making of the British Empire*. Cambridge: Cambridge University Press, 1988.

Bose, Sugata, ed. *South Asia and World Capitalism*. Delhi: Oxford University Press, 1990.

Chatterjee, Partha, ed. *Texts of Power: Emerging Disciplines in Colonial Bengal*. Minneapolis: University of Minnesota Press, 1995.

Cooper, Frederick. "Modernity." *Colonialism in Question: Theory, Knowledge, History*. Berkeley: University of California Press, 2005.

Goldstone, Jack. "The Problem of the 'Early Modern' World." *Journal of the Economic and Social History of the Orient* 41, no. 3 (1998), 249–83.

Kaviraj, Sudipta. "Outline of a Revisionist Theory of Modernity." *European Journal of Sociology* 46, no. 3 (2005), 497–526.

Latour, Bruno. *We Have Never Been Modern*. Cambridge, Mass.: Harvard University Press, 1993.

Marshall, P. J. *The Eighteenth Century in Indian History: Evolution or Revolution?* New Delhi: Oxford University Press, 2003.

Pollock, Sheldon. "Deep Orientalism? Notes on Sanskrit and Power beyond the Raj." *Orientalism and the Post-colonial Predicament*, ed. Carol Breckenridge and Peter van der Veer. Philadelphia: University of Pennsylvania Press, 1993.

———. "Literary Culture and Manuscript Culture in Precolonial India." *History of the Book and Literary Cultures*, ed. Simon Eliot, Andrew Nash, and Ian Willison. London: British Library, 2006.

———, ed. *Literary Cultures in History: Reconstructions from South Asia*. Berkeley: University of California Press, 2003.

Prakash, Gyan. *Another Reason: Science and the Imagination of Modern India*. Princeton, N.J.: Princeton University Press, 1999.

Rao, Velcheru Narayana, et al. *Symbols of Substance: Court and State in Nāyaka Period Tamil Nadu*. Delhi: Oxford University Press, 1992.

Rao, Velcheru Narayana, et al. *Textures of Time: Writing History in South India, 1600–1800*. Delhi: Permanent Black, 2001.

Richards, John. "Early Modern India and World History." *Journal of World History* 8, no. 2 (1997), 197–209.

Starn, Randolph. "The Early Modern Muddle." *Journal of Early Modern History* 6, no. 3 (2002), 296–307.

Subrahmanyam, Sanjay, ed. *Merchants, Markets and the State in Early Modern India*. Delhi: Oxford University Press, 1990.

———. *Penumbral Visions: Making Polities in Early Modern South India*. Ann Arbor: University of Michigan Press, 2001.

Tavakoli-Targhi, Mohamad. *Refashioning Iran: Orientalism, Occidentalism, and Historiography*. New York: Palgrave, 2001.

Washbrook, David. "From Comparative Sociology to Global History: Britain and India in the Pre-history of Modernity." *Journal of Economic and Social History of the Orient* 40, no. 4 (1997), 410–43.

———. "Orients and Occidents: Colonial Discourse Theory and the Historiography of the British Empire." *The Oxford History of the British Empire*, vol. 5, *Historiography*, ed. Robin W. Winks. Oxford: Oxford University Press, 1999.

Weber, Max. "Vorbemerkung." *Die protestantische Ethik und der Geist des Kapitalismus.* Tübingen: Mohr, 1934.

Part I

Communication, Knowledge, and Power

CHAPTER ONE

The Languages of Science in Early Modern India

SHELDON POLLOCK

An important factor in the modernization of the production and dissemination of knowledge in Europe was the transformation, beginning in the seventeenth century, of the vernaculars into languages of science and the eventual displacement of long-dominant Latin. By contrast, although South Asia had known a history of vernacularization in the domain of expressive textuality (*kāvya*, "literature") astonishingly comparable to that of Europe, Sanskrit persisted as the exclusive medium of communication outside the Persianate cultural sphere for many areas of science, systematic thought, and scholarship more generally until the consolidation of colonial rule in the nineteenth century. This is a puzzling and arguably a consequential difference in the histories of their respective modernities.

The problem of the relationship between knowledge forms and language choice has a long history in India, beginning with the multiple linguistic preferences shown by Buddhists until Sanskrit gained ascendancy in the early centuries of the Common Era. I address some of this premodern history elsewhere.[1] Here I want to situate the problem of language and science more narrowly conceived within the context of the collaborative research project in which I first formulated it, and that has something to do with the descriptor "early modern" in my title. I then reflect briefly on what we might mean by the category *science* (or *systematic knowledge* or *learning*) in this period and in its relationship to the complex "question of the language" with its two kinds of concerns, epistemological and social.[2] After delineating the boundaries of language choice in a number of specific intellectual disciplines and vernaculars, I look more closely at one tradition, that of Brajbhasha. I then review some of the presuppositions in Sanskrit language philosophy that may have militated against the vernacularization of intellectual discourse. A useful orientation here, which summarizes the dominant position of early modern Sanskrit intellectuals, is offered by *mīmāṃsā* (discourse analysis and scriptural hermeneutics), in particular the work of Khaṇḍadeva, the discipline's foremost exponent in mid-seventeenth-century Varanasi. I end by drawing and weighing some contrasts with the case of Europe.

It bears remarking at once how thoroughly the question of the medium of intellectual discourse in early modern India has been ignored in scholarship. Thanks to the work of Frits Staal and others, we may understand something of the discursive styles of the "Sanskrit of science."[3] But we still understand next to nothing of its ideology or sociology, let alone how this might compare to other cultural formations contemporaneous with it. These are obviously vast and complex issues, and it is not possible in this brief space to offer more than a brisk and tentative sketch.

Knowledge Systems on the Eve of Colonialism

The collaborative research project of this name that forms the context for the thematic of the languages of science aims to investigate the substance and social life of Sanskrit learning from about 1550 to 1750 across four geographical areas and seven intellectual disciplines.[4] As for the time boundaries, the endpoint is set by the consolidation of colonial domination in our spatial foci (Bengal 1764; Tanjavur 1799; Varanasi 1803; Maharashtra in the course of the following decade). Somewhat more arbitrary is the starting point. It was certainly not meant to be hard and fast, and it has become clear that different knowledge systems followed different historical rhythms. But in many ways the work of the logician Raghunātha Śiromaṇi in the north and the polymath Appayya Dīkṣita in the south (both fl. ca. 1550) marked something of an intellectual and historical rupture that we are only now beginning to understand.[5] The spatial boundaries are similarly somewhat flexible, but to the degree possible attention is being concentrated on trying to understand the varying conditions of intellectual production in what are, in sociopolitical terms, very different regional complexes (Delhi/Varanasi, Tanjavur/Madurai, Mithila/Navadvip, and Maharashtra). In addition to these time-space limits, the project restricts itself to seven disciplines: *vyākaraṇa* (language analysis), *mīmāṃsā, nyāya* (logic and epistemology), *dharmaśāstra* (law and moral philosophy, broadly speaking), *alaṅkāraśāstra* (poetics), *āyurveda* (life science), and *jyotiḥśāstra* (astral science). These have been selected for their centrality to Sanskrit culture (language and discourse analysis), for their comparative and historical value (life and astral sciences), or for the new vitality the system seems to have demonstrated during these centuries (logic and epistemology).[6]

The Eve of Colonialism project is at once self-contained and preparatory to a comparative history, first with Indo-Persian and vernacular scholarship of the sort offered in this volume, and second, more grandly, with European and other Asian systems of thought.[7] It was largely a matter of pragmatic method, intuition, and professional orientation that the project was originally organized according to language, first Sanskrit and eventually Persian and vernacular. The decision to concentrate initially on Sanskrit was made also because it appeared that the Indian knowledge systems of the period were in fact concentrated in Sanskrit. But is that impression more than an appearance? Was science in the period 1550–1750 in fact restricted to production in the Sanskrit language (outside the Persianate sphere, that is), and if so, why was it restricted and with what consequences? More generally, has language choice in India (or anywhere else) ever been pertinent to the production of science, systematic thought, and scholarship, and if so, how and to what degree?

Science and Language in Premodern India

Before the problem of the relationship of language and science can even be raised we need to ask what is meant by *science*. This is no easy question to answer, however, for the intellectual history of premodern South Asia, or indeed for that of the West. As recently as 1993 European scholars were bemoaning the fact that there existed "no critical discussion of the changing meaning of the word 'science'" in the West; in fact an important recent collection on science and language in Europe over the past four centuries evinces astonishing indifference to the historical semantics of the term that defines the book's very problematic.[8] The situation is hardly less acute in South Asian scholarship. *Science, systematic knowledge, scholarship, learning* (as well as *rule* and even *scripture*) would all be legitimately translated by the Sanskrit word *śāstra*. But what exactly is *śāstra*, and how does it relate to other, kindred concepts, such as *jñāna* (and *vijñāna*) and *vidyā* (all variously translated as knowledge, learning, scholarship . . . and science)? The English word *science* points to no natural kind but is a worrisomely pliable signifier, indeed almost a talisman (witness *Christian science* or *creation science* or *political science*), and clearly it is no straightforward matter to map onto it the congeries of terms and texts and intellectual practices we find in India during the two or three centuries before colonialism.

At the same time we must address a certain circularity, for traditional India, that presents itself in the very formulation of the central problem of this essay. If, from a long-term perspective, *science*, whether as *jñāna* in the sense of comprehension or *śāstra* in the sense of system, is simply *knowledge*—Sanskrit *veda* (from the root *vid*, "to know")—then science can have been expressed *only* in the Sanskrit language. This is surely one implication of the discourse on the *vidyāsthānas*; these fourteen (later eighteen) "knowledge sources" or disciplines, which were held to exhaust the realm of systematic thought, all derive their truth from their relationship to Vedic revelation. As the *Yājñvalkyasmṛti* expresses it, "No *śāstra* exists other than the Veda-*śāstra*; every *śāstra* springs from it."[9] Accordingly throughout much of Indian history new—or, ipso facto, counter—*śāstra* (or *jñāna* or *vidyā*) required new or counter language, beginning with the *śāstra* comprised of the teachings of the Buddha, composed originally in Gandhari and other local languages in the north and Pali in the south.

This apparently general cultural presupposition finds an echo in the widespread commitment to a postulate of Sanskrit language ideology: correct language is required for the correct communication of reality (science). This idea is at least as old as the seventh century, when Kumārila, the great scholar of *mīmāṃsā*, argued "The scriptures of the Śākyas [Buddhists] and Jains are composed in overwhelmingly corrupt language [*asādhuśabdabhūyiṣṭha*]—with words of the Magadha or Dakshinatya languages or their even more dialectal forms [*tadapabhraṁśa*]. And because of their false composition [*asannibandhanatva*], they cannot be considered science [*śāstratvaṃ na pratīyate*]. . . . When their words are false [*asatyaśabda*] how could their doctrines ever be true [*arthasatyatā*]? . . . That the Veda, on the other hand, is an autonomous source of true knowledge is vouchsafed by its very form [*rūpād eva*]."[10] Kumārila is entirely typical in his view on the relationship between "correct" language, Sanskrit, and truth, and in his conviction that only Sanskrit can articulate reality and thus be the sole medium for science. Even the Indian Buddhists eventually agreed after all, adopting Sanskrit for all their writings from the first or second century onward. And this position was one *mīmāṃsakas* such as Dinakara Bhaṭṭa (fl. 1625) were still endorsing a millennium later: "The remembered Vedic text [*smṛti*] that restricts usage to grammatically correct language [i.e., Sanskrit]—the one that enjoins us to 'Use only correct words, not incorrect ones' [*sādhūn evābhibhāṣeta nāsādhūn*]—derives its authority from the extant Vedic text [*śruti*] requiring one to speak the truth and to avoid lies."[11]

A language ideology of this sort is not, to be sure, peculiar to Sanskrit intellectuals: for Derrida, only Greek can really speak philosophy, for Heidegger, only German. But Sanskrit intellectuals based their view on a far more explicitly enunciated theory, one that I examine in what follows. Some continuing energies from their various postulates and the quest for an ever more perfect fit between language and things—for an ever more Sanskritic Sanskrit—may also have conditioned one of the most far-reaching developments in early modern intellectual life: the fashioning of a new idiolect by *navyanyāya* (new logic), beginning around the fourteenth century, that was to profoundly influence discursive style across disciplines and regions. Indeed exploiting to an extreme degree linguistic capacities with which Sanskrit is especially well endowed (in particular nominal compounding), this philosophical register would make the transition to science and scholarship in vernacular languages even more difficult than language ideology already had. Sanskrit scientific thought had long been not only thought in Sanskrit but thought about Sanskrit, about the nature of this particular language and its attributes. (It is, for example, no easy thing to discuss *mīmāṃsā*'s concern with deontic verbal morphemes [*vidhi liṅ*] or possessive qualifiers [e.g., *matup*] in languages that lack them.) This was the tendency that *navyanyāya*, with its invention of a new philosophical vocabulary—far vaster than, say, the poststructuralist gallicization of English—exaggerated to the point of untranslatability, even unintelligibility.[12] And there are other elements of language ideology, in addition to the linkage between language that is correct or true (*sādhu* or *sat*) and the truth itself (*satya*), that I address separately below.

Let us be more empirical for a moment, however, and examine the language practices of science understood as broadly as possible. Were there forms of systematic knowledge that were never communicated in vernacular texts prior to the colonial age?

Consider first the Indian *vidyātraya* of *pada, vākya,* and *pramāṇa,* the "triple science" of words, sentences, and grounds of knowledge, which, whatever its status in earlier times, had by the seventeenth century become an actual ideal of intellectual perfection. (Every scholar now claimed for himself the sonorous title *padavākyapramāṇapārāvārapārīṇadhurīṇa,* "able to cross to the further shore of the ocean of grammar, hermeneutics, and epistemology.") No synthetic work on the question of language medium in these disciplines has ever been done, but an informal survey suggests strongly that access to them was attainable only through Sanskrit. Both *nyāya,* the *pramāṇaśāstra* (along with the larger questions of epistemology), and *mīmāṃsā,* the *vākyaśāstra,* were entirely untouched by vernacularization. I have been unable to locate a single premodern work in either field in any regional language, except for the occasional and very late, almost certainly colonial-era, translation.

The vernacular history of grammar and the related disciplines of poetics, metrics, and lexicography, is somewhat anomalous, and it also presents a significant, and puzzling, unevenness between north and south India. Philology (to use that term as the general disciplinary rubric of these arts) swept across most of south India more or less simultaneously. The Kannada tradition commenced in the late ninth century with an important text encompassing grammar and poetics, the *Kavirājamārga* of Śrīvijaya, which was quickly followed by elementary grammatical (and lexicographical and prosodical) works leading to one of the most sophisticated descriptions of a vernacular language in the premodern world, the *Śabdamaṇidarpaṇa* of Kēśirāja (mid-thirteenth century).[13] This philological activity continued into the seventeenth century with the *Śabdānuśāsana* of Bhaṭṭa Akalaṅka Deva (a grammar, written in Sanskrit, of the classical idiom of Kannada, which had become obsolete by the thirteenth or fourteenth century), but then mysteriously vanished. Developments in Tamil are more or less contemporaneous with Kannada; leaving aside the undatable *Tolkāppiam*,[14] these include the grammar *Naṉṉūl* by Pavaṇanti (early thirteenth century), the more strictly poetics texts *Vīracōḻiyakkārikai* (ca. 1063–69) and *Taṇṭiyalaṅkāra* (somewhat earlier), and a plethora of dictionaries produced continuously from around the eighth or ninth century into the eighteenth. Telugu philology begins only slightly later, with the appearance of important grammatical works from the thirteenth century onward (*Āndhrabhāṣābhūṣaṇamu* of Kētana, thirteenth century, and *Āndhraśabdacintāmaṇi* ascribed to the eleventh-century poet Nannaya but more likely authored by Appakavi in the last quarter of the sixteenth century).[15]

Wholly different is the situation in the north, where vernacular languages without exception remained untouched by formal grammaticization until the coming of the new colonial order of knowledge. A striking instance of this negative dynamic is Marathi. The language was conceptually objectified by the late tenth century and became the vehicle for expressive literature by the thirteenth. Four centuries later it was continually being adduced by Maharashtra-born scholars when glossing Sanskrit texts (a good example is the great *Mahābhārata* commentator Nīlakaṇṭha Caturdhara, fl. 1675), a sure sign of its primacy among their readership. Yet systematic reflection on Marathi grammar (and lexicon and prosody) is, with one exiguous exception, entirely absent before the coming of European science—a fact made doubly paradoxical by the fact that it was in Maharashtra, where Marathi is the dominant language, that the cultivation of Sanskrit grammatical studies attained the greatest brilliance in early modern India.[16] The same holds for poetics, which found no vernacular expression in the north except (admittedly a big exception) in the Brajbhasha appropriation of Sanskrit *alaṅkāraśāstra*.[17]

The almost total—and in some regions total—linguistic monopolization by Sanskrit over the three primary disciplines of grammar, hermeneutics, and logic and epistemology tallies with the evidence from many other areas of systematic knowledge. Again this question awaits detailed study, but some first observations are likely to be borne out by further work. In law (*dharmaśāstra*) vernacular works are exceedingly rare; there may well be more than the *Vijñāneśvariyamu*, a Telugu adaptation by Kētana of the celebrated Sanskrit work produced in Kannada country in the twelfth century, but that is all I have ever encountered.[18] In the field of life science (*āyurveda*), to take a second example, matters are somewhat less clear, but Sanskrit certainly appears to have maintained a statistical dominance in some areas until the second half of the eighteenth century. At which point, for reasons that await explanation, medical authors began to produce their discourses in more than one language, but this remained an occasional practice.[19]

Vernacular philosophical and religious poetry might seem to offer counterevidence to the overall pattern, for the genre is not only common but sometimes foundational to a regional tradition. Again Marathi offers an interesting case, with the (possibly) thirteenth-century *Vivekasindhu* of Mukundarāja presenting a remarkably precocious example of vernacular Advaita-vedantic exposition, and the near contemporary work of Jñāneśvara, the *Bhāvārthadīpikā*, providing an equally precocious example of vernacular philosophical and poetic commentary.[20] Similarly Śrīvaiṣṇava theology was composed in a new Sanskrit-Tamil register (*maṇipravāḷa*) in Tamil country, and Vīraśaiva theology in Kannada (and sometimes Telugu) in the Deccan. And yet these kinds of works do not really constitute an exception to the general rule of the language of science and its broader norms that, with the Hinduization of Sanskrit in the present age, we are apt to forget: the vehicle of organized, systematic *laukika*, or this-worldly, knowledge before colonialism was Sanskrit, while the regional languages, at least in their incarnation as literary idioms, were in the first instance the voice of *alaukika*, or other-worldly, wisdom (a situation closely paralleled by Latin and the European vernaculars).[21] To make this distinction is not to value information over imagination or to unjustly narrow the scope of the *śāstra*; it is simply to describe a historical division of language labor that was highly influential. By and large, *systematic knowledge* remained the preserve of Sanskrit, the literary and spiritual the preserve of the vernaculars, outside the Persianate world. Indeed that is precisely how the Persianate world understood the situation: Mīrzā Khān, in his remarkable overview of Brajbhasha, *Tuḥfat al Hind* (ca. 1675), defines "Sahāskirt" as the language in which "books on various sciences and arts are mostly composed."[22]

The general tendencies in learning and language suggested by these data are fully corroborated for a language tradition that I want to look at in a little more detail, Brajbhasha, the language that supplemented, and then effectively replaced, Sanskrit as the transregional literary code in north India during the early modern era.

The Language of Braj beyond the Literary

Brajbhasha is an important and especially good case to study for the problematic of language and science.[23] Although the history of nonliterary Old Hindi has never been written—all the important survey works entirely ignore such materials—the resources for doing so exist in abundance and are comparatively well ordered. These include the various manuscript catalogues compiled as a result of intensive searches in the early part of the twentieth century, including the three-volume manuscript catalogue published by the Nagari Pracharini Sabha that lists according to genre nearly forty-five hundred works (culled from a five-volume *Khoj* series).[24] While it is admittedly hazardous to draw large conclusions from one survey of manuscripts, however systematically prepared—let alone historical conclusions, since the majority of the manuscripts are undated—it does seem significant that upwards of *70 percent* of these are texts we would broadly classify as expressive, imaginative, literary, and religious. Of the remaining quarter, the greater part (five hundred or so) deal with practical arts: *jyotiṣ* (astrology), *śakun* (augury), *śālihotra* (veterinary science), *sāmudrikaśāstra* (physiognomy), and the like; religious practices, including works on *karmavipāk* (karma theory), *māhātmya* or *vrat* (sacred topography, religious vows), *stotra* (hymnody), *tantra, mantra, yantra,* or *indrajāl* (mystical and magical arts), and gnomic wisdom (versions of the Sanskrit classics *Hitopadeśa* and *Pañcatantra*).[25] Works that concern themselves with *darśana* (the philosophical viewpoints) are conspicuous for their rarity.[26] The only areas of growth for Brajbhasha scientific textuality in the early modern period are *āyurved* (forty-eight manuscripts) and the adjacent field of *kāmaśāstra,* or erotology (numerous examples of *Kokaśāstra* manuscripts). Once again specific exceptions tend to prove a general rule.

Brajbhasha shows a remarkable and relatively early development of a science of poetics (which is as noted strikingly absent in every other north Indian vernacular tradition). The two foundational works of Keśavdās, *Kavipriyā* and *Rasikapriyā* (ca. 1600), were preceded by a certain kind of philological interest absent elsewhere in north India (indicated by, among other texts, the *Mānmañjarī*, a thesaurus composed by Nandadās ca. 1550) and succeeded by attempts toward a more fully systematized discipline (as visible in the works of Cintāmaṇi, fl. 1650, and Bikhārīdās, fl. 1730).[27] But again grammatical analysis remains completely nonexistent. Some works of spiritual reflection were composed in Brajbhasha prose, including a *guruśiṣyasaṃvād* (teacher-student dialogue) titled *Siddāntabodh* by Jaswant Singh, king of Jodhpur (1667; what appear to be comparable texts are noted in Hindi manuscript catalogues).[28] A tradition of expository prose in the form of commentaries began with Indrajit, king of Orchha (ca. 1600), who commented on two of the *Śatakas* of Bhartṛhari; especially noteworthy are commentaries, something on the order of fifty, on the works of Keśavdās. As indicated by Indrajit, Jaswant Singh, and many others (including Rāyasiṃha, king of Bīkāner ca. 1600, to whom is attributed a Rajasthani commentary on an astronomical text, Śrīpati's *Jyotiṣaratnamālā*), courtly notables played a prominent role in the creation of a vernacular scholarly idiom.[29] This merits further scrutiny, as indeed does premodern vernacular literary commentary itself, especially from a comparative perspective. (In Kannada and Telugu, for example, virtually none exists before the modern period.)

Science did find expression in Brajbhasha, then, but in a highly restricted sense. Something of this constrained character of vernacular knowledge production is illustrated by the career of one of the more interesting seventeenth-century scholars, Kavīndrācārya Sarasvatī (ca. 1600–1675).[30] A Maharashtrian cleric, Kavīndra, according to François Bernier (and there can be little doubt that the reference is to him), was Dārā Shikoh's chief Sanskrit scholar, "one of the most celebrated pandits in all the *Indies*," and later Bernier's constant companion over a period of three years. He was a familiar at the court of the Mughal emperor Shāh Jahān, who conferred on him the title "Hoard of All Knowledge" and provided him with a rich annuity enabling him to assemble one of the most celebrated Sanskrit libraries of the day. (Many of the manuscripts, recopied expressly for Kavīndra's collection, are today to be found in the Anup Sanskrit Library Bikaner, the Library of the Maharaja of Jammu, and the Sarasvati Bhavan, Varanasi.) Kavīndra's extant work in Sanskrit consists largely of commentaries on Vedic and classical texts, but one could argue that, historically viewed, his more remarkable contribution—less for its intellectual originality than for its sociolinguistic symbolism—was to Brajbhasha.

Indeed the very fact that he wrote in Braj is remarkable. So far as I can tell—a provisional claim that sounds too extreme to be true, though it is borne out by materials currently available to me—he is the single Sanskrit scholar in the intellectually vibrant world of seventeenth-century Varanasi to have written in the vernacular.[31] But his relationship to the vernacular was conflicted. His most important work is the *Bhāṣāyogavāsiṣṭhasār* (also known as *Jñānsār*), a version of the anonymous Sanskrit *Laghuyogavāsiṣṭhasāra*, which he prepared in 1656–57. In the introduction to this text Kavīndra celebrates his learning in the Sanskrit knowledge systems: "the four Vedas and their meanings; the six *vedāṅgas*, on which he has given lectures; *nyāya, vedānta, mīmāṃsā, vaiśeṣika, sāṃkhya, pātañjala*, on which he has cleared up all doubts and confusions. He has taught *nyāya* and so on repeatedly, and written many works on *sāhitya*." Then he adds, "He lived first on the banks of the Godāvarī, and then came to live in Kāśī. He is a Ṛgvedin of the Āśvalāyana *śākhā* [school]—and he has composed the *Jñānsār* in the vernacular."[32] Kavīndra's celebrating his Sanskrit learning in the introduction to a vernacular text implies less pride in his multilingualism, as one might suppose, than condescension toward the *bhāṣā*. This is confirmed elsewhere in his oeuvre, where a clear note of unease in writing in the vernacular can be heard. He actually uses the term *lāj* (shame) in the *Kavīndrakalpalatā*, a collection of his *bhāṣākavitā*, or vernacular poetry:

> One feels ashamed to compose in the vernacular
> It was only for the sake of others that this book was written.
>
> bhāṣā karat āvati hai lāj
> *kīnai gramth parāe kāj.*[33]

Whatever we may make of this vernacular anxiety, however, what is not in doubt is that for Kavīndra, Brajbhasha was a language of poetry, not science; nothing of the vast scholarship he claimed was ever transmuted into the language, with the sole exception of the text in hand, a work, as he calls it, of "Upanishadic" wisdom comparable to the other kinds of theological poems mentioned earlier.[34]

What the case of Kavīndra and Brajbhasha more generally suggests are the clear and untranscendable limits of vernacular textualization in the early modern period. Aside from poetics, which was crucial for the constitution of the "illustrious vernaculars" as such, the central concerns of the Sanskrit thought-world—and these constitute the central concerns of science and scholarly thought of precolonial India outside the world of Persian—remained almost entirely locked in the Sanskrit language. In linguistic philosophy, hermeneutics, logic and epistemology, jurisprudence and moral reflection, or other disciplines (and the situation seems only marginally more favorable in life science and astral science), no original work whatsoever seems to have been composed in Brajbhasha. In fact not one of the standard Sanskrit works—the classical foundational text, commentary, exegesis, or exposition (*sūtra, vṛtti, bhāṣya, vārtika*), or any of the great independent (*prakaraṇa*) treatises—appears ever to have been made available in translation before the colonial period.

Sanskrit Language Ideology and the Character of Early Modern Science

The exclusion of the vernacular from the realm of scientific discourse has deep roots, I suggested earlier, in a complex language ideology. Sometimes this theory is formulated by way of a simple typology, articulated already in the prevernacular world in Bhoja's early eleventh-century treatise on literature, *Śṛṅgāraprakāśa*: "Words with unitary meaning constitute a unit of discourse [*vākyam*]. There are three species of such discourse: Sanskrit, Prakrit, and Apabhramsha. As for Sanskrit discourse, it is of three types: relating to revelation, to the seers, and to the world. . . . Discourse relating to the world has two subtypes: *kāvya* [literature] and *śāstra* [systematic thought]." The world of written discourse as a whole is here radically restricted to nonregional languages. Sanskrit occupies the domain of science, to the exclusion of all others; Prakrit and Apabhramsha, which Bhoja goes on to describe solely in sociolinguistic terms, are shown to be restricted in their usage entirely to poetry.[35] As the *Tuḥfat al Hind* again shows, this tripartite division was tenacious and remained alive more than half a millennium after Bhoja, but with this change: that Brajbhasha (Bhakha) replaces Apabhramsha as the third language of literature.[36]

More instructive than this kind of typological presentation, which carries a second-order pragmatic dimension (as if simply reporting what the world of textual production consisted of), are the philosophical arguments that have a primary force in buttressing constraints on the production of science in the vernacular. Central here is the episteme mentioned earlier that links grammatical correctness and truth, the axiom of intrinsic Sanskrit veracity—and intrinsic vernacular mendacity. But a range of other, more abstract tenets of Sanskrit language philosophy also enters into the mix. One was the old notion found in *vyākaraṇa* (language analysis and grammar) that non-Sanskrit language is able to exercise *śakti* (signifying power) only by the mediation of the original Sanskrit from which the vernacular was believed to derive and which was somehow thought (but in a way never explained) to be recognized in the process of communication. Whatever is sayable in the vernacular, this implies, has already been said, and said more clearly, in Sanskrit.

Counterarguments were raised against this position in the early modern era, such as those of the important linguistic philosopher Kauṇḍa (or Koṇḍa) Bhaṭṭa (fl. 1625), nephew of the celebrated grammarian Bhaṭṭoji Dīkṣita, in his *Bṛhadvaiyākaraṇabhūṣaṇa* (which exists in an abridged version as well, the *Sāra*), a commentary cum exposition of his uncle's *Vaiyākaraṇamatonmajjana*.[37] There are a number of important new (or newly clarified) ideas that Kauṇḍa offers; note in particular his view that it is precisely Sanskrit's cosmopolitan presence that in the eyes of previous writers endowed it alone with the capacity of the direct signification:

> [According to the "new logician," against whom the "new grammarian" Kauṇḍa is arguing,] signifying power is found only in Sanskrit. It cannot exist in vernacular words even though the putative communicative exchange in the vernacular may be identical to what is found in Sanskrit. This is so because the vernaculars vary across regions [whereas Sanskrit words are thought to be everywhere the same]. . . . However, given the absence of any decisive argument one way or the other, we must conclude that vernacular language, too, possesses the power of signifying directly. Nor would this lead to any lack of parsimony [i.e., the need to postulate multiple words—which is to say, multiple spellings of a single word—that all directly express the same meaning] since it is impossible to avoid attributing signifying power to Marathi [*mahārāṣṭrabhāṣā*] no less than Sanskrit. This is so because Marathi, too, remains self-identical in every single region. [*Sāra*: Like Sanskrit the vernacular of Maharashtra and all others are everywhere one and the same.][38] Thus, because there is no conclusive evidence for exclusion in the case of other languages, the rejection of signifying power with respect to any single one of the vernaculars is itself refuted. Indeed, even in the case of Sanskrit conclusive evidence for exclusion is absent. [*Sāra*: If by "conclusive evidence" were meant acceptance by the learned everywhere that a given form is correct, then even in the case of Sanskrit there might be incorrect words, since the word *śava* is used as a verb of motion among the Kambojas (i.e., in part of today's Afghanistan), and as a noun meaning corpse in Āryāvarta (Domain of the Aryans, i.e., India), according to the *Mahābhāṣya* (Great Commentary on Pāṇini's Grammar).]

One might argue further that it is not the fact of a language's being Sanskrit or a vernacular that determines whether or not it has signifying power, but rather its orthographic stability, which [in the case of the vernaculars] is everywhere variable. But this would hardly differ from the case of Sanskrit synonyms: *ghaṭa* and *kalaśa* [are spelled differently but mean the exact same thing, "pot"]. Given this, the one [Sanskrit] cannot entail that signifying power in the other [the vernacular] is a false attribution. [*Sāra*: Moreover, even if one were to agree that the vernaculars are marked by variation and argue that it is orthography that defines a word as such, one could reply as follows: The variable orthography in the vernacular is like the variability in Sanskrit with respect to synonyms (that is, various spellings of a single vernacular word all mean the same thing, just as various spellings in Sanskrit in the case of synonyms all mean the same thing); what is the difference between the two that allows us to count the latter as correct and the former as incorrect?] It is precisely because non-Sanskrit language can have signifying power that the *Kāvyaprakāśa* (Light on Literature) quotes a *Prakrit* verse to illustrate a case of aesthetic implication of the expressed meaning.[39]

It is this radically modernist position represented by Kauṇḍa Bhaṭṭa that came under attack from the widely influential Varanasi intellectual Kamalākara Bhaṭṭa (also, let us note, a Maharashtrian Brahman writer). Kamalākara reiterates the old position in *mīmāṃsā* language philosophy (though tinged in fact with *navyanyāya*) when arguing that the very capacity of vernacular language to produce meaning is a pure illusion, since authentic meaning presupposes language that does not change—that is, Sanskrit:

> The new intellectuals [*navya*] hold that [inherently expressive] words and sentences must exist in dialect, that is, in vernacular-language texts, as well as in [newly coined] technical terms and proper names, because these actually do communicate verbal knowledge. These thinkers, however, fail to grasp the logic in the argument that "a multiplicity of equally expressive speech forms cannot be logically posited" [PMS 1.3.26]. Nor do they understand that, by thereby rendering grammar itself irrelevant and accepting as valid words and meanings in use among the *mlecchas* [the uncivilized, those who stand outside of Sanskrit culture], they are destroying the Veda. There cannot exist in dialectal words such as *gāvya* [instead of Skt. *gauḥ*, "cow"] the expressive power conferred by divine will, because these dialectal words have no stable form [whereas the words stamped by God's will, i.e., Sanskrit words, are invariable]. . . . In short, [if one accepts direct signification in vernacular words] one would have to attribute the power of

> signification to the sounds of seashells and bells. By the same token, the vernacular can be said to possess real words only in one of two ways: either by the illusion of their being expressive in themselves, or through the presence of the grammatically correct Sanskrit words that they imply. Words are actually changeless and eternal, because the phonemes of which they are composed are such.[40]

Another, related axiom is the *mīmāṃsā* postulate of the natural and uncreated (*autpattika*) connection of signifier and signified, along with its theory of reference, whereby all substantives are believed to refer to class properties (*ākṛti*), or indeed universals (*jāti*), and not individuals (*vyakti*), which they connote only secondarily, and each signified is believed to have only one signifier.[41] We cannot scrutinize these theorems here, but what they imply for vernacular knowledge should be obvious: in a world of nonarbitrary and singular language it is impossible for any language but Sanskrit to make scientific or other sense; non-Sanskrit languages would not be referring to the universally real since they would be using false words, and if they were using real words (what are called *tatsamas*, or vernacular words identical to Sanskrit) they would be completely redundant.

Other old but still functioning components of Sanskrit language ideology persisted; these may have been bent in the early modern period, but they were not broken. Consider first the discussion of the well-known *pikanemādhikaraṇa* by Khaṇḍadeva in his remarkable comprehensive treatise on *mīmāṃsā*, the *Mīmāṃsākaustubha*.[42] The larger context of this topic (the *smṛtipāda*, or Section on the Authority of Tradition), to characterize it generally, is the grounds for the authority claimed by various Sanskrit knowledge systems per se. The specific question at issue in the topic concerns the words *pika* and *nema*, non-Sanskrit words present (or held to be present) in Vedic texts and yet having no currency among *ārya*s themselves, but only among *mleccha*s: Are the latter competent to explain the meaning of their own language, or must the signification of such words be determined by the application of Sanskrit knowledge techniques, especially etymology?[43] To be sure, Khaṇḍadeva accepts the *mīmāṃsā* tenet: the communicative practices of the *mleccha*s can be shown to be beginningless, for words such as *pika* and *nema* cannot be proven to be corrupted either phonologically or semantically (unlike other lexemes, such as *pīlu*, that are current among both *ārya*s and *mleccha*s but in radically different senses, and where, therefore, the suspicion of corruption among the latter cannot be removed).[44] "This leads us to assume that their linguistic usages do express meaning. Accordingly, their practices, too, [no less than those of the *āryas*,] should be authoritative in determining the signification of words."

It is to Khaṇḍadeva's *pūrvapakṣa*, or prima facie argument, however, that I call special attention. *Mīmāṃsā* is celebrated among pandits for avoiding the straw man and mounting the strongest arguments possible against its own tenets (since, as Bhoja says [ŚP 742.3], the stronger his adversary, the more ennobled becomes the victor). There is little reason to doubt that the following position as formulated by Khaṇḍadeva, constructed only to be rejected though it may be, seemed entirely reasonable in the seventeenth-century Sanskrit thought-world:

> Lacking education [*abhiyoga*] the *mlecchas* are observed to corrupt [*viplu-*] language by using incorrect [*asādhu*] speech items, and so they have no competence to determine the real phonetics of words [*śabdatattvāvadhāraṇa*]. By the same token, neither have they competence to determine their semantics [*tadarthāvadhāraṇa*], because of their mistaken use of words like *pīlu* and so on. One cannot argue that since we do not find any corruption in words such as *pika* that it should be possible to accept the meaning attributed to them by *mlecchas*. For those words, too, are in fact phonologically corrupted [*apabhraṣṭa*], insofar as only the stems [and not the full inflections] are used. What the *mlecchas* are therefore employing are words *similar to* the Sanskrit words used in the Veda, not those very same Vedic words themselves. And we cannot, on the basis of mere similarity, conjecture the meaning of the words *pika* and so on [as found in] Sanskrit texts from the meaning of the words known to *mlecchas*. Were one to base oneself on mere similarity, one could wind up assuming that, for example, the word *śālā* [room] expresses the same meaning as *mālā* [garland]. In his *Tantravārtika* Kumārila considered at length the difficulties of trying to conjecture, by means of similarity or the interpolation of additional phonemes, the Sanskrit words that lie at the origin of words used in the Āndhra and Draviḍa languages and thus their capacity to signify what the original Sanskrit words signify. He showed accordingly how just for those two languages it is impossible to determine the words and meanings in any systematic way.[45] This is a fortiori the case with respect to languages of those even more remote than the Āndhras and the Draviḍas, such as the Pārasi [Persians] and the Romakas ["people of Rome," i.e., Constantinople or Istanbul? Or the French, or the Portuguese?]. Accordingly, the knowledge of *mlecchas* has as little authority in the determination of linguistic meaning as it does in the determination of *dharma* and *adharma*.[46]

What is perhaps most remarkable here, amid the many older arguments, is the fact that the question of whether Persians and Europeans were competent to understand their own languages was still being seriously discussed in the 1660s.

Elsewhere in his work, too, what Khaṇḍadeva chooses to recover from early discussions suggests that his general attitude toward language and sociality retains many traces of the archaic. Here is one example:

> The following objection has been raised: It may be granted that the [beginningless] communicative practice of their ancestors is authoritative for the *mleccha*s [which would validate their own linguistic competence], but since they are disallowed from hearing the language of the Veda, and *ārya*s are prohibited from speaking with them or learning their speech, there is no possibility for *ārya*s to come to know the meanings familiar to the *mlecchas*. But this objection has no force. *Mleccha*s might have learned Sanskrit from bilingual *ārya*s [*dvaibhāṣika*] who violated the prohibition, and these *mleccha*s might have taught to *ārya*s the meanings of words known only to them. Thus there is no insurmountable obstacle in the *ārya*s' acquiring the requisite linguistic knowledge.[47]

On matters of true knowledge, communication outside the domain of Sanskrit was clearly still viewed as transgressive and exceptional in the *imaginaire* of Sanskrit scholarship. As far as the vernacular in particular is concerned, Khaṇḍadeva does acknowledge a communicative space for it, but it is tellingly narrow. When considering the injunction noted earlier to employ only correct Sanskrit (*sādhūn evābhibhāṣeta,* "One should use only grammatically correct words"), he argues, in what appears to be an open-minded way, that the rule has reference only to the domain of sacrificial activity; it does not constitute a general moral principle and thus does not militate against use of the vernacular—that is, the degenerated (*apabhraṣṭa*) Sanskrit words thought to be the source of the vernaculars—in other contexts: "For these degenerated Sanskrit words are used by learned men of all regions [*sakaladeśīyāḥ śiṣṭāḥ*] in their everyday activities as well as in chanting the name and virtues of God [*hari*]." His general conclusion is that there is no primary human end (*puruṣārtha*) attaching to the prohibition on ungrammaticality (or dialectism, or vernacularity, *asādhubhāṣaṇa*): "While ungrammaticality can impair a sacrifice it cannot impair other Vedic activity nor pose a direct threat to human welfare [*puruṣasya pratyavāya*]." This would seem to open the door to a wide range of vernacular practices, but it is surely significant that Khaṇḍadeva restricts this to *vyavahārakāla* and *saṃkīrtana*, the pragmatic and devotional, activities outside the realm of science, learning, and scholarship. In general his position on language is as inflexible as other *mīmāṃsakas* of his day, such as Dinakara Bhaṭṭa, with whom Khaṇḍadeva directly agrees on the question of Persian when he states:

> However, there does indeed exist a prohibition of a general moral scope [*puruṣārtha*; rather than one restricted to ritual, *kratvartha*] applying to words of barbarian [*bārbara*] and other languages, since there is a scriptural prohibition against learning them at all: "One should not learn a *mleccha* language [*na mlecchabhāṣāṃ śikṣeta*]." With regard to this statement there are no grounds such as primary context [as there is in the case of another scriptural prohibition, "One is not to barbarize" (*na mlecchitavai*)] for setting aside the conventional meaning of the word *mleccha* [which he elsewhere identifies as Pārasika and Romaka] [and interpreting the word as referring more narrowly to ungrammatical Sanskrit]. Thus the prohibition on barbarian and other languages only is purely of a general moral sort, whereas the prohibition on other language [i.e., *apabhraṣṭa* Sanskrit, as expressed in *na mlecchitavai*] relates to sacrificial activity and that only.[48]

The actual degree of Sanskrit-Persian intercommunication in the period 1550–1750, like so many other questions raised here, awaits systematic study.[49] We do know that, whereas intellectual intercourse among astronomers may have been relatively relaxed and some scholars, such as the Jain Siddhicandra, celebrated their skills in Persian (*yāvanībhāṣā*), other sources substantiate Khaṇḍadeva on the resistance among most Sanskrit intellectuals (Jains aside) to the use of Persian.[50] Among Kashmiri Brahmans there even emerged a new caste division between the *kārkun* (bureaucrats) who learned Persian and entered the service of the sultans, and the *bhāṣbhaṭas* (language scholars) who maintained a Sanskrit cultural identity. In the description of Maharashtra in the contemporaneous *Viśvaguṇādarśacampū* of Veṅkaṭādhvarin, scorn is heaped on those who, at the time of life they should be practicing Vedic recitation, do nothing but learn Persian. But also derided are those (Tengalai Śrīvaiṣṇavas are intended, though Kavīndra might just as well have been included) "who senselessly bother with vernacular texts [*bhāṣāprabandha*] when the Veda, source of all human values, is at hand. You don't run off to a cowherd's hut for a glass of milk when standing on the shore of the milk ocean."[51]

To be sure, at precisely the same moment others appeared to be speaking in favor of a *bhāṣā* competence even on the most transcendent plane. Nīlakaṇṭha Caturdhara, for example, the celebrated editor of and commentator on the

Mahābhārata, argued in his *Śivatāṇḍavatantraṭīkā* not only that tantric texts should be numbered among the fourteen knowledge sources and so be adjudged Vedic in origin and hence true knowledge, but that the power of their mantras even when composed in the vernacular (*bhāṣā*) was undiminished:

> Their actual sequence of phonemes may not be Vedic, but their meanings are Vedic, and it is precisely this that gives them their efficacy. And it is perfectly possible that Vyāsa, Śabara,[52] and others were able to set out the meaning of Vedic texts in vernacular as well as in Sanskrit language, and to compose texts through the power of their asceticism. The sequence of phonemes arranged by them could have likewise the entire efficacy of [Vedic] mantras. Therefore, the Vedic origins of . . . the vernacular mantras is proved beyond doubt. It is precisely as a result of the differences [from Vedic mantras] in the sequence of their phonemes that both higher and lower castes, as appropriate, have the right to pronounce the phonemes.[53]

Yet there is an archaic exception to this, as it were, modernist innovation that is almost too obvious for comment: for knowledge to be true it must have Vedic affiliation, and even to claim vernacular truth meant to set forth the claim, as Nīlakaṇṭha of course does here, in Sanskrit.

The Case of Europe

I noted at the beginning of this essay the remarkable asymmetry between literary and scientific vernacularization in India and Europe. It is especially the parallel in literary language change and the linkage often assumed between the development of scientific and literary discourse that make the apparent resistance to scientific vernacularization in India so puzzling. I have written about literary vernacularization elsewhere and need state here only that the commonalities, conceptual, social, and chronological, in the emergence of the vernaculars in the two regions are astonishing.[54] As for the vernacularization of scientific knowledge in western Europe, this commenced in the natural sciences in the mid-sixteenth century with Peletier writing in French on algebra (1554) and had gained powerful momentum by the time Galileo published his *Discorsi* in Italian (1638); in philosophy Bacon's *The Advancement of Learning* (1605) and Descartes's *Discours de la méthode* (1637) are among the most important early works.[55] Latin long retained its appeal, of course. Scientists from Copernicus and Kepler to Newton and Gauss continued to use the language (though philosophers had abandoned it entirely by the time of Kant) because of its supposed universality, stability, prestige, and demonstrated communicative capacity. But the trend toward science in the demotic idiom was irreversible.

Sometimes the choice of the vernacular was not in fact a choice but a matter of practical necessity; Peletier is said to have used French simply because he was ignorant of Latin. Sometimes the use of the vernacular was an attempt to achieve a certain new kind of diffusion of a national-popular sort, a goal pursued, it seems, by Descartes with his *Discours,* despite the substantial conceptual challenge of presenting a discourse on universal reason in a nonuniversal language.[56] The role of the new academies (the Académie française was established two years before the *Discours* was published, virtually the moment, half a world away, when Kamalākara Bhaṭṭa was arguing out the essential incoherence of the vernacular), and more largely, of the knowledge initiatives of the nascent nation-state, are pertinent factors here too; note that with Bacon science itself became a state enterprise.[57] Other motives for the vernacularization of science, as conceived by the agents themselves, include the confirmation by language choice of the idea of *translatio studiorum et imperii*; popular disclosure of useful information hitherto kept secret; the education of women and aristocratic officials. Pertinent also are the arguments, ever more forcefully made, that favored the supposed natural language, especially its facility and putative transparence, over the artificial classical, something already evident in Dante, who proclaimed in 1300 what no one in Europe had ever proclaimed before: *nobilior est vulgaris,* "More noble [than Latin] is the vernacular."[58]

Several hard questions are raised by thinking through the cases of Europe and India together. With respect to the vernacularization of literature as a cultural and political process, similar developments occurred more or less simultaneously in both Europe and India to produce, each autonomously, its own brand of modernity, on the one hand national, on the other, for want of a better term, *deshi.* But the vernacularization of scientific discourse never happened in precolonial India, certainly not for most of the core disciplines of the dominant intellectual order, and this needs to be explained.[59]

One's first impulse is to interpret this commitment to Sanskrit as obscurantism or blind traditionalism, a practical enactment of Sanskrit's archaic language ideology—in short, as failure. To be sure few of the factors identified for European scientific vernacularization were present in early modern South Asia. Sanskrit competence among intellectuals never deteriorated to the degree that made writing in the vernacular unavoidable. No national-popular projects, let alone institutions, that instrumentalized and rationalized cultural practices were ever developed. No polity ever sought to draw on culture to make its language the *compañera del imperio*. But these are again absences; is there a more positive interpretation?

Here I am put in mind of a remark made by the historical sociologist Shmuel Eisenstadt regarding an old text of Werner Sombart's, *Why Is There No Socialism in the United States?* For Eisenstadt it is just as reasonable or even more so to ask, instead, Why was there socialism in Europe? Similarly we might want to turn the tables of our assumptions and ask, not why India failed to vernacularize science but why Europe did, and conversely what intellectuals in South Asia sought to achieve by their choice to remain transregional. I have elsewhere sought to make sense of the continuing commitment to Sanskrit on the part of late precolonial intellectuals as an attempt to reinvigorate and sustain an old ecumenical cultural order in a changing world where a middle-class, national-cultural regime was not yet a historical possibility.[60] Perhaps, in accordance with the Eisenstadt principle, we ought to proceed even further against the obvious grain. Not only is it the case that few of the factors present in early modern Europe are relevant to India, but deeper or wiser promptings may also have been in play. If, unlike literature, systematic knowledge in general and science in particular are not idiographic (let alone ethnographic) but nomothetic, then the cultural nationalization of science and scientific language in early modern Europe turns out actually to have been a curious experiment—and indeed it has largely now been abandoned.[61] Modern supranational communication forms, whether transnational English or the abstract language of mathematics, constitute a Latin *redivivus*, and we now think of "German chemistry" or "French mathematics" not as science but as chapters in the history of science. Might therefore a conceptual "provincialization of Europe," as Dipesh Chakrabarty puts it, permit us to think of the Sanskrit domination of science as a good universalism, and thus not as a failure according to the norms of European modernity but, according to an Indian ethos, as a kind of civilizational achievement?

Abbreviations

AMKV	*Adhvaramīmāṃsākutūhalavṛtti*
BhYVS	*Bhāṣāyogavāsiṣṭhasāra*
BhD	*Bhāṭṭadinakara*
MK	*Mīmāṃsākaustubha*
NS	*Nyāyasudhā*

PMS	*Pūrvamīmāṃsāsūtra*
ŚD	*Śāstradīpikā*
ŚP	*Śṛṅgāraprakāśa*
TV	*Tantravārtika*
VGĀC	*Viśvaguṇādarśacampū*
VMP	*Vīramitrodaya Paribhāṣāprakāśa*

Notes

1. Pollock, *The Language of the Gods in the World of Men*, 39–74.

2. These two concerns are well described for Europe by Chartier and Corsi, *Sciences et langues en Europe*, 12.

3. See the learned and challenging account in Staal, "The Sanskrit of Science." The true question in the history of Indian science for Staal is not why it never vernacularized (an issue not in fact raised at all), but why India failed to invent an artificial language for science, except in linguistics, to which it remained confined. The issue of language medium is raised here (as it is elsewhere) only in passing.

4. Further information may be obtained at the project's website, http://www.columbia.edu/itc/mealac/pollock/sks. In addition to the materials available there, "Working Papers on Sanskrit Knowledge Systems on the Eve of Colonialism I" was published in the *Journal of Indian Philosophy* 30, no. 5 (2002); "Working Papers . . . II" in 33, no. 1 (2005), and "Theory and Method in Indian Intellectual History" in 36, nos. 5–6 (2008).

5. With respect to the latter, in addition to the papers by Y. Bronner and myself on the project website, see the summaries of the presentations of Bronner, M. Deshpande, L. McCrea, and C. Minkowski in American Oriental Society, Abstracts of Communications.

6. A preliminary synthesis is offered in Pollock, "Ends of Man at the End of Premodernity."

7. For an account of an experimental seminar, see Pollock, "Comparative Intellectual Histories of the Early Modern World."

8. See, respectively, Cunningham and Williams, "De-centering the 'Big Picture,'" 420 n., and Chartier and Corsi, *Sciences et langues en Europe*.

9. *Na vedaśāstrād anyat tu kiṃcic chāstraṃ hi vidyate | niḥsṛtaṃ sarvaśāstraṃ tu vedaśāstrāt sanātanāt* (cited in VMP, 20). For the *vidyāsthānas*, see Pollock, "The Idea of *Śāstra* in Traditional India." An influential enumeration is found in the *Viṣṇupurāṇa* (perhaps fifth century): "The four Vedas, the [six] *vedāṅgas* (language analysis, phonetics, etymology, metrics, astral science, ritual science), *mīmāṃsā, nyāya, purāṇa, dharmaśāstra* are the fourteen sciences. These number eighteen with the addition of *āyurveda, dhanurveda, gāndharva*, and *arthaśāstra*" (*Viṣṇupurāṇa* 3.6.28–29, ed. Rajendra Nath Sarma, Delhi, 1985).

10. TV 164, 1.3.12, lines 9–15 (slightly rearranging the verse and the prose that glosses it); 166, line 2; compare NS, 236.10. Kumārila curiously ignores the fact that the Buddhists had turned to Sanskrit for both scriptural and scholarly purposes some four to five centuries before his time.

11. BhD fol. 41v, lines 1–2 (see also ŚD, 47: *sādhūn śabdān satyaparyāyān*, "'correct language,' i.e., truthful"). Injunctions such as *nāsādhu vadet* ("One should not speak ungrammatically"), *sādhubhir bhāṣeta* ("One should use grammatical speech"), *na brāhmaṇena mlecchitavai* ("A Brahman must not barbarize his speech"), and *na mlecchabhāṣāṃ śikṣeta* ("One should not learn a *mleccha*'s language") are often discussed together, as in the TV and NS on the *vyākaraṇādhikaraṇa* (PMS 1.3.24–29). As we see below, early modern *mīmāṃsāka*s like Khaṇḍadeva discriminate among different realms of application of these *vidhi*s.

12. A consideration of the place of *navyanyāya* terminology in the early modern *mīmāṃsā* is offered in McCrea, "Novelty of Form and Novelty of Substance in Seventeenth-century Mīmāṃsā," and a useful general account is in Staal, "The Sanskrit of Science," 79–88. For the ridicule the *navyanyāya* style earned in some quarters of the seventeenth-century intelligentsia, see *Viśvaguṇādarśacampū* v. 555bc: *paraṃ vāco vaśyān katipayapadaughān vidadhataḥ / sabhāyāṃ vācāṭāḥ śrutikaṭu raṭanto ghaṭapaṭān*, "The [logicians] use a few terms—but use them in a flood—that are entirely dependent on language itself [i.e., metalinguistic?], and in the halls of debate stridently bang their pots and flap their cloths" (the two items adduced as standard examples in *navyanyāya* syllogisms; my thanks to A. Wezler for correcting an earlier oversight of mine).

13. On the former, see Pollock, "India in the Vernacular Millennium"; the latter is examined in detail in Pollock, *The Language of the Gods in the World of Men*, 283–329.

14. This is so for several reasons. The work itself is multilayered; the date of the literature it refers to is itself undetermined; the date of a grammar need not be contemporaneous with the language it describes, as Akalaṅka Deva's work shows; its commentaries do not appear until the twelfth century. See Swamy, "The Date of the Tolkāppiam"; Takahashi, *Tamil Love Poetry and Poetics*, 15–29.

15. On the dating of the *Āndhraśabdacintāmaṇi*, see Rao, "Multiple Literary Cultures in Telugu." The history of Malayalam stands apart; see Freeman, "Rubies and Coral."

16. The exception is a brief account of Marathi morphology in the *Pañcavārtik* of Bhīṣmācārya sometime in the fourteenth century. On the vernacular glossators, the old essay by Printz, "Bhāṣā-Wörter in Nīlakaṇṭha's Bhāratabhāvadīpa usw," remains useful. The north-south difference in grammaticization is discussed in Pollock, *The Language of the Gods in the World of Men*, chapter 9.

17. See Busch, this volume.

18. This work is complemented by what appears to be one of the earliest vernacular texts on polity, the *Beddanīti* (perhaps as early as the fourteenth century; see Wagoner, "*Iqtā* and *Nāyaṃkara*"), but except in literary texts the tradition of vernacular political thought seems not to have been continued.

19. We thus find one Vyāsa Keśavarāma composing a bilingual Gujarati-Sanskrit

medical glossary, while Mahārāja Pratāpasiṃha of Jaipur wrote in Marwari and then translated his own work into Sanskrit verse and Hindi prose (Dominic Wujastyk, personal communication); for the situation in Brajbhasha, see below. In the Siddha tradition of Tamil Nadu, oral transmission was the rule.

20. On the *Vivekasindhu*, see Tulpule, *Classical Marāṭhī Literature*, 316.

21. To be sure in virtually every case in South (and Southeast) Asia the inaugural use of the vernaculars was entirely pragmatic—in the business end of inscriptions—and such usage did leave later textual traces in some regional traditions. See Pollock, *The Language of the Gods in the World of Men*, especially 121.

22. Literally, "the language in which composition is done of *a'lum* and *fanun*" (sciences and painting, music, crafts, etc., i.e., *kalā*). See Ziauddin, *A Grammar of the Braj Bhākhā*, 34 (53 of the Persian text).

23. I owe a number of references in this section to Allison Busch and profited greatly from discussions with her on the issues raised here.

24. Pandey, *Hastalikhit Hindī Granthasūcī*. These findings are largely confirmed by the two-volume manuscript catalogue of Varma et al.'s *Hastalikhit Hindī Granthoṁ kī Vivarṇātmak Sūcī*. No works at all in the *bhāṣā* are listed for *vyākaraṇa, mīmāṃsā, nyāya* (with the exception of two recent *ṭīkās* on the last), or any other philosophical system save *pātañjalayoga* (two or three manuscripts); *āyurveda* and *jyotiḥśāstra* are more substantially represented, but their numbers remain small.

25. There is also listed a *Rājanīticandrikā* (vol. 3.3420, 3421), but I have been unable to examine the manuscript. Note also the *Devīdāsa kṛta Rājanīti* and *Nathurāma kṛta Rājanīti*, two works on the syllabus of the Brajbhāṣā Pāṭhaśālā discussed by Mallison (this volume).

26. Only *vaidika* works are found: *Caturvedasatśāstramata* of one Balirām "Bali" (vol. 1.30, unpublished); Sundaradās's *Jñānsamudra* (*Advaitasiddhāntanirūpaṇ*) (verse, often printed); the anonymous *Bodhadarpaṇ* (an exegesis of the *Puruṣasūkta*) (vol. 1.42); *Vedāntaratnamañjuṣā* of one Puruṣottamācārya (1.52); *Sāṃkhyaśāstra*, anon. (1.56) (all unpublished).

27. See Busch, *Poetry of Kings*, chapter 3. As she notes, it is a measure of the underdevelopment of our knowledge that several texts of Cintāmaṇi, the most important Brajbhasha poetician of the seventeenth century, remain unpublished or virtually inaccessible.

28. The *Siddāntabodh* is available in *Jasvant Siṃh Granthāvalī*, edited by Mishra. (For other, comparable texts, see Varma et al., *Hastalikhit Hindī Granthoṁ kī Vivarṇātmak Sūcī*.) The fact that, in the case of another work of the king's, the *Ānandavilāsa*, a Sanskrit translation was prepared contemporaneously (32) raises in a pointed way questions about language, communication, and intellectual community of the epoch about which at present we know next to nothing.

29. For Indrajit, see McGregor, "The Progress of Hindi," and for the full exposition, McGregor, *The Language of Indrajit of Orcha*; for Keśavdās and his commentators, Busch, *Poetry of Kings*, chapter 3; for Rāyasiṁha, Pingree, *From Astral Omens to Astrology*, 93.

30. Details are in Pollock, "The Death of Sanskrit," 407–8; see also Pollock, "New Intellectuals in Seventeenth-century India," 20–21.

31. A collection of Vaiṣṇava *bhajans* titled *Kīrtanapraṇālīpadasaṃgraha* is ascribed to a Jagannātha, and a "Jagannātha Kavirai" is mentioned as a composer of *dhrupads* in the late seventeenth-century *Anūpasangītaratnākara* (Delvoye, "Les chants *dhrupad*," 169). (The *Kīrtanapraṇālīpadasaṃgraha* exists in a single unpublished manuscript, once in the temple library in Kankroli and now reportedly in Baroda and inaccessible to scholars.)

32. BhYVS vv. 3–4.

33. Divākar, *Kavīndracandrikā*, 34, citing the *Kavīndrakalpalatā*. In the citation from the *Samarasāra* (an unpublished work on astral science; in Divākar, *Kavīndracandrikā*, 34), *samarasāra bhāṣā racyo, chamiyo budh aparādh*, we may have instead merely the conventional apologia.

34. Note too that among the more than two thousand manuscripts in his library only two or three, on *vaidya*, are in the vernacular (see Sastry, *Kavīndrācārya's List*).

35. ŚP 165. The Jain canon, in Prakrit, was obviously not considered *śāstra* by Bhoja; Prakrit was rarely used by Jains (or anyone else) for scholarly purposes after the second or third century. (A work like the *Mahārthamañjarī* of Maheśvarānanda from twelfth-century Madurai, which uses Maharashtri Prakrit for its *kārikās*, is a self-acknowledged anomaly.) Apabhramsha figures occasionally in tantric philosophical texts but typically only for *saṃgrahaślokas* (in, e.g., Abhinavagupta's *Tantrasāra*).

36. "The people of India have a number of languages, but those in which books and poetical works may be composed . . . are three," and he goes on to list Sanskrit, Prakrit, and Brajbhasha (Ziauddin, *A Grammar of the Braj Bhākhā by Mīrzā Khān*, 34).

37. Some of the following discussion is adapted from Pollock, "New Intellectuals in Seventeenth-century India," 27–29. Even while defending the autonomous expressivity of Marathi, Kauṇḍa wrote not a single line in the language.

38. The *Prabhā* commentary adds: "That is, all vernaculars produce meaning in one form only. None of them varies across regions, for when it does become truly transformed, it turns into another language."

39. Which demonstrates that other important authorities hold *śakti* to exist in *bhāṣā* (*Bṛhadvaiyākaraṇabhūṣaṇa* 218, 220; *Vaiyākaraṇabhūṣaṇasāra*, 341–42 = Benares Sanskrit Series edition, 248–49: "Thus, because there is no conclusive evidence for exclusion in the case of [lit., with] other languages"—that is, because just as in the case of Sanskrit, so in the case of the vernaculars the learned use one and the same form everywhere—"the rejection of signifying power with respect to any single one [of the vernaculars] is itself refuted" (*bhāṣāntarair vinigamanāvirahān naikatra śaktir iti parāstam*). The Prakrit citation from the *Kāvyaprakāśa*, an important early twelfth-century text on poetics, is chapter 2, v. 6 (*māe gharovaaraṇam*). This notwithstanding, it is more likely that Kauṇḍa Bhaṭṭa held Māhārāṣṭrī and Marathi to be related, rather than that he meant by *mahārāṣṭrabhāṣā* the Prakrit (which for a thousand years had been called *māhārāṣṭrī*). The presence of signifying power in non-Sanskrit is asserted

by another *navya* grammarian of the preceding generation, Annam Bhaṭṭa (according to his subcommentary on the *Mahābhāṣya* cited in Coward and Kunjuni Raja, *The Philosophy of the Grammarians*, 237).

40. *Mīmāṃsākutūhalam* 77 ("dialect," *apabhraṁśa*; "vernacular-language texts," *bhāṣāprabandha*; "proper names and technical terms," *saṅketaśabda*). By and large this is the dominant position across disciplines, from logic (see Mahādeva, *Nyāyakaustubha Śabdapariccheda*, 549) to literary criticism (*Alaṅkārakaustubha* of Kavikarṇapūra [ca. 1600, Navadvip], 30–31). Supporters of the new grammarians seem few and far between, though consider the following comments of Gāgā Bhaṭṭa in his commentary on the *Candrāloka* v. 4. With regard to Mammaṭa's by then canonical definition of poetry as *nirdoṣā . . . vāk*, or "faultless language," Gāgā remarks, "Some people hold that, even though faultless usage is absent from vernacular verse and the like (*bhāṣāślokādau*) given the presence there of phonological and morphological solecisms and so on (*cyutasaṃskṛtitva-*), people still apply the word 'literature' to it, and accordingly 'faultlessness' should be taken not as a defining property (*viśeṣaṇa*) of literature but as a secondary property (*upalakṣaṇa*)."

41. See PMS 1.3.26 (*anyāyaś cānekaśabdatvam*). It was precisely a proposition in European scholasticism comparable to the *autpattikasaṃbandha* that Descartes, the first great French philosophical vernacularizer, challenged with his proto-Saussurean declaration in *Le Monde*, "Les paroles, n'ayant aucune resemblance avec les choses qu'elles signifient" (quoted in Chartier and Corsi, *Sciences et langues en Europe*, 109).

42. MK 79–84, lines 1–16 (the *Kaustubha* was evidently prized by Kavīndra as well, who acquired a copy for inclusion in his library; see *Kavīndra's List* no. 368); see PMS 1.3.10.

43. The words in question, which are said to mean "cuckoo" and "half," respectively, are non-Indo-Aryan, perhaps Munda, though the argument could be and has been extended to non-Sanskrit as such.

44. Kumārila had argued that, with respect to a word like *pīlu* (meaning a type of tree in Sanskrit and elephant or ivory staff in some indeterminate but almost certainly non-Dravidian language), *ārya* usage, based on learning, is primary and authoritative, and *mleccha* usage is secondary and erroneous (TV 143–44, 1.3.9). Khaṇḍadeva addresses the question on 58–59 and concurs with Kumārila.

45. Kumārila's rather convoluted discussion of Draviḍa and other non-Sanskrit languages is found in TV 150–51. The *pūrvapakṣa* seems to claim that Dravidian dialectal pronunciations (*apabhāṣaṇa*) are mere copies (*pratirūpa*) of Sanskrit words, used with different (i.e., erroneous) meanings; if *āryas* were to try to restore the Sanskrit for such words, to make them accord with meanings current among Tamil users—if for instance Tamil *pā[m]p[u]* (snake) were to be derived from Sanskrit *pāpa* (evil) because snakes are wicked (lines 24–25)—such a procedure would consist of entirely arbitrary conjecture (*svacchandakalpanā*). The meaning of the putative original Sanskrit word can therefore be truly determined only on the basis of etymology. In his conclusion, as I read him, Kumārila demurs: "The corruptions in the vernaculars are so deep

that it is impossible to distinguish" the correct Sanskrit words and meanings from which they derive (*deśabhāṣāpabhraṃśapadāni hi viplutibhūyiṣṭhāni na śakyante vivektum*, TV 151, line 23). Note that Kumārila also refers to "Pārasika, Bārbara, Yavana, and Raumaka [*sic*] languages"; the seventeenth-century understanding of these terms, however, is likely to have been quite different.

46. MK 79, lines 15–80, line 3 (*pūrvapakṣa*). The *siddhānta* is found on 82, lines 10–23. As late as the early decades of the eighteenth century the south Indian *mīmāṃsāka* Vāsudeva Dīkṣita felt it necessary to exclude from the domain of solecism (largely *tadbhavas*) such Tamil words as *ayyā* and *appā*. These are not to be considered *asādhu* because they do not "share a similar form" with a correct word. *Tadbhavas* are produced by a failure to generate the correct Sanskrit form, and they convey meaning only by prompting recollection of that form, to which they bear a resemblance (incorrect *gāvi* leading to correct *gauḥ*). *Appā* and the like, however, are simply "a separate species" (*vijātīya*) of words (AMKV 1.3.24).

47. MK 82, lines 4–9 (see also TV 152, lines 5–6).

48. MK 132, lines 14–18 (discussed further in Pollock, "The *Bhāṭṭadinakara* of Dinakara Bhaṭṭa (1.3)").

49. Audrey Truschke's forthcoming dissertation (Columbia University), "Cosmopolitan Encounters: Sanskrit and Persian at the Mughal Court," promises to provide the first detailed account of the question for the early Mughal era.

50. On the astronomers, see Minkowski, "Astronomers and Their Reasons"; on Siddhicandra, see Pollock, "The Death of Sanskrit," 406.

51. See Kachru, *Kashmiri Literature*, 25, n.4; VGĀC 230, vv. 134. See also v. 89, where Brahmans of Kāśī who consort with Yavanas (Muslims) are criticized (see also vv. 96, 97).

52. The former is the compiler of the *Mahābhārata* and the *purāṇas*, the latter the author of a celebrated commentary on the PMS.

53. *Śivatāṇḍavatantraṭīkā* 2v–3r. I thank Christopher Minkowski of Oxford University, who provided me with his transcription of a manuscript of this work in his possession.

54. Pollock, "Cosmopolitan and Vernacular in History."

55. Note however that there were large-scale translation programs since the late Middle Ages. Nicole Oresme's French translation of Aristotle's *Ethics* of 1370 was the first complete version of an authentic Aristotelian work in any modern language. This of course intensified over the centuries: Tesauro's *La filosofia morale* (1670) saw twenty-seven editions over the course of the following century and translation into other vernaculars—as well as Latin (*The Cambridge History of Seventeenth-Century Philosophy*, 1282). Even earlier is Gossouin of Metz's *Image du Monde* (Lorraine, 1246), probably the oldest encyclopedic treatise written in a European vernacular. Such initiatives are entirely absent in the Indian context.

56. The issue is raised and explored in Derrida, "Languages and Institutions of Philosophy."

57. Gaukroger, *Francis Bacon and the Transformation of Early-modern Philosophy,* 160.

58. *De vulgari eloquentia* 1.4. See also the essay by Pantin, "Latin et langues vernaculaires," from whom I adopt a number of ideas in this paragraph. As she points out, there was no clear and invariant line of progression (for example, most of Galileo's students reverted to Latin), and no good explanations are available to account for this indirect route of the vernacular's eventual conquest. Even as French, Italian, and English became the principal vehicles of scientific expression, anomalies continue to be found, such as Latin treatises produced for local aristocratic environments and vernacular treatises destined for Europe-wide dissemination.

59. Colonialism and globalization have changed the rules for game of vernacular language science. I cannot address that here, but among the useful resources are Naregal, *Language Politics, Elites, and the Public Sphere* (the place of Marathi in nineteenth-century education in Maharashtra) and Minault, "Delhi College and Urdu" (the place of Urdu, with important remarks on the Vernacular Translation Society).

60. Pollock, "New Intellectuals in Seventeenth-century India," 30–31.

61. This was recognized to some degree from the start by European vernacular intellectuals like Bacon. The Latin translation of his *Advancement* (which he commissioned in 1607–8) was, he said, a book that "will live, and be a citizen of the world, as English books are not. . . . My end of putting it into Latin was to have it read everywhere." Similarly regarding the Latin translation of his *Essays*: "For I doe conceive, that the Latine Volume of them, (being the Universall Language) may last, as long as Bookes last" (quoted in Kiernan, *The Advancement of Learning,* liv).

References

PRIMARY SOURCES

Bhoja. *Śṛṅgāraprakāśa.* Vol. 1. Ed. V. Raghavan. Cambridge, Mass.: Harvard University Press, 1998.

Dinakara Bhaṭṭa. *Bhāṭṭadinakara.* Ms. shelf list *mīmāṃsā* mss. no. 30. Sarasvati Bhandar, Fort, Ramnagar, Varanasi.

Gāgā Bhaṭṭa. *The Rākāgama Commentary on the Chandrāloka of Jayadeva.* Ed. Anantarama Sastri Vetala. Varanasi: Chowkhamba Sanskrit Series Office, 1938.

Jasvant Siṃh Granthāvalī. Ed. Vishvanathprasad Mishra. Varanasi: Nagaripracarini Sabha, 1972.

Kamalākara Bhaṭṭa. *Mīmāṃsākutūhalam.* Ed. P. N. Pattabhirama Sastry. Varanasi: Sampurnanand Sanskrit University, 1987.

Kavikarṇapūra. *Alaṅkārakaustubha.* Ed. Shivaprasad Bhattacharya et al. Delhi: Parimal Publications, 1993.

Kavīndrācārya Sarasvati. *Bhāṣāyogavāsiṣṭhasāra*. Ed. V. G. Rahurkar. Pune: Bharatavani-Prakasanamala, 1969.

Khaṇḍadeva. *Mīmāṃsākaustubha*. Ed. Chinnaswami Sastri. Varanasi: Chowkhamba Sanskrit Series Office, 1923.

Kumārila. *Tantravārtika*. Vol. 2 of *Mīmāṃsādarśana*. Pune: Anandasrama, 1970.

Mahādeva Punatāmbekara. *Nyāyakaustubha Śabdapariccheda*. Ed. V. Subrahmanya Sastri. Thanjavur: Saraswati Mahal Library, 1982.

Pārthasārathi Miśra. *Śāstradīpikā*. Ed. Dharmadatta Jha. Varanasi: Krshnadasa Akadami, 1988.

Someśvara. *Nyāyasudhā*. Ed. Mukund Shastri. Varanasi: Chowkhamba Sanskrit Book Depot, 1901.

Vāsudeva Dīkṣita. *Adhvaramīmāṃsākutūhalavṛtti*. 4 vols., ed. P. N. Pattabhirama Sastri. New Delhi: Lal Bahadur Shastri Vidyapeeth, 1968–72.

Veṅkaṭādhvarin. *Viśvaguṇādarśacampū*. Ed. Jatashanakara Pathak. Varanasi: Chowkhamba Vidyabhawan, 1963.

SECONDARY SOURCES

American Oriental Society. *Abstracts of Communications Presented at the Two Hundred and Nineteenth Meeting*. New Haven, Conn.: American Oriental Society, 2009.

Busch, Allison. *Poetry of Kings: Classical Hindi Literature of Mughal India*. New York: Oxford University Press, forthcoming.

Chartier, Roger, and Pietro Corsi. *Sciences et langues en Europe*. Paris: Ecole des hautes études en sciences sociales, 1996.

Coward, Harold G., and Raja K. Kunjuni. *The Philosophy of the Grammarians*. Princeton, N.J.: Princeton University Press, 1990.

Cunningham, Andrew, and Perry Williams. "De-centering the 'Big Picture': *The Origins of Modern Science* and the Modern Origins of Science." *British Journal for the History of Science* 26 (1993), 407–32.

Delvoye, Françoise. "Les chants *dhrupad* en langue Braj." *Littératures médiévales de l'Inde du nord*, ed. Françoise Mallison. Paris: École française d'Extrême-Orient, 1991.

Derrida, Jacques. "Languages and Institutions of Philosophy." *Recherches Sémiotiques/Semiotic Inquiry* 4, no. 2 (1984), 91–154.

Divākar, Kṛṣṇa. *Kavīndracandrikā: Kavīndrācārya Sarasvatī Hindī Abhinandana Grantha*. Pune: Maharashtra Rashtrabhasha Sabha, 1966.

Freeman, Rich. "Rubies and Coral: The Lapidary Crafting of Language in Kerala." *Journal of Asian Studies* 57 (1998), 38–65.

Garber, Daniel, and Michael Ayers, eds. *The Cambridge History of Seventeenth-century Philosophy*. Cambridge: Cambridge University Press, 2003.

Gaukroger, Stephen. *Francis Bacon and the Transformation of Early-modern Philosophy*. Cambridge: Cambridge University Press, 2001.

Kachru, B. *Kashmiri Literature*. Wiesbaden: Harrassowitz, 1981.
Kiernan, Michael, ed. *The Advancement of Learning*. Oxford: Clarendon Press, 2000.
McCrea, Lawrence. "Novelty of Form and Novelty of Substance in Seventeenth-century Mīmāṃsā." *Journal of Indian Philosophy* 30, no. 5 (2003), 481–94.
McGregor, Stuart. *The Language of Indrajit of Orcha*. Cambridge: Cambridge University Press, 1968.
———. "The Progress of Hindi, Part 1: The Development of a Transregional Idiom." *Literary Cultures in History: Reconstructions from South Asia*, ed. Sheldon Pollock. Berkeley: University of California Press, 2003.
Minault, Gail. "Delhi College and Urdu." *Annual of Urdu Studies* 14 (1999), 119–34.
Minkowski, Christopher. "Astronomers and Their Reasons: Working Paper on *Jyotiḥ-śāstra*." *Journal of Indian Philosophy* 30, no. 5 (2003), 495–514.
Naregal, Veena. *Language Politics, Elites, and the Public Sphere: Western India under Colonialism*. New Delhi: Permanent Black, 2001.
Pandey, Sudhakar. *Hastalikhit Hindī Granthasūcī*. 3 vols. Varanasi: Nagariparcharini Sabha, 1989–93.
Pantin, Isabelle. "Latin et langues vernaculaires dans la littérature scientifique européenne au début de l'époque moderne (1550–1635)." *Sciences et langues en Europe*, ed. Roger Chartier and Pietro Corsi. Paris: Ecole des hautes études en sciences sociales, 1996.
Pingree, David. *From Astral Omens to Astrology: From Babylon to Bīkāner*. Rome: Istituto Italiano per l'Africa e l'Oriente, 1997.
Pollock, Sheldon, ed. "The *Bhāṭṭadinakara* of Dinakara Bhaṭṭa (1.3), a Seventeenth-century Treatise on Mīmāṃsā. Edited for the first time, with an Introduction." Unpublished manuscript.
———, ed. "Comparative Intellectual Histories of the Early Modern World." *International Association of Asian Studies Newsletter* (Leiden) 43 (spring 2007), 1–13.
———. "Cosmopolitan and Vernacular in History." *Cosmopolitanism*, ed. Carol Breckenridge et al. Durham, N.C.: Duke University Press, 2002.
———. "The Death of Sanskrit." *Comparative Studies in Society and History* 43, no. 2 (2001), 392–426.
———. *The Ends of Man at the End of Premodernity*. Amsterdam: Royal Netherlands Academy of Arts and Sciences, Stichting J. Gonda-Fonds, 2005.
———. "The Idea of *Śāstra* in Traditional India." *Śāstric Tradition in the Indian Arts*, ed. A. L. Dallapiccola and S. Zingel-Avé Lallemant. Stuttgart: Steiner, 1989.
———. "India in the Vernacular Millennium: Literary Culture and Polity, 1000–1500." "Early Modernities," ed. Shmuel Eisenstadt et al. Special issue of *Daedalus* 127, no. 3 (1998), 41–74.
———. *The Language of the Gods in the World of Men: Sanskrit, Culture, and Power in Premodern India*. Berkeley: University of California Press, 2006.
———, ed. *Literary Cultures in History: Reconstructions from South Asia*. Berkeley: University of California Press, 2003.

———. "New Intellectuals in Seventeenth-century India." *Indian Economic and Social History Review* 38, no. 1 (2001), 3–31.

———. "A New Philology: From Norm-bound Practice to Practice-bound Norm in Kannada Intellectual History." *South-Indian Horizons: Felicitation Volume for François Gros*, ed. Jean-Luc Chevillard. Pondicherry: Institut français de Pondichéry, 2004.

Printz, Wilhelm. "Bhāṣā-Wörter in Nīlakaṇṭha's Bhāratabhāvadīpa usw." *Kühns Zeitschrift* 44 (1911), 70–74.

Rao, Velcheru Narayana. "Multiple Literary Cultures in Telugu: Court, Temple, and Public." *Literary Cultures in History: Reconstructions from South Asia*, ed. Sheldon Pollock. Berkeley: University of California Press, 2003.

Sastry, R. Ananta Krishna, ed. *Kavīndrācārya's List*. Baroda: Central Library, 1921.

Staal, F. "The Sanskrit of Science." *Journal of Indian Philosophy* 23, no. 1 (1995), 73–127.

Swamy, B. G. L. "The Date of the Tolkāppiam: A Retrospect." *Annals of Oriental Research, Madras*, Silver Jubilee volume (1975), 292–317.

Takahashi, Takanobu. *Tamil Love Poetry and Poetics*. Leiden: Brill, 1995.

Tulpule, Shankar Gopal. *Classical Marāṭhī Literature: From the Beginning to A.D. 1818*. Wiesbaden: Harrassowitz, 1979.

Varma, Ramkumar, et al., eds. *Hastalikhit Hindī Granthoṁ kī Vivarṇātmak Sūcī*. 2 vols. Allahabad: Hindi Sahitya Sammelan, 1971–87.

Wagoner, Phillip B. "*Iqtā* and *Nāyaṁkara*: Military Service Tenures and Political Theory from Saljuq Iran to Vijayanagara South India." Paper presented at the 25th annual Conference on South Asia, Madison, Wisc., 18–20 October 1996.

Ziauddin, M. *A Grammar of the Braj Bhākhā by Mīrzā Khān*. Calcutta: Visvabharati Bookshop, 1935.

CHAPTER TWO

Bad Language and Good Language

Lexical Awareness in the Cultural Politics of Peninsular India, ca. 1300–1800

SUMIT GUHA

Colonialism did more than change the political structures of South Asian society. Long-enduring forms of systematic knowledge lost their validity or mutated into unrecognizable forms in order to survive. These domains of knowledge were not the timeless traditions of unchanging civilizations: each had its own history, one worth revisiting. It is in this context that Sheldon Pollock has long been investigating the changes that generated the South Asian literary and scientific world overtaken by colonialism at the end of the eighteenth century. This project culminated in a massive volume that ranged across numerous Asian languages and literatures over many centuries.[1]

Most of these literary domains were generated by the process of vernacularization that began about the beginning of the past millennium. In South Asia the second stage of a self-conscious canonical culture was well under way by the seventeenth century, when elements of a vernacular literary canon were assembled for the widely used vernacular Braj.[2] Pollock sees the origins of vernacularization in "the conscious decisions of writers to reshape the boundaries of their cultural universe by renouncing the larger world for a smaller place. . . . New local ways of making culture—with their wholly historical and factitious local identities—and, concomitantly, ordering society and polity came into being, replacing the older translocalism."[3] But the vernaculars were nonetheless largely limited to nonscientific domains of thought. As Pollock has persuasively argued in his contribution to the present volume, a major community of South Asian intellectuals before the nineteenth century decided that systematic knowledge was to be expressed in Sanskrit. The sense that writing in Sanskrit was the equivalent of addressing the world is evidenced in the writings of the Mahanubhava, a secretive Vaishnava community that pioneered Marathi prose and verse. Beginning in about 1350 they wrote in several enciphered scripts, which served to prevent outsiders from reading their texts. When a disciple suggested rendering two chapters of the scripture into Sanskrit verse, the idea was rejected because it would expose esoteric knowledge to the world. One key text was in fact translated into Sanskrit in the sixteenth century, but it never gained any status within the community. Oral transmission in Marathi and recording in cipher were therefore seen as devices for secrecy.[4] This should certainly give us pause before we link vernacular with the popular and Sanskrit with the esoteric.

Let us return to the issue of language choice. The use of a vernacular was a choice made by the author who had abandoned the older translocalism so as to inhabit a narrower textual space. The Mahanubhava pioneers of Marathi literature narrowed their space of intelligibility through their choice of language and then still further by encipherment. Later bearers of the Marathi vernacular tradition were not so concerned with secrecy and saw themselves as addressing a larger Marathi literary world, a world that by the seventeenth century extended from Tanjavur to Banaras. Marathi had become the language of a transregional elite, patronized by the Dakhan sultans and their rivals, the Bhosle kings, who founded a regime that loomed large in South Asian politics through the eighteenth century. These developments may have impacted the cultural politics of language choice; it is significant that it was a Maharashtrian Brahman, Kaundabhatta, who selected Marathi as his example since he believed that although it was a vernacular language it was nonetheless invariant across space and therefore suitable for scientific discourse.[5]

Marathi-using soldiers and administrators had in fact been spreading across the Dakhan from at least the late sixteenth century and retaining their language even as they ruled and fought in areas far removed from their linguistic homeland. Marathi was well established as an administrative language by the early sixteenth century. It then came into use for historical writing that extended beyond the retelling of traditional histories found in the *purāṇa*.[6] The rise of a Marathi-speaking gentry culminated in the establishment of the Bhosle kingdom (1674), which extended down the western spine of the peninsula as far south as the Kaveri delta. Could these phenomena have informed Kaundabhatta's argument?

The novel political formation created by Chatrapati Shivaji certainly seems to have stimulated a new deployment of the Marathi language as a tool of systematic description and understanding. A major Marathi history, usually titled the *Sabhāsad Bakhar*, both explained and analyzed the career of the first *chatrapati*. This text was completed in 1696–97. It explicitly presents itself as an effort to record and systematize the feats and strategies by which he had defied powerful enemies and established a new empire. The newness of the enterprise is exemplified by a chapter in *Sabhāsad* that provides a vivid description of the functional structure of both the central administration and the entire state. The Persian names for the different departments as well as their functions reflect the effort to accurately describe administrative reality rather than simply echo previous Sanskrit treatises, as for example the contemporary Sanskrit compilation known as the *Daṇḍanīti* does.[7]

In 1716–17 the analytic process went a step further and a historically grounded and highly realistic treatise on political management was written in Marathi by Ramacandra Pant Amatya. Ramacandra Pant served as a plenipotentiary in southwestern Maharashtra after Chatrapati Rajaram had to flee to Senji (Jinjee) following Aurangzeb's capture of the major fortresses in Maharashtra in 1689–90. In 1716 Ramacandra set out to write a scientific analysis of the practice of statecraft. The first two chapters create a dramatic contrast by shifting between descriptions of the disordered state of the kingdom in the first part of the eighteenth century and its earlier greatness. The author's mission is put into the mouth of King Sambhaji of Kolhapur:

> You are a servant of high achievements. When the honored father [Rajaram] now residing in heaven went to Karnataka he entrusted this kingdom to you and you whole-heartedly devoted yourself to its development. You are supremely accomplished in statecraft, its procedures and policies. The King has commanded the writing and distribution of this scientific treatise on politics [*yathāśāstra rājnīti*] so that the young prince may be trained in government, and also so that governors and police officials posted in other provinces may conduct themselves lawfully. Write so that the prince should also be instructed and the others should all take this to heart and so conduct themselves that the kingdom is well protected.[8]

The text makes no overt reference to the old Sanskrit tradition of *nīti* or *arthaśāstra*; it is very much lodged in the real world of the time. For example, chapter 5 opens with the following words: "Businessmen are the ornaments of the kingdom. It becomes populous and prosperous by their presence." The presence of a completely different class of businessmen is also noted: "Businessmen include Phirangi [originally Franks, here meaning the Portuguese], Ingrez [English], Valandez [Dutch], Dingmar [Danes] etc., hat-wearing peoples who also trade. But they are not like other merchants. Each of them has a king for a master. It is at their command and on their account that they come to these lands. Has anyone ever seen a king who was not hungry for land?"[9]

Thus the Marathi language had become the chosen vehicle for the systematic understanding of the politics and history of the Maratha *Svarāj*. But as Pollock points out, despite the brilliance of the Maharashtrian tradition in Sanskrit little significant work was done in the grammatical study of Marathi, in contrast to the precocious development of that science in Telugu and Kannada.[10] The Mahanubhava community was the exception to this rule. In the fourteenth century Bhishmacarya composed a Marathi grammar focused on the correct reading of the cryptic *sūtras* of the community.[11] Mainstream Marathi scholars' linguistic concerns were (as we shall see) mainly limited to the lexical. In this essay I seek to delineate a history of language awareness and language use in the Marathi-speaking world as a backdrop to a discussion of the three lexicons that emerged in the seventeenth and eighteenth centuries. I end with an example of the lexical choices consciously made by a Maratha man of letters in the late eighteenth century.

The Marathi language is attested from at least the eighth century CE, when the *Kuvalayamālā* used a grammatical marker to identify as *Marhaṭṭe* those who used (the past tense forms) *diṇṇle* (gave) and *gahille* (took).[12] Its first major literary efflorescence coincided with the rise of the Yadavas of Devagiri at the end of the twelfth century. This culminated in the famous *Jñāneśvarī* (or *Dnyānesvarī*), completed in 1290. This massive work is a paraphrase and commentary of the entire *Bhagavadgītā* in 1,965 verses.[13] As Tulpule and Feldhaus point out, "Such great literary achievements were made in this period that it has come to be known as the 'Golden Age' in the history of the Marathi language. This period saw the rise and development of the Varkaris and the Mahanubhavas, the two sects that produced the bulk of Old and Middle Marathi literature."[14] Cynthia Talbot has observed that the find-spots of early Telugu inscriptions fall within the territories of the Kakatiya dynasty of Warangal. In Maharashtra the increasing use of Marathi was associated with the Kakatiyas' contemporaries and rivals, the Yadavas of Devagiri.[15] As regional languages became politically important, royal patrons began to sponsor their formal study. Tulpule cites the Hoysala inscription of 1290 found at Maillagi in the Mysore district of Karnataka. It provided a stipend for instruction in Nagara, Kannada, Tigul, and Are, that is, Sanskrit, Kannada, Telugu, and Marathi.[16] The urbane high language comes first, but interestingly Marathi is also seen as a noble (*āre/ārya*) tongue.

This sense of linguistic pride was naturally even stronger in the Maratha homeland. The *Jñāneśvarī* was created in a cultural setting where Sanskrit was the unquestioned high language, with Marathi immediately below it. As a dominant tongue Marathi used as an adjective carried positive connotations of frankness and honesty. So when Arjuna asks Krishna to unequivocally indicate the noblest course open to him (*Gītā*, 3, 2), the extensive Marathi commentary has him say, *maj viveku sāṃgāvā Marhāṭā jo*, "Instruct me in (what would be) the Maratha way."[17]

The Yadavas of Devagiri were defeated by Alauddin Khalji in 1294 and supplanted by governors sent by the sultans of Delhi in 1318. Their capital was renamed Daulatabad. The Turko-Afghan elite there rebelled against Sultan Muhammad Tughlaq in the 1340s and finally set up the Bahmani sultanate after his death. That sultanate disintegrated into five kingdoms at the end of the fifteenth century. Important works of Middle Marathi literature continued to be produced. Furthermore Marathi remained the language of administration and government at the local level, where hereditary officials maintained a tenacious grip on authority. As Krishnaji Ananta Sabhasad wrote in 1696–97, "Lands held by the Idalshahi, Nizamshahi, Mughalai were conquered [by Shivaji]. In those lands, the peasants had been until then completely in the hands of the hereditary headmen, accountants and district officers."[18]

These hereditary officials maintained records (and probably spoke) in variants of Middle Marathi. They may well have benefited from this linguistic segregation that rendered their records and accounts impenetrable to central authority and forced the finance minister to compound with them on a lump-sum tax payment which they would then collect from the peasants as they chose. A partition of domains between vernaculars and Sanskrit had begun to emerge by the twelfth century; take for example the Miraj copperplates of Calukya Virasatyashrayadeva, possibly from 1154. The first fifteen lines of obeisance to Shiva followed by a conventional tribute to the king are in Sanskrit. Then the text switches to a Kannada-influenced Marathi as it explains that the king camped near the shrine of Kopeshvaradeva after his victorious expedition to the south and made the grants enumerated in the plates.[19] Obviously this was because local officials and landowners who might challenge or encroach on the grant had to be able to read its details for themselves and they could not be presumed to know Sanskrit. After the Khalji conquests and the establishment of the southern sultanates Sanskrit had to deal with a powerful rival: a Persian adopted and developed by the conquering Turks. The sultans of the Deccan recruited top administrators from Iran and the lands along the Arabian Sea and used Persian at their courts and in most inscriptions. But Arabic was also known and patronized, a process encouraged by the ancient trading connections of southern Arabia with the Indian peninsula.[20] Where functionally necessary, Persian epigraphs were accompanied by regional language versions. The earliest bilingual inscription (Arabic-Marathi in this case) that I have found mentioned dates from 1377. Ziauddin Desai writes that another epigraph indicates "that it was the local or regional language which was considered the medium of communicating official orders—*qaulnama* [solemn reassurance] in this case—to the members of the public, apparently in non-urban areas."[21] But the dominant high language of the new urban centers was Persian, while the emerging vernacular Dakhani was their lingua franca.

Both the script and the register of the new Dakhani language proclaimed its distinctive affiliation. Various labels began to be applied to it. Thus a close disciple of the famous Marathi *bhakta* poet Tukaram (whose poem is included among Tukaram's own), in depicting the modern age of decay (*kaliyuga*) pointed to the use of *avindhavāṇī*, "the speech of those who have unpierced ears" (i.e., Muslims), even by Brahmans, as one of its features.[22] In the 1650s Jayarama Pindye claimed to compose freely in twelve languages, including *dākṣinātya yāvani* (southern Yavani). *Yavana* was by then a common term for Muslim, and Jayarama clearly recognized that the southern Muslims had a language distinct from Persian, which he simply termed *yāvani*.[23] The ecumenical Ibrahim Adilshah of Bijapur (r. 1580–1627) passed into southern folklore with the title *jagadguru* because of his affinity with all his subjects. But he too recognized the established linguistic distinctions even as he denied religious distinction. In the *Navaras Rāgamālā*, a collection of short lyric poems attributed to him, one poem declared *atai*s (bards), *dhāḍi*s (heralds), and *guṇijans* (artistic performers) to be like the three eyes of Shiva. They have different languages but one sentiment (to express): "Where is the Turk, where is the Brahman?" (*bhākā nyārī nyarī bhāva ek / kahāñ turak kahāñ brāhman*).[24]

The adoption of Persian as the language of central administration meant that intermediary translators would be needed. Anyone who dealt regularly with the central authorities would have had to learn some of the technical terminology of administration. But these new registers of speech began to leak out beyond the narrow channels of government communication and significantly changed popular speech. This is evidenced by its appearance in popular religious poetry.

The immensely popular Eknath (1533–99) and Tukaram (1608–50) both composed devotional poems that played upon these formats.[25] Such texts presuppose that the audience has at least a passive knowledge of specialized official vocabulary as well as a repertoire of certain formulas of greeting and petition that are embedded in official language and produced in appropriate contexts. Poets such as Eknath clearly had a keen awareness of linguistic diversity and linguistic change. It is not surprising then that the awareness of language difference begins with the sense that registers vary and vocabularies change. Many words in the much loved *Jñāneśvarī* (completed ca. 1290) had become incomprehensible by the sixteenth century, resulting in the production of lexicons specific to that text, including one written by Eknath himself.[26] Nor was this the only such work. Another lexicon was prepared in 1747, this one explaining all its obscure words, not in the sequence of their appearance in the text as a commentary might have done, but in alphabetical order.[27]

The devotional poets may of course have wished to display their linguistic virtuosity as well as their devotion, but there are many examples of how deeply these official processes imprinted ordinary rustic Marathi. This can be seen from a deposition by a Dalit (Mahar) claimant to the hereditary headman's post in 1608 (with Perso-Arabic derivates highlighted):

> *pascamvādi Jāv Pātil* ***bin*** *Bāda Pātil Khandge* ***takrīr*** *kelī aisije* ***mauze mazkurī kul kharāb*** *hote yāvari āplā paṇjā Kota Patil khālilekadun [****khalisekadun?****] etañ gāvi rāhilā hotā govīce [sic]* ***mokāsāi*** *panjā yās bolile je gaṃva khora paḍilā sadab [?] paḍilā āhe tumhās pātilkī* ***mirās*** *karūn deū gāṃva* ***māmūr*** *karṇe.*

> The respondent Jav Patil son of Bada Patil Khandge stated as follows: The aforesaid village was completely desolate, our grandfather Kota Patil came from the [office of] state lands and lived in the village. Then the fief-holder [of the village] told my grandfather: "The village is ruined. I will give you the headmanship as a hereditary possession and you should make the village prosperous again."[28]

Language use was marked by a tension between hybridization tending toward assimilation and distinction tending toward establishing identity. Seventeenth-century men of letters were certainly aware of this tension, and it played a role in the cultural politics of the day. So we get dictionary making as a response. On the other hand it afforded great opportunities for the display of virtuosity in double entendre and puns. In actual speech—and we should remember that many of the documents we have were either recited or dictated—the result could be a mixed idiom.

Early exercises in polyglossia such as Eknath's devotional poems distinctly tie language use with social roles. These poems are skillfully composed so that they can be read at two levels: the mundanely stereotypical and the spiritual. Poems in which the speakers are regular members of the village (or "insiders") are in Marathi; poems in which the speakers are itinerant outsiders are in

Dakhani and other versions of Hindi. These include a conjurer cum snake charmer, a Muslim Ethiopian mendicant (*Hāpasi*), a follower of Guru Nanak, a migrant herdsman, and a naked *phakīr*, who opens his chant:

> Good is the company of the holy
> drunk on the *bhang* of self-knowledge
> always blissfully intoxicated
> such is the naked *phakīr*.[29]

Already therefore by the end of the sixteenth century use of the Marathi language marks the distinction between those who belong to the village and those who transit through it. Versions of Dakhani or Hindustani mark outsiders. Such identification of languages with social and political identities developed further in the seventeenth and eighteenth centuries, in part because of the rise of a transregional Maratha imperial formation.

The Mughal Empire in North India was effectively refounded by Akbar in the mid-sixteenth century. By the end of that century his armies had crossed the Narmada and conquered two of the five sultanates and were pressing hard on the Nizamshahis of Ahmednagar. This seems to have led to widespread employment of Maratha horsemen on two distinct marchlands: the southern one, opened by the destruction of Vijayanagara in 1565, and the northern, created by the advance of Mughal power. Maratha light cavalry had long been employed in southern armies. Malik Ambar, regent of Ahmednagar during its long sustained resistance to the Mughals, was a major employer of Maratha soldiers, and many commanders gained experience under him. These included Shahaji Bhosle, father of the future *chatrapati*. Malik Ambar died in 1626; by that time the political significance of Maratha gentry and soldiers was manifested in the increasing volume of bilingual Persian-Marathi edicts, orders, and other official documents issuing from both the Nizamshi and Adilshahi courts. It is also perhaps significant that documents issued by the great Maratha families (Bhosle, Nimbalkar, Ghorpade, Kharate, Ghatge) were exclusively in (heavily Persianized) Marathi. An early survey by the leading scholar G. H. Khare found no examples of Persian or bilingual documents sent out by them.[30] There is considerable evidence that Shahaji Bhonsle considered the project of setting up an independent kingdom for himself. In the last days of the Nizamshahi sultanate he assumed the title of chief minister and took custody of an infant heir to that dynasty. But the scheme was foiled with some effort by the Mughal emperor Shahjahan, and Shahaji joined the Adilshahi court at Bijapur. He became a senior commander and was sent southward, where the Adilshahi and Qutbshahi sultanates were expanding into the wreck of the Vijayanagara Empire. Shahaji spent the last twenty years of his life there in efforts to build an autonomous appanage. The older lands around Pune were retained as a *jāgīr*, and his senior wife and her young son, Shivaji, were left in charge of them. Shahaji died in 1664.[31]

We may draw on the previously cited work by Ramacandra Pant as a source for the actual practice of cultural politics under the Bhosle kings. Chapter 3 of his treatise discusses the qualities needed in a king and the qualities needed in his chief attendants. The policies of the Maratha kings break into the text, which begins, "An honorable name is gained from the *ślokas, subhāṣitas* and poems of great [Sanskrit] poets." But then realism prevails and he adds, "Similarly, the previous kings have also patronized and kept *bhāṭs* [vernacular bards] in their entourage. Therefore assay the qualities of poets and bards and maintain them. But never call these people in when state business is being transacted."[32]

Bhāṭs (and the associated *cāraṇas*) were itinerant bards, especially influential among warrior communities of northern and central India.[33] The experienced minister was thus harking back at least as far as the cultural politics of Shahaji Bhosle, Shivaji's father. The *Rādhāmādhavavilāsa campū*, an important multilingual text from Shahaji's court, gives us some glimpses of the setting and its politics. It specifically names and cites poets of the various vernacular languages. Some of these seem real; others are almost certainly mouthpieces for the author. That includes Subuddhirav, a *bhāṭ* from Ghatampur in the north, whose farewell verse compares Shahaji to Krishna when he sheltered the people by holding up Govardhana mountain and then in an extravagant pun declares that the Creator has now given the Sisodiya (Rajput) lineage one who looks like its head (*sīso dīso diyā*).[34] Another poem was actually composed in Dingal (here referred to as *uttarabhāṣā*, "northern language") by Jayarama. It depicts the king of Amber (who would be the Kacchawaha lineage rival to the Sisodiya Rana of Mewad) learning about Shahaji's greatness from the bards and anxious to deliver supplies and gifts to Shahaji should he come that way. This is of course an entirely imaginary scene, but it does indicate that the *bhāṭs* were viewed as the creators of a transregional reputation, something increasingly important for aspirant potentates as a subcontinental political system was being generated under Mughal hegemony through the seventeenth century.

It is possible to date the *Campū* only by internal clues: the poet depicts himself meeting Shahaji's oldest son by Jijabai, Shambhuraja. The description of this prince is prefaced by a clause describing him as the elder brother of Shivaraja, who, fiery as the sun, was a flame to the forest of all the powerful

Yavanas of the *kaliyuga* (*Campū* 232). While some historians have argued that the older prince was alive until 1663, most scholars believe that he was killed at the siege of Kanakagiri in 1654.[35] Thus 1660 would seem a likely upper date for the completion of this text.

Not too much is known of life at the Bhosle court beyond what is depicted in the *Campū* itself. It is of course dangerous to assume that the *Campū* is a faithful journalistic record; it is first of all a work of the literary imagination, an offering to a lordly patron and an exercise in poetic skill. A clear example of the selective record of events is the fact that while a grandiloquent description of the Kanakagiri campaign is given, the poet never mentions that Prince Shambhuji was probably killed in that siege. At least one poem, the Marathi *Bhujaṃgaprayāga* (no. 43), was almost certainly composed after this event. This mentions Shahaji's son Shivaji as *ṭīkecā dhaṇī* (anointed heir), a status that he would have obtained only after the death of his older brother. It also mentions the youngest son, Ekoji.

The major patron is a historically attested figure; so are many of the lesser figures around him. We can therefore connect some features of what is deemed praiseworthy in a specific milieu to specific persons from the text itself and see the work as an exercise in poetic virtuosity that may perhaps help develop a new definition of the cultural politics of India in an important phase of transition and integration. As we shall see, the *Campū* complements the works recently discussed by Allison Busch, who has studied the development of aesthetic theory in Braj poetry. Here we have (among other things) the transference of popular Krishna-Radha poems into Sanskrit. Indeed Jayarama quotes learned aesthetes seeking to know the identity of the man of letters who had transcreated into Sanskrit poems what they had hitherto known in the *deśa bhāṣā*.[36]

In the sixth canto we are told that the fame of this work spread to the ends of the earth and that astonished connoisseurs said, "It is currently heard that the maker of this work that draws on the dalliance of Radha and Madhava in the *deśa bhāṣā* is a king of poets named Jayarama. Who indeed is this king of poets Jayarama? The lector then said that the poet was at this time close to a maharaja. Then they asked again 'Who is this maharaja?'" This affords the poet the opportunity to introduce several verse and prose passages of laudatory description of Shahaji of the Bhosle lineage.[37] Among the many virtues of this sustainer of the earth was being as heroic as Partha (Arjuna), as generous as Vikramarka, and as learned as Bhoja. As he was controlling the earth, hundreds and thousands of scholars and poets from the ends of the earth flocked to his court. The poet Jayarama, surnamed Pindye, who was skilled in twelve languages, heard of his generosity and was eager to see him. The waning of the Sanskrit cosmopolis is suggested by the fact that this learned Sanskrit scholar heard of Shahaji from traveling *bhāṭs* who were returning to their northern homes.

So the text returns to the court setting. We are told that when the opportunity arose the vernacular poets, thinking that vernacular composition was exceedingly difficult, all clamored to pose themes for Jayarama in the hope of baffling him. Whereupon he recited a verse: "The Sanskrit lion advances roaring to prize from the elephant the pearl-like word; while [others] sit like monkeys [*śākhāmṛga*] hidden in the many-branched languages."[38] Then follow sixty poems in different vernaculars. Many are represented as composed extempore, others at the request of different hearers or occasioned by incidents or special occasions; a few are simply introduced without contextual explanation. Poems 61 and 62 are again Sanskrit, the latter reporting that the poet requested leave from Shahaji to make a pilgrimage to Kashi, Prayaga, and Gaya to spread the raja's fame and also to take his very aged parents along to Kashi. He was promptly told to take what wealth he desired and proceed to Kashi. A short Marathi poem ends the collection.[39]

Inferences from the content and arrangement of this eleventh chapter are necessarily speculative. As Busch has pointed out, its separation is itself a marker of the lower status of the languages themselves. Can we infer anything from the ordering? The first twenty are in *bhāṣā*, the northern lingua franca close to Braj or Bundeli.[40] These are in fact simply described as *bhākhā* (*bhāṣā*), for example in poem 7, where the questioner is described as a poet named Kehari who is the lion (*kehari*) of the *bhākhā* forest. Marathi appears in poem 21. It begins in *bhākhā* and celebrates the defeat of the Qutbshahi general Mir Jumla; the king's jester then demands that it be completed in "Marhata." (This name is absent from the list of languages, where it is probably described as Prakrit.)[41] But poem 22 is again in *bhākhā*, and poem 23 is labeled *rekhtā*.

What may we learn about the cultural politics of Jayarama himself? At one level he presents himself as recapturing a certain cultural space from the vernaculars by moving the popular songs of Radha and Krishna into what he sees as the superior domain of Sanskrit. In that sense he is moving in the opposite direction from the *rīti* poets in Braja. But for Jayarama the really dangerous rivals to Sanskrit do not seem to be the vernaculars. Even though he claims to have the ability to versify in both Yavani and Dakshinatya Yavani, his text manifests an unmistakable hostility to the Yavanas and languages associated with them. This appears in Marathi poem 33, which is put in the mouth of a *hejib* (envoy) sent by Shahaji, who announces to the (generic) Yavana, "*Yavanīṃ* [Yavan-ness] has spread on earth, *muñja* grass and Vedic speech have drooped languidly; but now the Saheb who is the very image of fame has made my tongue his messenger [*he jīb hejib kelī*]." The stage was clearly being set for a hoped-for reintroduction of Sanskrit into statecraft.

Court panegyric and flattery were one thing; the public world of politics was another. The latter was formally addressed by public documents authenticated by a seal. Both Shahaji and his senior wife, Jijabai, had Persian seals. On the other hand, in common with many aristocratic Marathas, Shahaji's orders were not bilingual but written exclusively in Marathi. A significant break comes with Shivaji because as far as we know he never even had a Persian seal. Right from his teens his seal was in Sanskrit and compared itself to the first sliver of the new moon as it began to grow and shed a pleasing light.[42] (The metaphor of the growing moon is itself not without significance when deployed by the teenage ruler of a small principality.) And it is Shivaji who is hailed in the *Campū* as the fire that consumed the forest of the all-powerful Yavanas in the *kaliyuga*.

Initially however Shivaji's documents follow the pattern established by his father: they use the Perso-Marathi forms established since at least the late fifteenth century. But one strand of the cultural politics of the period was then selected by Shivaji for further development soon after his coronation at Chatrapati in 1674. He commissioned one of his officials to prepare a new comprehensive *kośa* intended to replace the unnecessary (or *atyartha*, "overvalued") Yavana terms of state with educated language (*vibudha bhāṣā*). Educated language, not surprisingly, is Sanskrit. This command led to the production of the *Rājavyavahārakośa*, the thesaurus of state usage completed around 1677.[43]

It contains ten chapters, each of which deals with a particular class of terms. The first is the *Rājavarga*, which naturally opens with the king: "Know *pādaśāha* to be *rājā*; for *sāheba* use *svāmī*. *Antahpuram* is what is termed *duruni* in Yavana language. *Yuvarājā* is *valīahada*; for *shāhazādā* use *rājaputra* for *pesvā pradhāna*," and so on. The poet does not hesitate to coin clumsy neologisms, such as "*vṛttāntalekhaka*" for "*vākānavīsa*," "*bahudeśeṣv adhikṛta*" for "*kāramulkī*" (roving plenipotentiary), which appear in this first section. Interestingly terms for the sun and moon, rain, lightning, and astrologer occur here too, as do equivalents for *pīr, paigambar, qalandar*, and *mullā* (holy man, prophet, religious mendicant, and religious scholar, respectively). A certain coyness about the realities of courtly life is manifest in that the wine pourer (text has *sakhī*) is rendered simply as *dātā* (literally "giver"). Sometimes terms of Sanskrit origin are also rejected, apparently only because they were thought to be current in Yavani. Thus *kalāvantī* (artistically accomplished courtesan) is replaced by *veśyā*. Similarly in the fourth chapter, *guptaśastra* replaces *guptī*, and in the fifth *kareṇu* replaces *hattiṇī*, *tejasvī* for *tejī*. The second chapter deals with the departments of state and is therefore the *Kāryasthānavarga*, opening therefore with a replacement for the common *kārakhāna* (department or workplace). This chapter is largely concerned with words for luxuries and items of value. Enjoyment and consumption are the subject of the third chapter (*Bhogyavarga*). It offers replacements for the words for kitchen and cook and evident neologisms such as *pānakaādi-rasa-sthānam* for *sharbatakhānā*. The leather water bag becomes either *skandhabhasrikā* or *meśabhasrikā* (literally "shoulder hide" or "sheep hide"). Chapter 4 deals with weapons and arsenals. It offers, for example, *asi* for *phiraṃga* and *khaḍga* for *saifa* (sword), and more pedantically *kaṭhārikā* for *kaṭār* (broad dagger) and *bhalla* for *bhālā* (spear). The fifth chapter is titled *Caturaṅgavarga* (Section Concerning the Four Parts of an Army), even though the fourth part, the chariot corps, had not existed for centuries. Many special terms relating to elephants are found here. *Biruda* replaces the Sanskrit *cihna*, and the neologism *vīrakaṃkaṇa* replaces *toḍara* (bracelet given in reward for bravery) and *valganaha* (lively, bounding) for *turakī* (from Turk?). Terms for senior ministers and military commanders fall into the sixth chapter, which is inappropriately titled *Sāmantavarga* (Feudatory Chapter). Among the *tadbhava* terms that have been Sanskritized, we find *nāyikvāḍi* (*pattināyika*, infantry commander). The dense column formation *golah* is renamed *senāgulma*; *āsarā* (refuge) is re-Sanskritized to *āśraya*; *bhālekarī* (spearman) is renamed *kauntika*. The next chapter deals with forts. Some older Sanskrit fortification terms are offered in the place of Perso-Arabic ones, such as *dvīpa* for *jaṃjīrā* (*jazirāh*); some are clear neologisms, such as *prākārapāla* for *taṭ sarnaubat* (commander of a section of rampart). The text then offers *vivara* as an alternative to the perfectly good Sanskrit *suranga*. The tadbhava *kudalī* (spade) becomes *kuddāla*. The eighth chapter deals with writing in the broad sense of record keeping and the terminology thereof. It opens by offering *lekhyaśālā* for *daphtar* (office) and *lekhaka* (recorder) for both *daphtardār* and *navisinda*. Occasionally we find literal translations, such as *urobalam* for *sināzorī* (physically overawing).

Once again *tadbhava* terms are sought to be replaced with Sanskrit; thus we find *vibhajanam* for *vāṇtinī* (sharing) and *satyam* for *saccā* (true). The verb *nāgvineṃ* (to denude of all possessions) is superseded by the alternative *sarvasvaharaṇa* and *gaṇatī* (count) by *gaṇanā*. The ninth and tenth chapters deal with rural administrative terms and commerce, respectively. Chapter 9 offers *jāṅgala* for the common *jaṃgaḷa*, and common *tadbhava* terms for goldsmiths and leather, copper, and bronze workers are re-Sanskritized in the tenth chapter. Even *telī* has to be replaced by *tailī* in conformity with the Sanskrit vowel progression from *tila* (sesame seed).

While often pedantic, this lexicon was not merely an academic exercise. The last years of Shivaji's reign as well as that of his successor, Sambhaji, are marked by a notable Sanskritization of official vocabulary. Many official titles were Sanskritized to a considerable degree, and we find significantly more Sanskrit words in official documents. This continued with the succession of Rajaram (1689) and the desperate guerrilla struggle of the ensuing years, when every ideological appeal was thrown into the scales, with routine invocation of jihad by the Mughals and appeals such as this from the Maratha ruler: *svāmīce rājya mhanaje deva-brāhmaṇācī bhūmi. Yā rājyācī abhivṛddhi vhāvi āni Mahārāṣṭradharma rahāvā*, "That the Lord [Rajaram] holds this kingdom is equivalent to the Gods and Brahmans holding it. This kingdom must be sustained for the dharma pertinent to Maharashtra to survive."

We also have a return to a stronger emphasis under Rajaram and Tarabai on the ethnic Maratha character of the kingdom. In a letter—likely one of many sent in the desperate year 1690—Rajaram wrote to Baji Sarjerao Jedhe, *he Marāṣṭa rājya āhe*, "This is a Maratha kingdom."[44] Writing in 1696–97 the experienced minister Krishnaji Ananta Sabhasad nostalgically read ethnic assertion into Shivaji's coronation as *chatrapati* in 1674: "In this epoch all the great kings have been barbarian [*mleccha*]; now a Marāṣṭ *pādshāh* became *chatrapati*. This was no ordinary event."[45] In fact the copious contemporary documentation surviving from that event suggests that it was designed to be much more pan-Indian and Sanskritic than Marathi in character. But by the beginning of the eighteenth century *Mahārāṣṭradharma* was invoked in various contexts, without requiring further definition.

It is interesting that the *peśvās* who took effective control of the Maratha state in the early eighteenth century, while lavishly patronizing the traditions of Sanskrit learning, did not promote it seriously in the sphere of government and diplomacy. Some Sanskrit correspondence continued, as for example in a letter sent with two emissaries to Jodhpur in 1736. But the text is a word-for-word translation of a Marathi official text, with all the conventions of that genre. It also bears a great formal resemblance to Rajasthani letters in the same collection. I surmise that scribes in all three languages were modeling themselves on well-established Persian epistolary conventions. The letter ends with the conventional Marathi protocol "Why should I write much?," but in Sanskrit.[46] Meanwhile, back in Maharashtra, the language of the administrative documents of the era reflects, if anything, the strong legacy of sultanate and Mughal statecraft and eighteenth-century Hindustani usage.

The junior branch of the Bhosle line established itself at Tanjavur, ruling there until the kingdom was finally annexed by the British in the early nineteenth century. Maratha soldiers and officials continued to use and patronize the language. But distance began to have its inevitable effect, and later generations began to introduce the Kannada and Tamil of the region into their speech. This resulted in efforts to strengthen Marathi. At least two works intended to sustain this effort have been located and edited by Tulpule. These are the anonymous *Akārādī Prākṛta bhāṣecā nighaṇṭu* (published as *Tañjāvarī Kośa*) and Ramkavi's verse *Bhāṣāprakāśa*. It is noteworthy that the compilers remain within what we might term the mainstream of Marathi language study; that is, they focus on vocabulary and its corruption rather than on other aspects of linguistic science. Tulpule, their modern editor, initially dated them a century apart but later revised his dating of the *Bhāṣāprakāśa* to place them both in the earlier period of Maratha rule at Tanjavur. Both authors clearly address an expatriate community that is increasingly losing touch with its ancestral tongue.

The *Tañjāvarī Kośa* assumes that its users know a number of Tamil (which it terms "Drāviḍa") and Kannada words, so we get such explanations as "'*pimpri'*—this is the tree called *icci* in Dravida" and definitions such as for *kulkarṇi*, "writer of village accounts; in Dravida termed *kudikanakapille*." In some cases the functional explanation is omitted and the Tamil equivalent deemed sufficient, as in "*Kaikaḍa* in Dravida called *Koravara*—that caste of people." The compiler also occasionally offers Sanskrit equivalents, for example, *gahum* for *godhuma*; unlike the *Rājavyavahārakośa*, however, words of Persian and Hindustani derivation are accepted without comment and sometimes offered as explanations. For instance, we find "*gaṃja*—*nakhrā*. In Hindustani termed '*majjākh*'" (jest) or simply "*nirop*—*hokum*" (leave to depart, command). Sometimes both the word and the explanation are Perso-Arabic derivates, as *pherist*—*sadarband* (classified listing) or *phauja*—*laśkar* (army). The compiler was sometimes shaky as to linguistic origins; for example, we find the Hindi *gora* (fair-skinned) marked as a "Tilingi" (Telugu) word. Grammar appears in the marking of verbs as distinct from nouns and occasionally in the presentation of different conjugations of the same verb.[47]

The *Bhāṣāprakāśa* of Ramkavi was written in the same setting and possibly in the same period as the *Kośa*. Unlike the latter, it is in verse. The vocabulary is provided by topic rather than in alphabetical order. Ramkavi assumes that his readers have lost touch with Marathi to the extent of having to be informed of the gender of nouns in order to conjugate them appropriately. The second canto of the text therefore explains how the author will explain genders within the constraints of the metrical form adopted. But it turns out that he finds it sufficient to explain words that do not conform to Sanskrit. So in explaining masculine nouns ending in i/ī he lists some Marathi words and then adds, "These many are thus; the rest as in Sanskrit [*sama Saṃskritīm*]." The target audience is therefore assumed to know Sanskrit. That the text is addressed to an educated audience is suggested in the opening canto itself, which the author explicitly addresses to *deśasthas* living outside the *deśa*. "*Deśa*" refers to the plateau region east of the Sahyadri mountains, so *deśastha* might simply mean people from there. On the other hand the term also refers specifically to the Brahmans from the area, who became very prominent as officials in Tamil and Andhra lands. The text is evidently designed more for men of letters at the Tanjavur court than for ordinary folk.

The keen lexical awareness that informed the educated Marathi speaker is strikingly manifested in the writings of the important Maratha statesman Nana Phadnavis (d. 1798). One was a letter (ca. 1776?) to King George III protesting against the intrigues and duplicity of the English in Bombay; another is an autobiographical fragment written toward the end of his life. Excluding proper names, the first passage is 160 words long; of these, 72 words are Perso-Arabic. The personal memoir is focused on the way divine protection safeguarded him at various times in his career and saved him from death in various crises. In this *dhārmika* as opposed to *laukika* mood, Nana employs a chaste Marathi with a significant infusion of Sanskrit. Only seven Perso-Arabic derivatives appear out of approximately eighteen hundred words.[48] These occur mainly in the description of the battle of Panipat. It is evident that a well-educated writer would be conscious of having different registers of language available and would draw on them depending on mood and context. Linguistic awareness extended well beyond a few lexicographical pedants.

IIIIIIIIIIII

I MAKE NO CLAIM that this brief survey of language awareness in Marathi literary culture through six centuries has touched upon more than a small part of the enormous volume of the surviving literature. Nonetheless it does suggest that Marathi litterateurs were aware of the distinctions among the many languages spoken in western India and had a strong sense of the identity of their own. They were also aware of how the language was changing over time and sought to retain contact with earlier forms by annotating and compiling specialized glossaries. But the major focus of the linguistic purists seems to have been on vocabulary rather than other aspects of descriptive linguistics, despite their awareness of the ancient and rigorous traditions of Sanskrit grammar. The immediate answer would appear to be that scholars were addressing the fringes of their own speech community and were not concerned with non-native speakers. It was thus left to Westerners, first the Portuguese and later the Serampore mission community, to address these structural features of the Marathi language.

Notes

I am indebted to Professor Sheldon Pollock for several close readings of successive drafts, to Dr. Allison Busch for an important observation, to Dr. Imre Bangha for a clarification, and to Dr. Gail Omvedt for spotting a significant error. The usual disclaimer holds.

1. Pollock, introduction this volume. The "massive volume" in question is *Literary Cultures in History*.
2. Busch, this volume.
3. Pollock, "Cosmopolitan and Vernacular in History," 592.
4. Tulpule, "Mahānubhava sāhitya samālocanā," 436–40.
5. Pollock, "The Languages of Science in Early Modern India," 23–25.
6. Guha, "Speaking Historically."
7. Names in B. Kulkarni, *Sabhāsad*, 86–89, contrast to Bendrey, *Keśava Paṇḍita's Daṇḍanīti*.
8. Banhatti, *Ājñāpatra*, 57–68, quote on 68.
9. Banhatti, *Ājñāpatra*, 90.
10. Pollock, "Languages of Science," 10–11. Govinda Malhar Kulkarni, in "Bhāṣika Vāṃmaya," 88–89, comments on the paucity of grammatical studies before the arrival of Europeans in India.
11. Tulpule, *Mahānubhava pantha*, 247–51.
12. Tulpule, *Marāṭhī vāṃmayacā itihāsa*, 43.
13. P. N. Kulkarni, *Jñāneśvarī*.
14. Tulpule, introduction to Tulpule and Feldhaus, *Dictionary of Old Marathi*, xi.
15. Talbot, *Precolonial India*, 30–38; Altekar, "Yadavas of Seunadesa," 569–71.
16. Tulpule, "Marāṭhīcā janmakāla," 41.
17. P. N. Kulkarni, *Jñāneśvarī*, 50. Compare the deployment of the medieval ethnonym "Frank" as a personal adjective in the compounds *frankpledge* and *francfief* and in verbal forms such as *enfranchise*.
18. B. Kulkarni, *Sabhāsad*, 25.
19. Tulpule, *Prācīna Marāṭhī Korīva Lekha*, 334–40.
20. Khan, *Arabian Poets of Golconda*.

21. Desai, "Epigraphy," 100.

22. Gosavi, *Śrīsakaḷasantagāthā*, 2:1023.

23. Pindye, *Campū*, 227.

24. Ranade, *Navaras Rāgamālā*, 101.

25. For examples, see Guha, "Transitions and Translations," 26–27. An oversight caused me to credit Tukaram with a poem written by an anonymous disciple (cited here in n. 21). I am indebted to Dr. Gail Omvedt for the correction.

26. Tulpule, introduction to *Bhāṣāprakāśa*, 1–3.

27. Tulpule, introduction to *Tañjāvarī Kośa*, 7.

28. Rajavade, *Sādhaneṃ*, 3:150.

29. Gosavi, *Śrīsakalasantagāthā*, 2:533–74.

30. Khare, "Śivakālīna lekhanapaddhati," 77.

31. Sardesai, *Marāṭhī Riyāsat*, 1:47–126.

32. Banhatti ed. *Ājñāpatra*, 74.

33. See the essays in Ratnavat and Sharma, *Cāraṇa sāhitya paramparā*, especially Bahura, "Maukhika va cāraṇa sāhitya paramparā," 53–58.

34. Pindye, *Campū*, 268–70.

35. Sardesai, *Riyāsat*, 1: 116, editorial note 125–26.

36. Pindye, *Campū*, 226.

37. I am indebted to Dr. Sanghamitra Basu for guiding me through the thickets of page-long *samāsa* in these parts of the text.

38. Pindye, *Campū*, 242–45.

39. Pindye, *Campū*, 246–79.

40. In "Transitions and Translations" I over-hastily identified the language as Bundeli. Dr. Imre Bangha (personal communication) recently pointed out to me that distinguishing seventeenth-century Braj from Bundeli is difficult.

41. Pindye, *Campū*, 227.

42. Khare, "Śivakālīna lekhanapaddhati," 65–96.

43. D. L. Vyasa, *Rājavyavahārakośa* [1676–77].

44. Pagdi, *Hindavī Svarāj*, 17.

45. B. Kulkarni, *Sabhāsad*, 76.

46. A. Vyasa and Khare, *Kāhī patreṃ*, 80–90.

47. Tulpule, *Tañjāvarī Kośa*, 17–18 and *passim*.

48. The letter to King George is cited in the introduction to *Phārśī Mrāṭhī kośa*, 7; the memoir is reprinted in Tulpule, *Prācīna Marāṭhī Gadya*, 96–102.

References

Altekar, A. S. "Yadavas of Seunadesa."1960. *Early History of the Deccan, Part II*, ed. Ghulam Yazdani. Delhi: Oriental Books Reprint Corporation, 1982.

Banhatti, S. N., ed. *Ājñāpatra*. Pune and Nagpur: Suvicar Prakashan Mandala, 1986.

Bendrey, V. S., ed. *Keśava Paṇḍita's Daṇḍanīti prakaraṇam or Criminal Jurisprudence (XVIIth Century)*. Pune: Bharata Itihasa Samshodhaka Mandala, 1943.

Busch, Allison. "The Anxiety of Literary Innovation: The Practice of Literary Science in the Hindi/Riti Tradition." *Comparative Studies of South Asia, Africa and the Middle East* 24, no. 2 (2004), 43–59.

Desai, Ziauddin A. "Epigraphy as a Source Material for History." *Proceedings of Seminar on Medieval Inscriptions*. Aligarh: Centre of Advanced Study, Department of History, Aligarh Muslim University, 1974.

Gosavi, R. R. *Śrīsakaḷasantagāthā*. 2 vols. Pune: Sarathi Prakashan, 2000.

Guha, Sumit. "Speaking Historically: The Changing Voices of Historical Narration in Western India, 1400–1900." *American Historical Review* 109 (2004), 1084–103.

———. "Transitions and Translations: Regional Power and Vernacular Identity in the Dakhan c. 1500–1800." *Comparative Studies of South Asia, Africa and the Middle East* 24, no. 2 (2004), 23–32.

Joshi, Shankara N., and Ganesh H. Khare, eds. *Śivacaritrasāhitya*, vol. 3. Pune: Bharata Itihasa Samshodhaka Mandala, 1930.

Khan, M. A. Muid. *The Arabian Poets of Golconda*. Bombay: University of Bombay, 1963.

Khare, Ganesh H. "Śivakālina rājapatrāñci lekhanapaddhati." *Śivajī Nibandhāvalī* vol. 2. Pune: Shiva Charitra Karyalaya, 1930.

Kulkarni, Bhimrao, ed. *Sabhāsad bakhar*. Pune: Anmol Prakashan, 1987.

Kulkarni, Govinda M. "Bhāṣika Vāṃmaya." *Marāṭhī Vāṃmayacā Itihāsa: Khaṇḍa Cauthā*. Pune: Maharashtra Sahitya Parishad, 1999.

Kulkarni, Madhava T. Introduction to *Phārśī Mrāṭhī kośa*. Pune: Varada Books, 1996.

Kulkarni, P. N., ed. *Eknāthasaṃsodhanapūrva āṇi Upalabdha Pratitīla Adya Hastlikhitācyā Ādhare sampādilelī Jñāneśvarī*. Kolhapur: Shivaji University Prakashan, 1993.

Pagdi, Setumadhavarao. *Hindavi Svarāj āni Mogal*. Pune: Venus Prakashan, 1966.

Pindye, Jayarama. *Rādhāmādhavavilāsacampū*. Ed. Visvanatha K. Rajavade. Pune: Varada Books, 1996 reprint [1922].

Pollock, Sheldon. "Cosmopolitan and Vernacular in History." *Public Culture* 12, no. 3 (2000), 591–625.

———. "Forms of Knowledge in Early Modern South Asia: Introduction." *Comparative Studies of South Asia, Africa and the Middle East* 24, no. 2 (2005), 19–21.

———, ed. *Literary Cultures in History*. Berkeley: University of California Press, 2003.

Rajavade, Visvanatha K. *Marāthyāñcyā Itihāsacīn Sādhaneṃ*. 5 vols. Ed. P. N. Deshpande. Dhule: I. Vi. Ka. Rajvade Samshodhana Mandala, 2002.

Ramkavi. *Bhāṣāprakāśa*. Ed. S. G. Tupule. Pune: V. H. Gole, 1962.

Ranade, Pandhrinatha, ed. *Ibrāhim Ādilshahā dvitīya likhita Navaras Rāgamālā*. Dhule: I. V. K. Rajvade Samshodhana Mandala, 1987.

Ratnavat, Shyamsingh, and Krishnagopal Sharma, eds. *Cāraṇa Sāhitya Paramparā: Professor Virendra Svarūp Bhatnāgar Abhinandana Grantha*. Jaipur: Rajasthana Visvavidyalaya, 2001.

Sardesai, Govind Sakharam. *Marāṭhī Riyāsat.* New enlarged and corrected edition, edited by S. M. Gange. 8 vols. Mumbai: Popular Prakashan, 1988.

Talbot, Cynthia. *Precolonial India in Practice: Society, Region and Identity in Medieval Andhra.* New York: Oxford University Press, 2001.

Tulpule, Sankara G. *Mahānubhava pantha āṇi tyāce vāṃmaya.* Pune: Venus Prakashan, 1976.

———, ed. Marāṭhī Bhāṣecā Tañjāvarī Kośa. Pune: Vhinasa Prakasana, 1973.

———. *Marāṭhī vāṃmayacā itihāsa–I.* Pune: Maharashtra Sahitya Parishad, 1984.

———, ed. *Prācīna Marāṭhī gadya.* Pune: Venus Prakashan, 1993.

———. *Prācīna Marāṭhī Korīva Lekha.* Pune: Pune Vidyapitha Prakashan, 1963.

Tulpule, Sankara G., and Anne Feldhaus. *A Dictionary of Old Marathi.* New York: Oxford University Press, 2000.

Vyasa, Akshayakirti, and Ganesh H. Khare. "Udepurcyā Vyāsagharānyakaḍīla kāhī patreṃ." *Bharata Itihasa Samshodhaka Mandala Traimasika* 33, nos. 1–2 (1952), 77–90.

Vyasa, Dhundhiraja Laksmana. "Rājavyavahārakośa." 1676–77. *Śivacaritrapradīpa,* ed. Kailash N. Sane. Pune: Bharata Itihasa Samshodhaka Mandala, 1925.

CHAPTER THREE

A New Imperial Idiom in the Sixteenth Century

Krishnadevaraya and His Political Theory of Vijayanagara

VELCHERU NARAYANA RAO, DAVID SHULMAN,
AND SANJAY SUBRAHMANYAM

It is pretty much a commonplace that when today's students of political science in India look to the history of political thinking in general, their principal focus is on the long tradition of Western political thought, from Plato and Aristotle, through the medieval Church, to Hobbes, Spinoza, and Locke, and finally to Hegel, Marx, and the twentieth-century thinkers.[1] When India is represented, it is only in terms of the thinkers who lived in colonial India, and who thus reacted overtly to Western political thinking, that is to say the tradition from Raja Rammohun Roy to Gandhi and perhaps Ambedkar. When we apply what the former thinkers wrote to the problems of statecraft and state formation in early modern South Asia the results can be quite disappointing, for the bulk of the Western thinkers after 1500 wrote in an institutional context that is much too distant in many respects from the South Asian context, and in their peculiar terms states such as the Mughal Empire are usually understood only as tyrannies or despotisms. At the same time, a political theory founded on the systematic opposition between church and state is of little conceptual use for historians of Asia, whether in India, Southeast Asia, Japan, or China. Further, the early modern European tradition of systematic intolerance and persecution of minorities (Jews and Muslims in the Iberian Peninsula, Catholics in the Netherlands, etc.) sets it apart from the situation in much of the rest of the world, be it the Ottoman Empire or China.

In this essay we take a rather different point of departure. Our notion is that in the thousand-odd years from 800 to 1800 CE South Asia produced a considerable variety of reflections on the question of statecraft, analogous to what Patricia Crone has recently analyzed for West and Central Asia as the tradition of Islamic political thought.[2] These reflections are in fact as diverse as the states in the region and range from grandiose imperial ideological statements to recipes for the survival of small kingdoms that are squeezed between massive rivals. However, despite this evident diversity one common thread that emerges is that the fundamental purpose of statecraft (which is to say, *nīti*) is to manage social conflict, and this is what really justifies the fact that taxes are claimed and taken from subjects. The idea of the divine right of kings, while present in both Islamic and other contexts in India, is never taken as a sufficient argument, save by very rare authors.

Yet this tradition has also been much misunderstood, especially in the past century and a half. In the second half of the nineteenth century most educated people in India who considered themselves to be modern would have agreed that India urgently needed moral reform. Raja Rammohun Roy's celebrated movement against widow burning in Bengal, as well as various measures taken by the colonial government, had precipitated a wave of concern among religious and moral reformers in the West against this and other "inhuman" Indian practices. Orientalist studies of India's exotic religious and literary texts and representations of the decadent lifestyle of elite Indians had both a direct and an indirect impact on British upper-class society. Many Western missionaries and liberal intellectuals were concerned that the colonial government in India was not taking a more active role in civilizing its subjects in the colony. However, stung by the bitter experience of the Mutiny of 1857 and its aftermath, when Indian sepoy soldiers and several heads of princely states rebelled against the rule of the East India Company primarily—or so the Britons perceived it—for religious reasons, the British colonial rulers were reluctant to interfere directly with the religious beliefs of their subjects. Since most of the social practices of Indians appeared to the British to be sanctioned by one religious belief or other, this meant to them a withdrawal of the government from almost any kind of significant social legislation. However, English was gaining ground as the language of power and creating fresh opportunities for employment. Already by the early decades of the nineteenth century there was in the making a group of English-educated Indians who were to form the basis—as Macaulay was notoriously to predict some three decades later—for "a class of persons Indian in blood and colour, but English in taste, in opinions, in morals and intellect."

Vennelakanti Subbarao (1784–1839), the translator of the *Sadr 'adālat* of the Madras presidency, a Telugu Niyogi Brahman who rose to the highest post a native could aspire to in the East India Company administration at the time, and who commanded competence in about half a dozen Indian languages in addition to English, was one of the most prominent of such persons. After being appointed a member of the Madras School Book Society he submitted a report in 1820 on the state of teaching, in which he wrote that children in schools were taught neither adequate grammar nor morals. Thus they came out of their schools with no real ability in using the language and were not trained to become upright members of their society. Addressing the need for teaching morals, he recommended that "tales extracted from different books composed chiefly of morals written in modern languages" be prescribed for study.[3]

The lack of adequate lines of communication between a society that already possessed a centuries-long set of continuous intellectual traditions and a new political power that had assumed the role of civilizing a group of uncivilized or partially civilized nations was never more striking than at this early point in colonial Indian history. For the traditionally educated Indian intellectual of the early nineteenth century whom the Company might have consulted, India had a sophisticated discipline termed *nīti*, beginning with the *Arthaśāstra*. Then there was a whole range of texts on *dharma*, beginning with Manu's *Dharmaśāstra* (dating perhaps from the early centuries CE) and continuing through the medieval period both in terms of a manuscript tradition and by way of extensive commentaries. But the British administrators and their native assistants in southern India were looking for "moral instruction." Of the two concepts in the Indian tradition that come close to the idea of morals, *dharma* and *nīti, dharma* was seen as somewhat unsuitable for moral instruction because it was too close to the religious world. Manu's celebrated *Dharmaśāstra* was also deeply embedded in the *varṇa* and *jāti* order and discussed legal matters relating to marriages, property rights, and so on. Law courts needed these texts to administer justice to Indians according to their indigenous laws. The story of Sir William Jones's efforts in this direction and Colebrooke's translation of Manu's code for use in the British courts is too well known to be repeated here.

At the same time it was easy enough to argue that there was a direct line of ascent between the medieval regional-language *nīti* texts and the *Arthaśāstra* of Kautilya, and thus to conclude that the regional-language texts were derivative and, if anything, bad copies of an original (however elusive that original was in purely philological terms) and therefore not particularly interesting. Another problem was that since the authors of *nīti* texts invariably claimed to be poets, literary scholars of the late nineteenth century and early twentieth, influenced by notions deriving from Western literary models, began by rejecting any formal literary merit in the texts and then showed no interest in analyzing them seriously for their content. Doubly neglected, the regional-language *nīti* texts were relegated to a sort of intellectual no-man's-land. Yet native schools still needed moral instruction, and in the absence of an Indian equivalent of the Ten Commandments or similar codes of virtue, teachers turned to *nīti* texts to fill the need.

The principal focus of this essay is a particularly salient example of *nīti* discourse from the early modern Telugu context. The author of this text is a rather celebrated figure in south Indian history, the Vijayanagara emperor Krishnadevaraya (r. 1509–29), and the text seems to come from about a decade into his rule. This was a busy time, though perhaps not unusually so, in the military life of Krishnadevaraya.[4] The early years of his reign had seen this ruler, son of the great Tuluva warlord Narasa Nayaka, and his younger wife Nagaladevi, preoccupied first with wars to the north and then with campaigns to the southeast against the Sambuvarayars. But from about 1514 a series of campaigns took him to the distant northeast, where he managed to extend his domains considerably, as far as the Godavari River and its delta. Inscriptions, such as an entire series on the second *prākāra* of the Tirumalai temple, inform us that this ruler, described by an elaborate title which includes the interesting epithet *Yavanarājya-sthāpanācārya*, "the lord who established the kingdom of the Muslims," set out in about 1514 from his capital of Vijayanagara on "an eastern expedition" and went on to capture not only Udayagiri, but such centers as Addanki, Vinukonda, and Nagarjunakonda, to say nothing of the great fort of Kondavidu, where we learn that he "laid siege to it, erected square sheds around the fort, demolished the rampart walls, occupied the citadel [and] captured alive Virabhadraraya, son of Prataparudra Gajapatideva."[5] He then made further substantial inroads into the kingdom of the Gajapati rulers of Orissa, captured a number of members of the Gajapati family as well as subordinate rulers (*pātra-sāmantas* and *manneyars*), but released them, and having given them "an assurance of safety for their lives" eventually returned to the city of Vijayanagara. A second expedition against Kalingadesa, the core Gajapati domain, is reported soon afterward, with the ruler making his way on this occasion through Bezwada (Vijayavada) to Kondapalli, which he occupied. After another series of extensive campaigns that took him as far as Simhadripotnuru and Rajamahendravara he eventually returned to his capital city by late 1516. In the first days of January 1517 the ruler made a triumphal visit to Tirupati, where he offered extensive gifts and grants of gold and jewelry to the god. He would continue to visit this great Vaishnava temple periodically, his last recorded visit being in February 1521.

At some point in the busy years of these northeastern campaigns, perhaps during the second of the expeditions just described, Krishnadevaraya began to compose his great Telugu work, the *Āmukta-mālyada* (The Woman Who Gives a Garland Already Worn), meant to recount the story of the Vaishnava goddess Goda or Antal from the southern town of Villiputtur.[6] Here is how the king himself recounts the circumstances of the text's composition:

> Some time ago, I was determined to conquer the Kalinga territory. On the way, I camped for a few days with my army in Vijayavada. Then I went to visit Andhra Vishnu, who lives in Srikakula. Observing the fast of Vishnu's day, in the fourth and final watch of that god's night,
>
> > Andhra Vishnu came to me in my dream.
> > His body was a radiant black, blacker than a rain cloud.
> > His eyes, wise and sparkling, put the lotus to shame.
> > He was clothed in the best golden silk, finer still
> > than the down on his eagle's wings.
> > The red of sunrise is pale compared to the ruby on his breast.

The god, it turns out, was familiar with the earlier literary works that the king had set down (but which are lost to us today): the story of Madalasa, that of Satyabhama, and works called the *Jñāna-cintāmaṇi* (The Gem of Wisdom) and *Rasa-mañjari* (The Pleasures of Poetry). But he still expressed disappointment that the king had composed nothing in Telugu, asking, "Is Telugu beyond you? Make a book in Telugu / now, for my delight." In the dream the god even specified the story to be told, that of "the girl who had given me a garland she wore first."

The king could hardly refuse such an imperious command, despite the fact that he was almost certainly more confident in his usage of Tulu, Kannada, and Sanskrit than of Telugu. He writes, "[On waking] I held court in the presence of my army men and subordinate kings, but dismissed them early to their homes. Then I called the scholars learned in many old texts, of various traditions, honoured them, and related my good dream. They were thrilled and astonished." The scholars urged him to compose the work, describing him as "sole sovereign in the fields of letters and wars."[7]

The work opens then with a stylized literary device of "objective" pretensions, in which the Vijayanagara ruler is described together with his family not by himself in an autobiographical mode, but in the words employed by his courtiers; this part of the work includes a section from Allasani Peddana's description of Krishnadevaraya in his *Manucaritra* and has at times misled commentators into attributing the *Āmukta-mālyada* to the court poet instead of the ruler. Elements of similar descriptions can also be found in inscriptions, as we see from one of the great Sanskrit epigraphs (in Telugu script) on the east wall of the third *prākāra* of the Tirumalai temple. Here the genealogy of Krishnadevaraya is traced back to the moon, then to Buddha, Pururavas, Ayu, and Nahusha. The first of the "Tuluva lineage" to be mentioned is Timmabhupati (and his wife Devaki), followed by his son Isvara and spouse Bukkamma, and then a certain Narasabhupala, to whom is attributed the conquest of the Ceras, Colas, and Pandyas, as well as of the Gajapatis and certain Muslim rulers.[8] This is none other than Krishnadevaraya's own father, Narasa Nayaka (d. 1503), kingmaker and warlord in the late fifteenth century. The inscription assures us that he "established his fame firmly over the country extending from the Ganges to Lanka and between the western and eastern ghats." We are now almost to the present time of the inscription, as we see from the following passage (lines 13–14), in which we learn that "by his queens Tippaji and Nagaladevi, like Dasaratha by Kausalya and Sumitra, the great king [Narasa] had two sons called Vira Narasimha and Krishnaraya, like unto Sri Rama and Lakshmana."[9] This is a rather modest introduction to Krishnadevaraya, one that seems to afford a far greater degree of glory to his older brother, who actually reigned a mere six-odd years. Indeed the verses that follow confirm this impression, detailing as they do the rule of Vira Narasimha, his conquests, his pilgrimages, and his sixteen *mahādāna*s. To be sure, Krishnadevaraya is the object of extensive praise in the following verses, now that he bears the earth on his shoulders. Here is how the king appears to those who composed the inscription in his honor:

> Filling the whole world with the camphor of his glory, and the renown acquired by his generosity, heroism and prowess, which have been praised by all the poets, terrible on the field of battle, putting in the shade the acts of Nriga, Nala, Nahusha, Nabhaga, Dundhumara, Mandhatru, Bharata, Bhagiratha, Dasaratha, Rama and others, protectors of the Brahmans and vanquisher of the sultans, the fever of the Gajapati elephants, well-versed in all the arts and sciences, with a face that overshadows the lotus, a second Bhoja, with deep insight into poetry, drama and rhetoric, and the true knowledge of righteous action.[10]

Here is a king who is not merely a warrior and a protector, patron to Brahmans and to the great spring festival, but a ruler who, unlike his father and brother, prides himself on being a connoisseur of poetry and theater, both Bhoja and Kalidasa rolled into one.

The Frame Text

This description of the warrior and poet king is what largely defines the tone of the *Āmukta-mālyada* itself. When we enter into the main body of the text we are transported directly to the Tamil country, beginning with a detailed and realistic description of the small town of Villiputtur, followed by a description of the principal personage in the story, a Brahman by the name of Vishnucitta. Throughout these sections the work shows a considerable preoccupation with a form of realism, as well as with providing detailed evidence that its royal author has a familiarity with life outside the palace, perhaps a sign of his days as the son of an up-and-coming warlord. A remarkable feature of this early part is a detailed version of the menu that Vishnucitta has available to feed guests in different seasons, and which takes us to the end of the first section (*āśvāsa*).

Yet we should not imagine that the whole text is simply dominated by the realistic mode. Instead the second *āśvāsa* opens with an elaborate and rather fantastic description of the city of Madhura (Madurai), far less effective however in poetical terms than that of Villiputtur, in the course of which a certain Pandya king (referred to as the Matsyadhvaja ruler) is introduced. He is described as a successful king, yet one day he decides to renounce his kingdom when he hears a Brahman singing a verse on the subject; this makes him the first of a whole series of renouncing kings in the text. At the moment when he has decided to retire from the affairs of state a debate is announced to resolve the vexed question of which of the various gods is best. In his heaven Vishnu is anxious to defend his own claim, and so instructs Vishnucitta to go to Madhura and defend his position as a Vaishnava, and also to convert the king. The text now shifts modes, as the *āśvāsa* ends with another meticulous and realistic description, this time of Vishnucitta's travel to the contest.

The third *āśvāsa* is a chapter focused on textual learning, with the assembled scholars in the courtly debate arguing in favor of one or the other position. The texts cited by them are once more very accurately quoted in what is effectively an authorial tour de force. As might be expected, Vishnucitta handily defeats the other pandits in the contest. He then tells them a story of two rival kings, who are cousins; one is a great expert in rituals, and the other in the path of knowledge. The two fight a battle, and the specialist in rituals is defeated and deprived of his kingdom. The winner, King Keshidhvaja, uses the occasion of his victory to start a royal ritual, but the sacrificial cow is snatched away and eaten by a tiger. The king is humiliated; he knows of no appropriate form of atonement, and so he has to go to his defeated rival, Khandikya, and submit to his superior expertise in matters of ritual. Khandikya could kill him in his moment of weakness, but decides not to. He instead gives him good advice; Keshidhvaja performs the ritual successfully and returns to give his rival his *guru-dakṣiṇā*. He offers to restore his cousin to his kingdom, but Khandikya decides he does not really want it back. Keshidhvaja also decides at this stage to retire, and gives his own kingdom over to his son. These renunciations are presented as the ideal acts for a ruler, and Khandikya himself (perhaps the nobler of the two) is presented as the champion of Srivaishnavism. The ideal devotee Vishnucitta, having recounted this story, returns to his hometown of Villiputtur, and the chapter closes.

In the fourth *āśvāsa* we return to Vishnu, who is now engrossed in conversation with his spouse, Lakshmi, to whom he singles out for praise two human beings, Vishnucitta and Yamunacarya. The first of the two is well known to us (and to the goddess) by now, but Lakshmi is curious to know more about the other character. In response to her questions Vishnu himself tells the story of Yamunacarya. This involves still another Pandya king, who becomes a fervent Virasaiva and takes it upon himself to humiliate Brahmans and Vaishnavas. But his wife is a Vaishnava, who, while loving and devoted to her husband, still keeps her own religious ways. Vishnu encourages a young Brahman bachelor named Yamuna to go to the king and tell him he wants a debate in his court on the superiority of Vishnu. Instead of directly asking the king, Yamuna approaches the queen to this end. Through her intervention the king agrees, but warns the young Brahman that if he is defeated he will himself have to become a Saiva. Again the debate is described with accuracy and detail. Yamuna naturally wins the debate, and the Pandya king offers him his sister in marriage. This is a chance for Yamuna's children to become the heirs to the throne, since succession in the kingdom is matrilineal, from uncle to nephew. Yamuna accepts the tempting offer, marries, and soon becomes lost in the pleasures of courtly life—a far cry from his earlier life as a Brahman bachelor.

In the autumn of a certain year Yamuna sets out on a *digvijaya* tour of world conquest. On the face of it, it would seem that his life is now ideal, but true Vaishnava devotees are worried about him, and the fact that he has partly lost his bearings. One of these true devotees, a Brahman named Srirama Mishra, offers a vegetable called *alarkaśākam* to the palace cooks to be used in the king's food. His act is based on the theory that the food one eats influences the thoughts one thinks; this particular vegetable will presumably clear up the clouds in Yamuna's mind. The king inevitably likes the vegetable and asks to see its supplier. In the course of the audience the Brahman recites an intriguing verse to describe a treasure that the king potentially has available to him. The poem in fact functions at two levels, for the treasure could be literal, or it could be the god himself, Ranganatha-Vishnu. At this point the king has a vision in which he sees Ranganatha and decides that he must henceforth look to his own salvation rather than pursue royal power. He renounces his kingdom and decides he must teach his own son political ethics (*rāja-nīti*) as he is leaving. This central—and deeply political—section of the work thus comes inserted inside two frames: Vishnu is recounting the story to Lakshmi, and inside his account, Yamunacarya is speaking to his son.

It is to this section that we shall return presently in greater detail, but the constant preoccupation in the work with the desire of the king to renounce his kingdom certainly calls for some immediate comment. At least one source, the account of the Portuguese trader Fernão Nunes (from the 1530s), offers us some insight into this question by recounting a story drawn from the Kannada oral tradition. In this retrospective view, in the summer months of 1509, as Vira Narasimha Raya lay dying, he ordered his minister Saluva Timma to blind Krishnadevaraya in order to ensure the clear succession of his own son. Nunes continues: "Salvatina [Saluva Timma] said that he would do so and departed, and sent to call for Crisnarao [Krishnaraya], and took him aside to a stable, and told him how his brother had bade him put out his eyes and make his son king. When he heard this, Crisnarao said that he did not seek to be king, nor to be anything in the kingdom, even though it should come to him by right; that his desire was to pass through this world as a *jogi*."[11]

The theme of the king as *jogi* is widespread in the period; we also encounter it with another contemporary monarch, Sultan Bahadur of Gujarat (r. 1526–37). But in the case of Vijayanagara it is for some reason particularly associated with Krishnadevaraya rather than with either of his brothers, who were also ruling monarchs. In a certain version it is even refracted in the later text, the *Rāyavācakamu* of the early seventeenth century, where the king is so frustrated by the power wielded by his ministers that he threatens to renounce his throne.[12]

In the fifth chapter of the *Āmukta-mālyada* we come at last to the story of Goda Devi, the tale that Krishnadevaraya was expressly commanded by Vishnu to tell. This story takes us back to Vishnucitta, who finds an abandoned baby girl in a mango grove under a *tulasi* plant and brings her up as his own adopted daughter, as he has no children of his own. The girl, who is Antal or Goda, grows up and falls in love with the god Vishnu. It would seem that this chapter draws on some earlier version from Villiputtur (perhaps the *Divyasūricaritra* or some other *purāṇic* text), but the circumstantial material is substantially truncated in this later Telugu version. The key part of the plot, which concerns the god's garland and the girl's wearing the garland before it is offered to Vishnu, is also much reduced. Instead there is a very long *viraha* section concerning the girl and the god, which largely takes place in the springtime. This incidentally allows the seasons, as is standard in all *mahākāvyas*, to play a central role in the text as a running line of development, for the earlier section on Yamunacarya has largely centered on the autumn. Of particular interest, though, is the fact that Yamuna is associated with the colder and darker season and Goda with the spring, as if the text were moving toward the more emotional and sensual pole of life. Vishnucitta eventually discovers what love is through his daughter, a feature of the story which is presented as somewhat comical. When Vishnucitta complains to the god in Villiputtur, Vishnu instructs him to take Goda to his temple in Srirangam.

At this point (we have by now spilled over into chapter 6) the plot takes on another twist, involving still another tale within a tale, for in the course of their conversation Vishnu himself tells Vishnucitta the story of the untouchable Maladasari and a certain Brahmarakshasa who wants to eat him up. Eventually the Brahmarakshasa realizes that he has gone down the wrong path and reforms his ways. The seventh chapter tells us how the Brahmarakshasa came to have the fate he did on account of his own greed and acquisitive nature, and also describes how he eventually becomes a normal human being through Vishnu's grace. At first Vishnucitta does not understand why he is being told this story and does not realize that the god is delicately hinting at his own acquisitive nature and reluctance to give up his daughter. Yet eventually he returns home and takes his daughter to the great temple at Srirangam. Here the god promptly falls in love with her, takes her into his palace, and also substitutes a magical image for the girl, which he returns to the devotee. But Vishnucitta, who is after all no fool, soon finds out he has been tricked and denounces the god for having stolen his daughter. Vishnu tells him to return home, where he miraculously finds his daughter. The devotee now realizes that he too has fallen short in his devotion, and that all his knowledge was really not enough for him to pass this test. At this point Vishnu sends Brahma, Rudra, Sarasvati, and Parvati to ask a contrite Vishnucitta for his daughter in marriage. He naturally agrees, and a wedding is performed in Srirangam, with which the text closes.

Underneath its layered complexity the structure of the *Āmukta-mālyada* text presents us with certain clear patterns. We can see it as a sort of protonovel, and the stories consistently are concerned with images of kings who renounce or are taught yogic dispassion. Despite its twists and turns the text remains well integrated, discursive, philosophical, and astonishingly personal in tone. A distinction is drawn—perhaps for the first time in southern India—

between the king as individual, with his individual inclinations and exigencies, and kingship as institution (which has to go on at all costs). A basis for stable kingship is elaborated around Srivaishnava and translocal values, with a yogic coloring, an aesthetic component (expressed through music), and strong themes of personal, nonascriptive loyalty. All this is expressed in a striking register of Telugu, one that is constantly creative yet at times harsh and reflective of the fact that its author is after all a non-native speaker, a royal transplant as it were from the fertile plains of the Tulu-speaking west coast to the Telugu cultural domain.

The complexity of the framing is also worth remarking upon, beginning with the external superframe, for Krishnadevaraya himself tells these stories in the first instance to the god Venkatesvara. At two points a double frame emerges within the superframe. The first of these is the *rāja-nīti* section, where Yamuna speaks to his son and Vishnu reports this to Lakshmi; in the second frame Vishnu tells Vishnucitta the story of Maladasari, within which Goda herself is said to have heard the story from Varaha at an earlier moment, when he lifted her up, as Bhudevi, from the ocean. In this second case, to render matters still more complex, there is also the explicit memory of the double frame. The recourse to this elaborate poet-listener (author-patron) frame allows for wide variation in meter, mood, and theme, in a manner that is arguably foreign to Sanskrit *kāvyas*.

On Royal Advice

It is notoriously easier to give advice than to take it. But what if you are giving advice to yourself? We can see the *rāja-nīti* section of the *Āmukta-mālyada* as standing at the confluence of two traditions, a fact that is rendered all the more complex by its remarkably intricate framing. On the one hand a long Sanskrit and vernacular tradition exists of texts of *nīti*, of political wisdom, that can be traced back to the *Arthaśāstra* of Kautilya, a text that was certainly known to Krishnadevaraya and those around him. In a Tamil context a work such as Tiruvalluvar's *Kuṟaḷ*—on which François Gros has written extensively—while embracing a variety of subjects and materials, also functions in an advice mode, even if its principal object is not to form the behavior of rulers.[13] Besides, closer to hand we find medieval texts from the Deccan, such as the *Mānasollāsa* of the twelfth-century Calukya king Someshvara III, which like the *Āmukta-mālyada* has the specific feature of possessing a royal author.[14] The medieval Deccan also produces a number of vernacular *nīti* texts, of which the best known is probably the Telugu text by Baddena, but of which examples could be multiplied.[15] Such works as these can certainly be seen to participate in a culture of political realism, and thus give the lie to those who have argued that precolonial politics in India was conceived along purely idealist lines. Yet the realism of these texts often dissolves into a series of clichés rather than demonstrating a capacity to react or adapt to the particular circumstances of one or the other polity in its historical context. At the same time, the genre of the "Mirror for Princes" is well known in the Indo-Islamic context, where a number of such texts exist both from the time of the sultanate of Delhi, under the later Mughals, and from the regional sultanates such as those of the Deccan.[16] Such texts, often written in Persian, are themselves at times influenced by Indic models such as the *Pañcatantra*, known in the Islamic world through its translation as the *Kalila wa Dimna*. Yet they also bear the clear imprint of the nontheological perspective on kingship that emerged in the Islamic lands in the aftermath of the Mongol conquests, when Muslim advisors and *wazīr*s struggled with the problem of how to advise *kāfir* rulers and princes on the matter of government without taking them into murky and controversial theological waters. The "Mirror for Princes" genre ranges wide and attempts to do everything from forming the prince's musical tastes to refining his table manners, but the core of the matter is usually politics, both in the sense of diplomatic relations between states and relations between a prince and his companions or between different elements in a courtly setting.

The *nīti* section of the *Āmukta-mālyada* appears to be aware of these different traditions, and even draws upon them quite explicitly.[17] Yet in contrast to the typical "Mirror for Princes" Krishnadevaraya offers a top-down, hands-on vision, partly rooted in pragmatic experience, partly creatively adapting the existing literature of *nīti* statecraft. This is no armchair pontificating but a largely practical synthesis reflecting the political, economic, and institutional changes of the early sixteenth century. Still, highly individualized statements that can be attributed directly to the book's author do alternate with verses that seem to be lifted from standard *nīti* texts about politics and kingship. Nonetheless we are left with the impression of a unique concoction of pragmatic wisdom, specific constraints, an inherited normative politics, and a meditative sensibility capable of formulating something entirely new. It may be useful in this context to recall, even if briefly, the political context in which the Vijayanagara ruler found himself.

Vijayanagara expansion over the peninsula begins in the later decades of the fourteenth century, and one of the significant moments is undoubtedly Kumara Kampana's southward campaign that takes him as far as the conquest of the sultanate of Ma'bar, based at Madurai. However, it seems likely that Vijayanagara rule in the fifteenth century was a relatively fragile affair once one moved away from the northern core of the kingdom, and we are aware that local rebellions and other forms of resistance occurred periodically, as for instance in the 1420s.[18] Our impression of Vijayanagara in the mid-fifteenth century, at the time of the rule of Devaraya, is still of a largely westerly kingdom, with contested frontiers and boundaries, as well as internal spaces that had been rather imperfectly integrated into the polity.

When Krishnadevaraya ascended the throne in 1509 (the earliest inscription bearing his name as ruler is dated to 26 July, and his coronation probably took place on 8 August) a number of crucial problems regarding political management remained to be resolved.[19] The first decade of the sixteenth century is generally thought to have been marked by internal rebellions because of the contested succession of his brother Vira Narasimha Raya, who had initially ruled as regent when the last king of the Saluva dynasty still held the throne. At the death of Saluva Narasimha Raya (r. 1456–91) his son Immadi Narasimha, or Dharmaraya, ascended the throne, but real power had increasingly fallen into the hands of a powerful warlord from the westerly Tulu region, Narasa Nayaka, and after the latter's death in 1503 into the hands of his son. It was this son, Narasimha Nayaka, who arranged the assassination in 1505 of Immadi Narasimha and then himself seized the throne as Vira Narasimha Raya. A contemporary Portuguese source dated 1506 informs us succinctly of these events: "The old king [Saluva Narasimha] had died quite some time ago, and after him, his son became king, but he was then killed too, and so there have been wars ever since."[20] These problems, both with regard to the neighboring kingdoms and to other sources of autonomous power within the bounds of the Vijayanagara state, seem to be central to the preoccupations of the *rāja-nīti* section. On what institutional basis could there be a lasting form of power for a fledgling dynasty, when the scale of kingship extended far beyond the confines of a compact region? Whom could one trust, and who had to be controlled?

Another way to formulate the dilemma that this king confronted is in terms of an enduring tension between local and translocal forces, a near universal problem in India in this period. There is a consistent effort to conceptualize some basis for a translocal polity that could extricate the state from its constant resubmergence in diffuse local contexts. A striking element in this conceptual effort lies in the king's own dynastic origins in one of the most marginal and recently conquered localities: the western coastal plain of Tulunad. A kind of upstart, whose own family inheritance dictated that he prove himself outside the family context, finds himself articulating, somewhat inchoately, a vision of transregional, highly personalized loyalties. Krishnadevaraya eschews the standard solution, namely the resort to *purāṇic* and *dharmaśāstric* normative language regarding an alliance between Brahmans and Kshatriyas. Instead his preference is markedly for yogic and renunciatory themes that at the same time are strongly and paradoxically allied to a Srivaishnava idiom rooted in the idea of *bhoga* (enjoyment), a theme to which we shall return below. We are incidentally aware that the Vijayanagara dynasty had switched its primary allegiance earlier, from Saivism to Vaishnavism, and here we see a bold attempt to imagine a royal order glued together with Vaishnava *bhakti* in both its renunciatory and sensual aspects.

Once a transregional state system is conceivable, its ruler runs up against its external boundaries. The *manyam* forest regions (especially the northern and northeastern frontiers but also implicitly to the southwest in Coorg and the Western Ghats) thus figure prominently in the *rāja-nīti* and require special treatment. External boundaries, however, coexist with the internal wilderness, as we see in a verse about the farmer marking off his field and then slowly making it free of stones and other impediments. But the text is also marked by a consistent suspicion, at times bordering on hostility or even contempt, for peoples like the Boyas and the Bhils, who could be found both at the border regions of the empire (in the northeast) and at the internal frontier. A prose passage within the *nīti* section thus advises the listener, "Allay the fears of the hill-folk, and bring them into your army. Since they are a small people, their loyalty or faithlessness, their enmity or friendship, their favour or disfavour, can all easily be managed." Another passage, this one in verse, runs as follows:

> Trying to clean up the forest-folk
> is like trying to wash a mud wall.
> There's no end to it. No point in getting angry.
> Make promises that you can keep and win them over.
> They'll be useful for invasions, or plundering an enemy land.
> It's irrational for a ruler to punish a thousand
> when a hundred are at fault. (257)

This then is *rāja-nīti* for building an empire, composed by a rather introspective, yet by now quite experienced king, who has been on the throne for perhaps a decade. In certain key respects the author departs quite some distance from conventional wisdom. For example, he recommends posting Brahmans as commanders of forts (*durga*), and the fact that this was practical advice is shown by studies of the prosopography of the notables of the empire in that time.[21]

> Make trustworthy Brahmans
> the commanders of your forts
> and give them just enough troops,
> to protect these strongholds,
> lest they become too threatening. (207)

Brahmans, in this view, have certain clear advantages over non-Brahmans, even though this caste is, theoretically at least, not to be associated with warrior functions (though numerous exceptions, both in the epics and earlier historical instances, could be found):

> The king will often benefit by putting a Brahman in charge,
> for he knows both the laws of Manu and his own dharma,
> and from fear of being mocked
> by Kshatriyas and Sudras,
> he will stand up to all difficulties. (217)

Beyond this, however, lies the Brahmans' relative freedom from local attachments. At the same time these Brahmans are clearly trained by now in military ways and engaged in worldly activities. This seems a clear precursor to the early eighteenth-century Maratha state and its so-called Brahman Raj under the Peshwas. In this latter case we see the rise to dominance of a secular Brahman elite, Persianized in its culture and largely made up of members of the Citpavan and Konkanastha subcastes. Between these two moments one can also cite the late seventeenth-century political conjuncture in the sultanate of Golkonda, when the Brahman brothers Akkanna and Madanna dominated the land-revenue administration and appropriated a series of military functions.[22] Arguably these later developments should be seen as variations on a theme that had already been set out by Krishnadevaraya.

But the king's leanings toward Brahmans do not extend as far as those learned men established in institutional settings that can become autonomous foci of power:

> If you are partial to learning
> and give lands and money away to the learned,
> mendicants, monks and men with matted hair
> will become swollen-headed.
> Famines, sickness and infant deaths will increase.

Just show devotion to the learned,
and if they resent their poverty—don't be concerned. (242)

This verse seems to be directed against *maṭhādhipatis*, the heads of sectarian *maṭhs*, and seems to reflect real tensions that existed in this context in fifteenth-century Vijayanagara.[23]

At the same time the text clearly reflects a state that is in movement, expanding outwardly. There is a strong sense of movement, incorporation, and the attempt to stabilize new areas and their peoples. With this comes a groundswell of economic expansion through external trade. Foreign traders are to be treated with respect, ports and harbors to be cared for. There is thus a feeling—which the Portuguese observers of the early sixteenth century confirm—that the kingdom was eager to foster external relations, quite distinct from the thalassophobia that could be found elsewhere in some parts of Asia in the same period.

Manage your ports well,
and let commerce increase
in horses, elephants, and gems,
pearls, and sandal-paste.
When drought, sickness and calamity
make foreigners seek refuge in your lands,
shelter them in keeping with their station.
Give out gardens, yards, and mines
to those whom you wish to favor. (245)

None of this is innocent, of course, in view of the close links between external trade and the import of strategic war animals. Fernão Nunes claimed in the 1530s that already in the late fifteenth century Saluva Narasimha had a monopolistic policy in this regard: "He caused horses to be brought from Hormuz and Aden into his kingdom, and thereby gave profit to the merchants, paying them for the horses just as they asked." Nunes adds in a later passage, devoted to the reign of Krishnadevaraya, "The king buys thirteen thousand horses of Ormuz, and country-breds, of which he chooses the best for his own stables, and he gives the rest to his captains, and gains much money by them."[24] A relatively simple verse in the *nīti* text confirms the claim of Nunes in an absolutely explicit manner:

Merchants who bring elephants and good horses
all the way from distant islands,
should be given villages and good houses and a favorable price

and honored with audiences,
so that they come to you in the normal course,
and the war-animals don't go to the enemy. (258)

At the same time the king must have an eye for investment in infrastructure, husbanding the resources (and especially the agrarian resources) at his disposal:

The extent of a country is the source of its wealth.
Make small ponds and channels,
so that even poor people can pay taxes in cash and kind.
Give them concessions,
and if you let them improve,
both your treasure and virtue will increase. (236)

Yet this is clearly a Srivaishnava-flavored form of legitimation, unafraid of enjoyment (*bhoga*), centrist, stable, and to an extent a precursor of the unabashedly hedonist view of the polity put forward in later Nayaka times by the smaller successor states spawned by Vijayanagara.[25]

The money you spend on elephants, horses, and their fodder,
and on salaries for your soldiers,
and for Brahmans, and temples,
and your own luxuries too,
this is money well spent. (262)

Asceticism has its rightful place, of course, but only in a sequential sense—after kingship has been renounced. There is no sense in denying oneself the pleasures of kingship while one is on the throne. This too is a form of realism, perhaps linked to the idea of kingship as display. For what is a king whose subjects do not see him as the First Consumer and leading precisely such a life of pleasure in the palace!

A king should enjoy life, each pleasure
in its season—massages, baths, good food,
ointments, rich clothes, and flowers.
The merit that comes from severe control of the body
he can attain by charity. (280)

There is at times even a blurring in this view between the king's own body and the body politic, as in a witty split-meaning verse, which can be read in two ways, as we have translated it using two columns (verse 270). We see in this verse a preoccupation with how the king treats his own body, a feature that even finds a reflection in the Portuguese accounts of the epoch, and reminiscent too of the preoccupation with the daily routine of the gods themselves in Vaishnava temples such as Tirupati. Thus we have the description of Domingos Paes from around 1510:

> The king is accustomed to drinking a quantity [*quartilho*] of sesame oil every day before daylight, and he anoints himself all over with the said oil; he covers his loins with a small cloth, and takes in his arms great weights made of earthenware, and then, taking a sword, he exercises himself with it till he has sweated out all the oil, and he then wrestles with one of his wrestlers. After this labour, he mounts a horse and gallops about the plain in one direction and another till dawn, for he does all this before daybreak. Then he goes to wash himself, and a Brahman washes him whom he holds sacred, and who is a great favourite of his and is a man of great wealth.[26]

We have already learned from an earlier passage in Paes's text that despite this rather rigorous fitness regime "the king is of medium height, and of fair complexion and good figure, rather fat than thin; he has on his face signs of smallpox." Besides being generally "cheerful of disposition and very merry" and though seen even by the Portuguese as a "most perfect" and just ruler, he is still regarded by those in his court as subject to fits of melancholy and changes of humor (*grandes súpitos*). These are characteristics that a later historiography in the same century would attribute to another ruler, the Mughal emperor Akbar (r. 1556–1605).

Whatever the symbolic import of the king's body, the reader of Paes is left in no doubt that this is an eminently practical and pragmatic kingship, one in which the king must play a constant and somewhat interventionist role rather than leaving all routine business to his subordinates. The king may exercise and attend to his body—he is certainly much given to pleasure—but he is equally a manager. We can see this from his idealized daily routine as described by Paes:

> Thence he goes to a building made in the shape of a porch without walls, which has many pillars hung with cloths right up to the top, and with the walls handsomely painted. . . . In such a building, he despatches his work with those men who bear office in his kingdom, and govern his cities, and his favourites talk with them. The greatest favourite is a man called Temersea [Timmarasu]; he commands the whole household, and to him all the great lords act as to the king. After the king has talked with these men on

subjects pleasing to him he bids enter the lords and captains who wait at the gate, and these at once enter to make their salaam to him.[27]

Pragmatic management with a set of record keepers is a feature of Vijayanagara that had been remarked even a century earlier by the Timurid ambassador 'Abd al-Razzaq Samarqandi in the 1440s. 'Abd al-Razzaq in particular focused on the office of the so-called *dannāyak,* whom he associates with the revenue administration (*dīwān-kh̲āna*), the place where he works being described as a pillared hall that is compared to the classic *chihil sutūn* (forty-pillared hall) of Sassanid times. Here one finds a number of scribes (*nawīsindagān*) who keep records, either on Indian coconut-palm leaf (*ba barg-i jauz-i hindī*) or, when the record is meant to be permanent and reliable (*daftar-i mu'tabar*), with white color on black stock.[28] But some things have changed between the mid-fifteenth century and the second decade of the sixteenth. The rather compact if impressive kingdom of 'Abd al-Razzaq's times has become a sprawling empire. The Vijayanagara military machine, with its emphasis on an effective cavalry, has begun by the early sixteenth century to pay heavy dividends, and it is possible to conceive of campaigns not only against the Bijapur sultans but the far less accessible Gajapatis in Orissa. At the same time excessive and unrealistic ambition is to be spurned: *Do your best, you don't have to achieve the impossible* seems to be the underlying message of *rāja-nīti.* Besides, even if the king keeps himself fit for warfare, there is really no point in exposing oneself to unnecessary risk.

> A king should never go himself on campaign.
> For his own comfort, he should find someone who is competent.
> A weak commander won't do. Even a good one needs full support—
> cash, elephants, and horses. Moreover, if you send a non-Brahman
> he'll turn against you. Give him—the Brahman commander—
> land equipped with forts and armies. (255)

It is true that Krishnadevaraya did not always follow this advice, as we see in the campaigns against Bijapur, but there is little evidence that he exposed himself in a constant and foolhardy fashion—as his son-in-law Aravidu Ramaraya did in the 1550s and 1560s, with notably disastrous consequences in 1565. The vision and the language in which it is expressed here is rather dispassionate, and all this is posed moreover in the form of a self-ironic, meditative vision:

> The king is nonviolent, though he kills.
> Chaste, though he has women.
> Truthful, though he lies.

Ever fasting, though he eats well.
A hero, though he uses trickery.
Rich, though he gives away.
Kingship is rather strange. (278)

As noted above, other sections of the text are far more conventional, dealing with the usual means with which one should treat neighboring kings (both the weak and the strong), administer justice, deal with spies and soldiers, and handle the delicate matter of rewards to subordinates. The potential for conflict between kings and ministers, which would be a staple of the histories and treatises produced by the *karaṇams*, the class from which the ministers themselves came, is already present as a theme here, though its resolution is rather more to the king's advantage. The following extended passage makes this clear enough.

Employ Brahmans who are learned in statecraft,
who fear the unethical and accept the king's authority,
who are between fifty and seventy,
from healthy families,
not too proud, willing to be ministers,
capable of discharging their duties well.
A king with such Brahmans for just a day
can strengthen the kingdom in all its departments.
If such ministers are not available,
a king must act on his own,
and do whatever he can.
If not, a bad minister can become
like a pearl as large as a pumpkin—
an ornament impossible to wear.
The minister will be out of control,
and the king will live under his thumb. (211–13)

The penultimate verse of this section of the text sums up this grounded, pragmatic, vision of kingship:

Don't assume that kingship
inevitably leads to wrong,
or that you can't escape it.
Texts don't ask the impossible.
They just tell you: do your best. (284)

Conclusion

It may be argued that the *rāja-nīti* section of the *Āmukta-mālyada* embodies a paradox since it gives pragmatic advice on rulership while at the same time is framed in a text in which the ideal that is insisted upon is renunciation. The paradox, however, is easily resolved, for the central thrust of the Srivaishnava political theology in which this advice comes embedded is that it is necessary to separate rulership as a social function—which is not merely desirable but essential for the world to cohere—from the personal salvation of the ruler. It is understood that at some moment or other in his life the king will wish to escape the bonds of his royal life and become a renouncer. The text does not suggest otherwise; on the contrary, it expresses its unbounded admiration for such rulers. But the view is clearly that renunciation is not a solution to the problems of kingship, only to the problems of the king as individual. We are not quite in the realm of the "two bodies" of the king developed by E. H. Kantorowicz, but what is obviously being proposed here is a view in which Srivaishnava this-worldliness plays a central role in protecting the king from renunciatory excess.

Of particular interest for us is that such a vision does not die out with Krishnadevaraya or even the Tuluva dynasty. We find echoes of it in later works, whether normative or not, concerning both kingship and the actions of kings. Three-quarters of a century later the great Mughal poet Abu'l Faiz ("Faizi") created a version of another exemplary story of kingship by transposing the tale of Nala and Damayanti (or Nal and Daman) into an Indo-Persian context.[29] Faizi's Nal is a great king, but one who loses his sense of balance, drawn as he is into the whirlwind of his passion (*junūn*) for Daman. Wisdom on political matters comes hard, and in Nal's case only at the price of extended exile and enormous personal suffering and transformation. Yet in Faizi's great didactic poem Nal ends his royal life after having ruled wisely for many years and hands his kingdom over to his son, with suitable words of advice (*nasīhat*). Curiously these words of advice echo other words that have been uttered in the same work, these by a renouncer king to Daman's father early in the text. The parallels between the *Āmukta-mālyada* and Faizi's *Nal-Daman* are striking but by no means conclusive evidence that the Mughal poet laureate had any knowledge of a south Indian precursor. They certainly do suggest however that the problems of kingship that were posed in Krishnadevaraya's great work had a significance that extended beyond his own lifetime and even that of his empire.[30] They offer us a glimpse into a political world in which the aesthetic and the pragmatic could be combined in ways that are as effective as they are surprising.

Appendix: A Fresh Translation of the Rāja-nīti *Section*

Rāja-nīti 4.204–85.

204

Never tire of protecting people.
When people in distress cry out to you,
listen to them, remove their suffering.
Don't give any responsibility to bad people.

205

The people [*rāṣṭramu*] seek the welfare of a king
who holds their well-being in his heart.
Don't discount this truth.
If all the people, from the Brahmans down,
want the same thing,
God, who moves inside them,
will surely give it.

206

A king should be obeyed.
Abhira pastoralists and Bhilla tribals obey a leader
who sends his order together with a string tied to an arrow.
So the command of a great king
should strike terror into everyone.

207

Make trustworthy Brahmans
the commanders of your forts
and give them just enough troops,
to protect these strongholds,
lest they become too threatening.

208

Don't give someone a high position
and then degrade him.
Nobody remembers that he began with nothing.
Instead he'll become resentful.

In keeping with his character, raise him up by steps,
and get work out of him in the ripeness of time.

209

These are the untrustworthy:
Those of low birth, or who live in a Boya village,
the uneducated, those who lie fearlessly,
the violent, those too-clever by half,
the foreigners, the unethical.
If you want to be king, reject all of these, even if they're Brahmans.

210

Avoid those raised in isolated villages.
Remember the story of the fallen Brahman
from one such place,
who killed the crane that had saved him,
just to eat for a day.

211

Employ Brahmans who are learned in statecraft,
who fear the unethical and accept the king's authority,
who are between fifty and seventy,
from healthy families,
not too proud, willing to be ministers,
capable of discharging their duties well.
A king with such Brahmans for just a day
can strengthen the kingdom in all its departments.

212–13

If such ministers are not available,
a king must act on his own,
and do whatever he can.
If not, a bad minister can become
like a pearl as large as a pumpkin—
an ornament impossible to wear.
The minister will be out of control,
and the king will live under his thumb.

214

For each task, appoint more than one person.
Then the work will be finished quickly.
If the numbers are reduced, it creates trouble,
but if they are increased, problems will be solved.

215

No job is done with money alone.
Many people should work at it with commitment.
If the king is generous, truthful and fair,
such qualities may attract the best to the job.

216

Though you have money in your treasury and horses,
without the right sort of men
those assets will be ruined
and the kingdom will fall into enemy hands.
Haven't we heard of such kingdoms?

217

The king will often benefit by putting a Brahman in charge,
for he knows both the laws of Manu and his own dharma,
and from fear of being mocked
by Kshatriyas and Sudras,
he will stand up to all difficulties.

218

Don't place important temples
in the hands of a tax-collector [*āyatikāḍu*].
Being greedy for money, he'll make up his shortfalls
by drawing on the temples.
He'll bring that income to the king's treasury.
That isn't right.
Appoint someone exclusively to that job,
and if he's a little corrupt
God will see to him.

219

Just as a farmer occupies new lands,
fences them in, and digs out rocks and roots
to prepare them for cultivation,
a king can occupy a country,
by giving battle to the enemy
or by taking over forts by bribes,
then seek out and uproot the enemies inside
one by one.

220

If you find a liar in your ranks,
don't condemn or expel him at once.
And if he lies all the time,
rather than discharge him,
move him to a minor post.

221

Farm out those villages
which lie by impenetrable hills and forests,
and are harassed by the nearby hunters
to discharged soldiers from elsewhere.
Let them fight it out,
and whatever happens will be to the good.

222

(Prose)
Allay the fears of the hill-folk, and bring them into your army. Since they are a small people, their loyalty or faithlessness, their enmity or friendship, their favor or disfavor, can all easily be managed. Here is an illustration.

223

A certain Bhil had gone with his bow and arrow
to eat in the house of another.
He was given milk and rice;
but seeing fiber on the boil in a pot,
and believing it was meat, grew enraged.
By the time his host walked him back,

the guest was ready to kill him.
Suspecting this, the host said:
"Let me go, for that fiber will become too soft."
Then the other understood his error and let him go.

224

Give them a little rice and milk,
and they'll never go back on their word.
But when they suspect the smallest deviation,
their imperfect minds cannot forgive it,
and they become your worst enemies.

225

Make friends with the forest-folk through honesty.
Honor the ambassador to make a friend
even of an enemy king.
Pay the foot-soldier punctually to earn his love,
and win over the horseman with gifts.

226

Give a good horse and a good elephant
only to a loyal warrior.
Take care of the stables yourself.
Don't delegate that task to someone else.

227

If one person gives good advice,
and another condemns it from jealousy,
don't contradict either.
When the court has concluded, go home;
and do what the first adviser said.
That is to the good.

228

They ally with outside enemies,
and foster trouble inside the kingdom,
they create problems for the king,
through neglect and bad policy

just to make the king dependent on them.
Such are bad ministers!

229

They make you give gifts to their friends.
They exclude others.
They make you promise and make you break your promise.
And when people say you're untrustworthy,
everyone believes it.

230

When the fire in your belly
grows weak because of phlegm
or other problems, medicine from outside
may heal it. Like this, a competent outsider
may counteract the machinations in your court.

231

If you ask me how:

232

The king—who is both knowledgeable and able—
must retain control of the treasury, the cavalry,
and the elephant stables. In this case,
impediments will fall away of their own.

233

If even a single morsel of food
falls short, the king's servants
become rebellious. A king has no real friends.
So don't cast such people off.
Try to manage them, without trusting them,
as best you can.

234

If a certain person doesn't shrink from doing a certain thing,
you can imagine all the rest. For example, Drupada in the *Mahābhārata*
wanted to perform a killing ritual. He asked a certain sage

to preside. The sage refused but said that his brother
would surely do it. Once he had seen this brother
pick up a fruit from an unclean place
when no one else would touch it.
You can infer people's character by such means.
You'll never know everything.

235

If a certain man bears enmity to you,
and if you defeat him in battle,
don't be merciless.
Just take his wealth, and leave him alone.
What can a snake do
when its fangs have been drawn?
Besides your enemy will trust you,
for the kindness you have shown.

236

The extent of a country is the source of its wealth.
Make small ponds and channels,
so that even poor people can pay taxes in cash and kind.
Give them concessions,
and if you let them improve,
both your treasure and virtue will increase.

237

The people have fled the country in distress,
but the officials don't call them back.
Instead they think, "Why not sell their cattle and grain,
and tear down their houses for firewood?"
A king supported by such jackals of the battlefield—
even if he conquers all seven parts of the world—
will gain nothing.

238

From his revenues a king
should spend one quarter for charity and enjoyment
and two parts for his army,
and save one part in the treasury.

His spies should watch his enemies,
his ministers, and the other pillars of the state.
As for thieves, kill them all.

239

Treat your watchmen well
and use them to discover thieves.
If the real robbers escape prison
and the innocent are punished instead,
your reputation will suffer.
It's like the king who impaled a fat merchant
because the real thief was too thin for the stake.

240

A king should understand three-fourths of a problem
by his own intelligence.
The remaining quarter he can think through
with the help of his good friends.
He should always be free from anger
and moderate in his punishments.
Such a king's rule will last long.

241

Like a bear asleep on a tree,
who still has one eye open,
a king should be ever watchful
of enemies within and without,
even in moments of enjoyment.

242

If you are partial to learning
and give lands and money away to the learned,
mendicants, monks and men with matted hair
will become swollen-headed.
Famines, sickness and infant deaths will increase.
Just show devotion to the learned
and if they resent their poverty—don't be concerned.

243

Before you punish a condemned man,
let him appeal for mercy thrice.
But if someone is a real threat to you,
use your weapon without delay.

244

Though a king might be a great warrior,
it's good to allow his soldiers to boast.
That may swell them up with pride,
and make them better fighters.

245

Manage your ports well,
and let commerce increase
in horses, elephants, and gems,
pearls, and sandal-paste.
When drought, sickness and calamity
make foreigners seek refuge in your lands,
shelter them in keeping with their station.
Give out gardens, yards, and mines
to those whom you wish to favor.

246

Await the right moment
to punish the evil-doer
like the waiting bowman who releases his arrow,
when the target comes into view.

247

Gradually advance into enemy territory,
building up your armies,
just as waters must build up
before they flow into a river.
If the enemy is too strong,
accept his honors and retreat.
But if your spies tell you of his weakness
advance and engulf him.

248

If a king takes money through torture,
takes counsel from the low,
cedes a part of his land to his enemies,
and suspects the loyalty of his friends—
give him secret gifts of gems and ornaments,
reassure him, and create divisions among your enemies.
But take care you don't become like him yourself.

249

If an enemy asks, give him half your kingdom.
Make a strong bond with him. Remove his enmity.
Fear of an enemy is worse than the fear of snakes.

250

Why waste words?
A king should uncover hidden enemies
in his kingdom and destroy them, so he can move
freely through his own country like a man among
beautiful women. Otherwise, what use is a kingdom?
Is it only to give you grief?

251

If a strong enemy is running from you,
don't reel him in right away.
Follow him and take him at the right moment,
as a fisherman who has a big fish on the line
plays him out until he turns back.

252

Harshness in punishment,
acting on slander,
refusing an offer of peace,
turning on a stranger who offers strategic information,
failure to conceal your actions from an enemy,
working with a known traitor,
alienating a faithful ally,
going along with bad advice out of politeness,

failure to punish indiscretion,
shying away from unpleasant events,
denying minimal respect to those who deserve it,
keeping bad company,
addictions,
and obstinacy—
all these a king must avoid.

253

If the three kinds of portent occur,[31]
the king should spend much money
on feeding Brahmans, rituals for the gods,
and fire rituals.

254

The king should encourage competition
among subordinates and soldiers.
That is how their qualities, good and bad,
will come out. They will be so obsessed
with winning the king's attention and honor
that they'll have no time for treacherous plots.

255

A king should never go himself on campaign.
For his own comfort, he should find someone who is competent.
A weak commander won't do. Even a good one needs full support—
cash, elephants, and horses. Moreover, if you send a non-Brahman
he'll turn against you. Give him—the Brahman commander—
land equipped with forts and armies.[32]

256

Make dense forests at the borders.
Thin out the forests in the middle of the kingdom
so that you won't have problems with bandits.

257

Trying to clean up the forest-folk
is like trying to wash a mud wall.
There's no end to it. No point in getting angry.

Make promises that you can keep and win them over.
They'll be useful for invasions, or plundering an enemy land.
It's irrational for a ruler to punish a thousand
when a hundred are at fault.

258

Merchants who bring elephants and good horses
all the way from distant islands,
should be given villages and good houses and a favorable price
and honored with audiences,
so that they come to you in the normal course,
and the war-animals don't go to the enemy.

259

Ambassadors from countries on the border
should be treated with familiarity.
A subordinate should make hard talk
in businesslike or warlike tones,
while the king affects a light air.

260

When you raise someone up in a post
don't share state secrets at once.
To show that he's in the corridors of power,
he may boast of what he's heard,
and both he and the secret will be ruined.

261

The Brahmans who are fortress commanders
should be your well-wishers, learned,
devoted to virtue, heroic, and known over generations.
These are the sort who should be given forts.
Stock the forts up with a lifetime's provisions,
even cheese from tiger's milk.
Grant lands to your subordinates,
even if it's just an ant's portion,
and never give them too much.
Let your incomes exceed your expenditures,
but never make the people suffer.

Watch over the weak kings around with spies,
like a meditating crane that suddenly snatches a fish.
Protect the common people,
but attack the pillars of the enemy's state.
Such a king can sleep with his hand on his heart.

262

The money you spend on elephants, horses, and their fodder,
and on salaries for your soldiers,
and for Brahmans, and temples,
and your own luxuries too,
this is money well spent.

263

Never boast in the face of an enemy.
You may win in battle, or lose,
or you may win only through attrition.
A king should mean business, not battle.

264

If you drive an enemy force into a corner
and it turns on you, you may win
or you may die.
Provoke the local population
to face those forts
where there are various dangerous devices.[33]

265

If someone advises a king well once or twice,
he becomes a trusted adviser.
Access to the king makes the man
the target of bribes.
Now he starts to give bad advice.
Spies should keep an eye on ministers
to know how they lead their lives.

266

If a neighboring king is headed for ruin,
help him along.

If he manages to recover, become his friend.
If he's your enemy, it's the king beyond
who can help against him.
When they fight, your borders will be safe.

267

Cause problems in neighboring lands.
Take their forts, but if you seize their women,
treat them like your own sisters.
Say no harsh words to ambassadors,
you may yet have to make peace with them.

268

If your neighboring land has sorcerers,
their water is poisoned, illness is rampant,
and there are dense forests and hills,
full of barbarians,
don't enter there even for a pile of gold
as large as Mount Meru.
Just send in a force to seize it.

269

There are those who talk smoothly,
just to get what they want.
They are fair-weather friends
who leave you in troubled times.
Such people are bad, and the king should be like a goldsmith
who can assay their real worth.

270

Assemble knowledgeable people.
Find out where there are precious metals.
Mine gold from the land.
For the good of the people,
tax them as little as possible.
If there is a rebellion somewhere
pound it into submission.
Have a soft heart
and take care of everyone.

Collect good doctors.
Learn the elements of your body.
Ingest gold.
Eat moderately in keeping
with your needs.
If you have an excess of wind,
manage it by massage.
Bathe in oil.
Adopt a holistic approach.

Keep people from violating
caste conventions
And take special care of Brahmins.
Settle areas that need population
and thin out overpopulated ones.
Clear the land of enemies.

Find remedies for graying hair
and skin blemishes.
Take special care of your teeth.
Be plump where you should be plump
and thin where you should be thin.
Cleanse the system to enhance your
glow.

Take scrupulous care of your body and the body politic
and both will endure.

271

In the early morning, after being woken up by courtiers who inquire about his sleep,
the king meets with physicians and Brahman astrologers.
In the next watch, he meets his ministers, subordinate kings, and accountants.
At midday, after wrestling and being massaged, he checks with his cooks and hunters.
In the afternoon, he worships the deities and meets temple managers and ascetics.
After dinner come the clowns, *purāṇic* scholars, and poets.
At dusk, after meeting his spies, he listens to musicians.
At night, after a good sleep, he plays with women.

272

Companions come in three kinds: those who always wish you well,
those who wish you well sometimes, and those who never wish you well.
Let me elaborate.

273

Those who always wish you well are your physicians, astrologers, pandits, poets, and priests.
Those who collect money for you sometimes wish you well.
Those who never wish you well are the ones whose property you have confiscated
and who are eager to recover it.
A king should manage all of them skillfully, according to context.

274

A king should identify a deserving person
before he asks and before anyone recommends him.
He should surprise with a sumptuous reward, the way the jack-fruit tree
yields plentiful fruit, like a dream coming true.

275

Between rituals for the gods
and ceremonies for dead ancestors,
the ceremonies are more important.
Send your ancestors to higher worlds
by giving to learned, upright, gentle devotees of Vishnu.

276

Charity is for protecting Brahmans.
Learning is for protecting yourself.
Surrender to Lord Narayana.
They say, *rājyānte narakaṃ dhruvam,*
"Hell is certain after being king."
There is no other hope.

277

Wives obey their husbands,
men and women avoid incest,
yogis control their passions,
the lower castes emulate the higher,
servants willingly serve their masters—
all because they fear punishment by the king.

278

The king is nonviolent, though he kills.
Chaste, though he has women.
Truthful, though he lies.
Ever fasting, though he eats well.
A hero, though he uses trickery.
Rich, though he gives away.
Kingship is rather strange.

279

These are the qualities of a spy:
He should live in the same city as the king.
He must know many languages.
He should know no other spies
and have no unusual external marks.
The king should reward him beyond his hopes.
Otherwise, no one will take the job.

280

A king should enjoy life, each pleasure
in its season—massages, baths, good food,
ointments, rich clothes, and flowers.
The merit that comes from severe control of the body
he can attain by charity.

281

A king's food has many flavors.
He should eat only in the afternoon
and on an empty stomach.
This is a healthy regimen.

282

If a king pays equal attention
to all the three goals in life,
but has a preference for *dharma* above all,
he'll be as happy as the farmer
whose rice fields are watered when
the dam with other fields is accidentally breached.

283

Wear a precious stone
to stand out in the crowd.
And, when you are dressed up,
above all wear the stone
that is appropriate to that day.

284

Don't assume that kingship
inevitably leads to wrong,
or that you can't escape it.
Texts don't ask the impossible.
They just tell you: do your best.

285

Manu, the first king, and Yama, lord of justice, are considered just because they strictly punish wrong-doers. An anointed king who takes his rule seriously and bears with the necessary pain is known in the Veda by various names: Virat, Samrat, and others. He is like a god. Until he resolves his people's problems, his life is unfulfilled. If sensual pleasure is the only goal, even bandits who seize other men's wives and rob wayfarers achieve it. Nor is it good to say that kingship is mostly a bother, so why get involved in protecting the earth? In the Krta Yuga, Kartavirya held up the earth with his thousand arms; he would turn up with the appropriate weapon—sword, club, or bow and arrow—wherever, whenever, however anyone was contemplating a foul act. Such absolute authority is no longer possible for us, in this final age, with our limited energies. Things have come to a point where not only kings but even Brahmans are not what they used to be. In those days, a single Brahman drank up the entire ocean.[34] Another Brahman created a counter-world to God's creation.[35] Another one used his Brahman staff to hold back God's own weapon.[36] So can today's Brahmans give up their Brahman duties, to the extent they can perform them, just because they're not up to what these earlier Brahmans could do? Do we cease to respect them for this same reason? So stay alert and do your best, without ignoring what you see and hear. Protect and punish, and leave what is beyond your own power to lotus-eyed Vishnu, the ultimate protector of those who surrender to him. Put the burden on him. If you behave without self-importance, all power will be within your grasp. A crowned king should act with *dharma* as his only goal. Even the gods stationed in the corners of the cosmos—Varuna, Kubera, the Wind, Fire, Indra and the others—were elevated to those posts because of *dharma*. So, my son, follow *dharma*, pay the three debts[37] and, honored by your equals, rule the kingdom.

Notes

1. One thinks of standard books such as Sabine, *A History of Political Theory.*

2. Crone, *Medieval Islamic Political Thought.*

3. Subbarao, *The Life of Vennelacunty Soobarow,* 65–75.

4. A vast, and largely repetitive, body of work exists on this monarch, to which we shall not refer in detail here. For significant exceptions, see Ramachandraiya, *Studies on Krsnadevaraya of Vijayanagara,* and Venkataramanayya, *Krishṇadevarāyalu.*

5. Vijayaraghavacharya, *Inscriptions of Krishnaraya's Time,* inscription 76, 172–75.

6. A recent scholar of European travel accounts has concluded from a handful of such facts that the text could not have been composed by Krishnadevaraya, but is instead a "fabrication" from the Madurai region, by a "brahmin of Tirumalai-Tirupati" (Rubiés, *Travel and Ethnology in the Renaissance,* 238–39). His view would inspire more confidence if Rubiés had read the entire text in its original version and not merely a summary and antiquated English translation of the *nīti* section.

7. Narayana Rao and Shulman, *Classical Telugu Poetry,* 252–65. Here, and elsewhere in the text, we have used the critical edition by Venkataraya Sastri, *Āmukta-mālyada nāmāntaramu Viṣṇucittīyamu.*

8. For various versions of this genealogy, see Hymavathi, *Vijayanagara,* 17–21.

9. Narasa Nayaka also had a third wife, Obambika, the mother of Acyutaraya and Rangaraya, but she is not mentioned here.

10. Inscription 65 from Tirumalai-Tirupati, in Vijayaraghavacharya, *Inscriptions of Krishnaraya's Time,* 148–57.

11. Sewell, *A Forgotten Empire,* 300–301.

12. This faintly ridiculous early seventeenth-century story about Krishnadevaraya, and his tantrums at his own lack of power, may be found in Wagoner, *Tidings of the King,* 100–109.

13. Tiruvalluvar, *Le Livre de l'Amour.*

14. See Someshvara, *Mānasollāsa.*

15. See Baddena, *Nītiśāstramuktāvaḷi.* Baddena is at times incorrectly termed the author of the *Sumati-śatakamu.* For a brief discussion of such texts, see Wagoner, *Tidings of the King,* 182, 197.

16. Darling, "'Do Justice, Do Justice,'" 3–19.

17. For an earlier translation, see Rangasvami Sarasvati, "Political Maxims of the Emperor-poet Krishnadeva Raya," 61–88; also the later rendition (with the Telugu text of the *rāja-nīti* section) in Nilakantha Sastri and Venkataramanayya, *Further Sources of Vijayanagara History.*

18. Some authors claim that the fifteenth century represented a high point of centralized rule; for this view, and more generally for Vijayanagara rule in the Tamil country, see Karashima, *A Concordance of Nayakas.*

19. For the succession dates of Krishnadevaraya and his coronation, see Sree Rama Sarma, *A History of Vijayanagar Empire,* 133.

20. For a discussion of these Portuguese materials, see Subrahmanyam, "Sobre uma carta de Vira Narasimha Raya," 677–83.

21. Talbot, "The Nayakas of Vijayanagara Andhra."

22. On these figures, see, most recently, Kruijtzer, "Madanna, Akkanna and the Brahmin Revolution," 231–67.

23. This may be an oblique reference to the complex relations between the Sringeri establishment and Vijayanagara, for which see Kulke, "Maharajas, Mahants and Historians," 120–43. But also see the more general view in Sarojini Devi, *Religion in Vijayanagara Empire.*

24. Sewell, *A Forgotten Empire,* 294, 362.

25. We have discussed these later polities at length in Narayana Rao, Shulman, and Subrahmanyam, *Symbols of Substance.*

26. Sewell, *A Forgotten Empire,* 241.

27. Sewell, *A Forgotten Empire,* 241–42.

28. Kamaluddin Abdul-Razzaq Samarqandi, "Mission to Calicut and Vijayanagar," 299–321.

29. For a discussion of this text, see Alam and Subrahmanyam, "Love, Passion and Reason in Faizi's *Nal-Daman*" and Alam and Subrahmanyam, "The Afterlife of a Mughal *Masnavī*."

30. For the afterlife of such texts, also see such important late seventeenth-century formulations on statecraft from the Maratha domains as Ramacandra Pant Amatya, *Ājñāpatra.*

31. *Divyam,* divine—unfavorable planetary conjunctions; *antarīkṣam,* celestial—comets; *bhauma,* terrestrial—strange creatures.

32. The Telugu is somewhat ambiguous, and the editor Vedamu Venkataraya Sastri reads: "You need the non-Brahman, too."

33. With the implication "And keep your own army in reserve."

34. Agastya.

35. Vishvamitra.

36. Vasistha.

37. To the gods, the sages, and your ancestors.

References

Alam, Muzaffar, and Sanjay Subrahmanyam. "The Afterlife of a Mughal *Masnavī*: The Tale of Nal and Daman in Urdu and Persian." *A Wilderness of Possibilities: Urdu Studies in Transnational Perspective,* ed. Kathryn Hansen and David Lelyveld. Delhi: Oxford University Press, 2005.

———. "Love, Passion and Reason in Faizi's *Nal-Daman*." *Love in South Asia: A Cultural History,* ed. Francesca Orsini. Cambridge: Cambridge University Press, 2006.

Baddena. *Nītiśāstramuktāvaḷi*. Ed. M. Ramakrishna Kavi. Tanuku: Śrī Narēndranādha Sāhityamaṇḍali, 1962.

Crone, Patricia. *Medieval Islamic Political Thought, c. 650–1250*. Edinburgh: University of Edinburgh Press, 2004.

Darling, Linda T. "'Do Justice, Do Justice, for That Is Paradise': Middle Eastern Advice for Indian Muslim Rulers." *Comparative Studies of South Asia, Africa, and the Middle East* 23, nos. 1–2 (2002), 3–19.

Hymavathi, Polavarapu. *Vijayanagara: The Life and Times of Tuluva Vira Narasimha Raya*. Madras: New Era Publications, 1994.

Kamaluddin Abdul-Razzaq Samarqandi. "Mission to Calicut and Vijayanagar." *A Century of Princes: Sources on Timurid History and Art*, by Wheeler M. Thackston. Cambridge, Mass.: Aga Khan Fund, 1989.

Karashima, Noboru. *A Concordance of Nayakas: The Vijayanagar Inscriptions in South India*. Delhi: Oxford University Press, 2002.

Kruijtzer, Gijs. "Madanna, Akkanna and the Brahmin Revolution: A Study of Mentality, Group Behaviour and Personality in Seventeenth-century India." *Journal of the Economic and Social History of the Orient* 45, no. 2 (2002), 231–67.

Kulke, Hermann. "Maharajas, Mahants and Historians: Reflections on the Historiography of Early Vijayanagara and Sringeri." *Vijayanagara—City and Empire: New Currents of Research*, ed. A. L. Dallapiccola and S. Zingel-Avé Lallemant. 2 vols. Stuttgart: Franz Steiner Verlag, 1985.

Narayana Rao, Velcheru, and David Shulman. *Classical Telugu Poetry: An Anthology*. Delhi: Oxford University Press, 2002.

Narayana Rao, Velcheru, David Shulman, and Sanjay Subrahmanyam. *Symbols of Substance: Court and State in Nayaka-period Tamilnadu*. Delhi: Oxford University Press, 1992.

Nilakantha Sastri, K. A., and N. Venkataramanayya. *Further Sources of Vijayanagara History*. 3 vols. Madras: University of Madras Press, 1946.

Ramacandra Pant Amatya. *Ājñāpatra*. Ed. Vilas Khole. Pune: Pratima Prakashana, 1988.

Ramachandraiya, Oruganti. *Studies on Krsnadevaraya of Vijayanagara*. Waltair: Andhra University, 1953.

Rangasvami Sarasvati, A. "Political Maxims of the Emperor-poet Krishnadeva Raya." *Journal of Indian History* 4, no. 3 (1926), 61–88.

Rubiés, Joan-Pau. *Travel and Ethnology in the Renaissance: South India through European Eyes, 1250–1625*. Cambridge: Cambridge University Press, 2000.

Sabine, George H. *A History of Political Theory*. 3rd ed. New York: Holt, Rinehart and Winston, 1965.

Sarojini Devi, K. *Religion in Vijayanagara Empire*. New Delhi: Sterling, 1990.

Sewell, Robert. *A Forgotten Empire—Vijayanagar: A Contribution to the History of India*. Delhi: National Book Trust, 1962.

Someshvara. *Mānasollāsa*. Ed. Gajanan K. Shrigondekar. 3 vols. Baroda: Oriental Institute, 1925–61.

Sree Rama Sarma, P. *A History of Vijayanagar Empire*. Hyderabad: Prabhakar, 1992.

Subbarao, Vennelakanti. *The Life of Vennelacunty Soobarow (Native of Ongole) as Written by Himself*. Madras: C. Foster, 1873.

Subrahmanyam, Sanjay. "Sobre uma carta de Vira Narasimha Raya, rei de Vijayanagara (1505–1509), a Dom Manuel I de Portugal (1495–1521)." *Professor Basilio Losada: Ensinar a pensar con liberdade e risco*, ed. Isabel de Riquer, Elena Losada, and Helena González. Barcelona: Universitat de Barcelona, 2000.

Talbot, Cynthia. "The Nayakas of Vijayanagara Andhra: A Preliminary Prosopography." *Structure and Society in Early South India: Essays in Honour of Noboru Karashima*, ed. Kenneth R. Hall. Delhi: Oxford University Press, 2001.

Tiruvalluvar. *Le Livre de l'Amour*. Trans. François Gros. Paris: Gallimard, 1992.

Venkataramanayya, N. *Kṛṣṇadevarāyalu*. Hyderabad: Archaeological Series, 1972.

Venkataraya Sastri, Vedamu. *Āmukta-mālyada nāmāntaramu Viṣṇucittīyamu*. Madras: Vedamu Venkatarayasastri, 1964.

Vijayaraghavacharya, V. *Inscriptions of Krishnaraya's Time, from 1509 AD to 1531 AD*. Delhi: Sri Satguru, 1984.

Wagoner, Phillip B. *Tidings of the King: A Translation and Ethnohistorical Analysis of the "Rāyavācakamu."* Honolulu: University of Hawaii Press, 1993.

Part II

Literary Consciousness, Practices, and Institutions in North India

CHAPTER FOUR

The Anxiety of Innovation

The Practice of Literary Science in the Hindi *Rīti* Tradition

ALLISON BUSCH

Keshavdas has described the various gestures of Radha and her lover according to his understanding. May master poets forgive his audacity.
—*Rasikpriyā* 6.57

Mughal-period North India saw the rise of many important new cultural patterns, one of which was a major expansion of the literary and intellectual ambitions of Hindi writers. The literary dialect of Brajbhasha, whose fame (and name) derived from a popular style of rustic, devotional lyrics centered on the Krishna legend, achieved new prominence in a range of courtly settings both Mughal and Rajput.[1] The traditional knowledge system of literary science (*alaṅkāraśāstra*), one of the core disciplines of Indian scholarship before the advent of colonialism and a cultural space long monopolized by Sanskrit intellectuals, emerged as a fertile site for the development of a new form of vernacular poetics and poetry. Rulers at dozens of courts from north India to the Dakhan commissioned Brajbhasha works on classical literary topics. So central was this genre, often termed the *rītigranth* (book of method), to the literary life of late precolonial India that modern historians of Hindi routinely term the entire period from 1650 to 1850 the *rītikāl* (period of method, i.e., neoclassicism).[2]

The Brajbhasha *rītigranths* straddle two different genres simultaneously: they are both theoretical works in the tradition of Sanskrit *alaṅkāraśāstra* and collections of poetry. This dual purpose explains the structure of the works: they consist of alternating sequences of definitions of Sanskrit poetics topics (*lakṣaṇ*) and verses that exemplify variations on those topics (*udāharaṇ*). Typical of the genre in both Sanskrit-derived content and classificatory style is the following eightfold analysis of female characters (*aṣṭa-nāyikābheda*) excerpted from Keshavdas's *Rasikpriyā* (Handbook for Poetry Connoisseurs, 1591), one of the earliest vernacular works on classical literary systems:

All female characters may be described in keeping with an eightfold system. These are called

1) whose lover is under her control
2) anxious
3) who has decorated her bed
4) stubborn
5) angry
6) whose lover has gone far away
7) whose lover did not keep the tryst
8) who goes out boldly to meet her lover

Know all these to be the eight types of female characters.[3]

After outlining his overarching system the poet proceeds to delineate each of the eight subtypes of female characters individually, pinpointing the specificities of each with example verses. An illustration of this complementary system of definitions and example poems is Keshavdas's elaboration of the anxious heroine (number 2 in his eightfold system):

A definition of the "anxious" [utkā]

Keshavdas says that the "anxious" is a woman whose lover does not show up for some reason, causing her heart to brim with sorrow.

An example of the sub-type of the "hidden anxious" [prachannā utkā]

Said the anxious woman to herself,
Is it some matter at home?
Or did his cowherd friends detain him?
Is he fasting today?
Did he fail to pay a debt?
Did he get into a fight?
Has he suddenly taken a religious turn?
Perhaps he is unwell?
Or his love for me is false?
Did he see the clouds and hesitate to come in the middle
  of the night?
Or is he testing my love?
Again today he hasn't come!
What could be the matter? (vv. 7.7–8)

Employing this style of classification and exemplification Braj *rītigranth* authors undertook the wholesale systematization of both vernacular poetics and poetry. Hundreds of these texts were produced in early modern north India. Among Indo-Muslims such vernacular works were a point of access into the general principles of Indian poetry, a world largely unavailable to them in the earlier Sanskrit texts; Braj poetry also provided an interesting diversion from works written in the primary Mughal court language, Persian. At Rajput courts the *rītigranth* was both an intellectual and an aesthetic enterprise: Braj writers relished the new literary possibilities of vernacular poetry, but they also began to encroach upon some of the intellectual terrain that had earlier been inhabited exclusively by Sanskrit writers.

In striking contrast to its extensive cultural reach in early modern times, the Hindi *rīti* tradition is today little studied and remains poorly understood. Modern readers tend to feel bewildered by the hypertaxonomical style of *rīti* authors, whose works catalogue dozens, even hundreds of types of *nāyikā*s or *alaṅkāras* (figures of speech), complete with subtypes. In contrast to the simplicity and naturalism celebrated as characteristic of premodern Hindi's better-known corpus, *bhakti* literature, the *rīti* poets' use of a high register of vernacular diction and preoccupation with time-worn themes from Sanskrit have come to be viewed as representative of the decadent and mannerist tendencies of a tired feudal age. Whereas *bhakti* poetry has been embraced by modern scholars, who seem especially to value its forms of demotic expressivity, *rīti* literature by virtue of its association with late medieval courtly life has been dismissed as stilted and reactionary. Whatever the provenance of such perceptions—perhaps they stem from a generalized post-Romantic distaste for courtly literature, or perhaps from now outdated theories about India's late precolonial decline—they fail utterly to account for or even to address the most interesting features of *rīti* literature, such as the cultural meaning(s) of neoclassicism.[4]

In what follows I demonstrate that the unfavorable reactions to *rīti* that predominate in modern Hindi scholarship were not even remotely shared by members of the literary public in early modern times. For *rīti* writers such as Keshavdas, forging connections to classical traditions was a sine qua non of vernacular literary and intellectual life. I explore the methodologies that underpinned the developing field of Brajbhasha *alaṅkāraśāstra,* with a special interest in perceptions about both the constraints and the new creative potential of this medium during a moment of intense growth for Hindi writing, as well as the responses of select Sanskrit intellectuals to the irrepressible popularity of the vernacular style. I conclude with some remarks about the reception contexts of the *rītigranth* genre, reflecting on the meaning and value of *rīti* literary systems for Hindi poets and scholars of the late precolonial world. If we suspend judgment and try to think outside the narrow constraints of modern literary biases, which deem courtly literature stilted and insincere or which expect of poetry or intellectual practices something other than what premodern Indians expected, we come much closer to being able to understand *rīti* literary trends from the perspective of their own cultural milieu.[5]

Vernacular Incompetence?

Both in early modern South Asia and elsewhere vernacular writers faced obstacles to literary success due to a fundamental distinction in status between local and prestige languages. In South Asia an otherwise common pattern of linguistic hierarchies was exacerbated by the potent belief of Sanskrit-using scholars that their language was the only one that mattered, intellectually, culturally, spiritually.[6] From the perspective of one firmly rooted in a Sanskrit worldview the movement between classical and vernacular languages was unidirectional, and that direction could only be downward. To be a vernacular writer was to exhibit both a linguistic and a scholarly failing. The hierarchies involved are implicit at the most basic lexical level. Vernaculars were by definition "corrupted" (*apabhraṣṭa*) languages, and their low status may be divined from the fact that they apparently did not even merit their own names; they were usually just called "language" (*bhāṣā*). The very word *Sanskrit*, in contrast, denotes (and connotes) the height of dignity: it means "perfected." Sanskrit was also widely venerated as the "language of the gods" (*devavāṇī, suravāṇī*). It was no simple matter to compete with a language that claimed not only perfection but divine status.

Traditional hierarchies concerning the inferior intellectual status of vernacular writing were doubtless enshrined in theory, but actual practice during the Mughal period reveals a far more complex picture in which the relational dynamics of Sanskrit and Brajbhasha were being renegotiated. This is particularly true in the disciplines of literature and literary science. Whereas vernacular writers had once been dismissed as inarguably inferior to their Sanskrit counterparts, Brajbhasha poets could now achieve spectacular recognition even in elegant courtly circles. For instance, Braj poetry is widely held to have influenced Jagannatha Panditaraja (d. ca. 1670), who is often hailed as the last great Sanskrit poet-intellectual before the vernacular wave eroded the once solid embankments of classical textual authority.[7] Cultural attitudes were shifting, and newfangled vernacular poets and thinkers felt emboldened to challenge the monopoly of Sanskrit.

Striking testimony to a new sense of the validity of vernacular scholarship is offered by the *Śṛṅgāramañjarī* (Bouquet of Passion, ca. 1670), an *alaṅkāraśāstra* text from the Dakhan attributed to Akbar Shah, son of Shah Raja, teacher to Sultan Abul Hasan Qutb Shah of Golkonda (r. 1672–87). Both the work's textual history and its content articulate something important about new perceptions of vernacular authority. In a telling reversal of the normal positions of source and target language, the work was first composed in Telugu (*āndhrabhāṣā*) and then translated into Sanskrit (*suravāṇī*). From Sanskrit the text soon made its way into Brajbhasha through yet another act of translation.[8] It is, to my knowledge, unprecedented in its citing of Brajbhasha authors as sources alongside Sanskrit literary authorities, as though they were newly perceived as intellectual equals. Two of the earliest Braj *rītigranths*, Keshavdas's *Rasikpriyā* and Sundar's *Sundarśṛṅgār* (Sundar's Love Poems, 1631), share the designation "principal text" (*pramukhagrantha*) with such illustrious Sanskrit works as Dhananjaya's *Daśarūpaka* (Ten Genres, late tenth century), Mammata's *Kāvyaprakāśa* (Light on Literature, mid-eleventh century), and Bhanudatta's *Rasamañjarī* (Bouquet of Emotion, ca. 1500).[9] Although Sundar is only mentioned, Keshavdas's theorization of the *premābhisārikā nāyikā* (lovelorn woman who ventures out to meet her lover) is actually discussed in the Sanskrit text. In the end Keshavdas's proposed new category is not endorsed, but this intellectual rebuff has nothing to do with the writer's choice of a vernacular medium.

That Brajbhasha could now function alongside Sanskrit as a transregional language of letters at a Dakhani court is a telling index of its new cultural status. Nor is this instance of its diffusion far from home particularly exceptional. Compositions in Braj and other dialects of Hindi were also routinely sponsored by the Maratha courts.[10] The name *Brajbhasha* (language of Braj) in wide use today marks the language's cultural and linguistic ties to the particular area of north India that was celebrated as the center of Krishna lore, but by the second half of the seventeenth century Brajbhasha had moved far beyond its original parameters, both geographically and expressively.

Although Brajbhasha writing was acquiring a new degree of circulation and intellectual cachet, it was not always readily embraced. There is, on the contrary, much evidence of profound ambivalence toward its literary and scholarly potentialities. The seventeenth-century intellectual Kavindracarya Sarasvati, for instance, spoke of his sense of shame (*lāj*) at writing in the vernacular, a sentiment that was echoed by other anxious vernacular-using pandits and poets of the day.[11] And yet for all this pandit's disclaimers, there are strong tensions between his professions of vernacular inferiority and the actual strength of his vernacular writerly persona. Kavindra may have expressed shame at using *bhāṣā*, but that hardly stopped him from doing so, and judging from his extant works he wrote in Braj and Sanskrit to an almost equal extent. In fact the very contours of Kavindracarya's life work appear to illustrate a newer pattern of vernacular-classical parity, hardly the older paradigm of vernacular inferiority.

The *Rādhāmādhavavilāsacampū* (Love-Play of Radha and Krishna, henceforth *Campū*) of Jayarama Pindye, Kavindracarya's fellow Maharashtrian and contemporary, is similarly contradictory in its unease about vernacularity while simultaneously endorsing it. At first glance Jayarama's *Campū* would appear to be a veritable paean to polyglossia: the work is composed in a combination of Sanskrit and eleven regional languages (*deśabhāṣā*), and the author boldly proclaims himself to be a master of poetry in all twelve.[12] But upon closer scrutiny the reader becomes confused about the relative status of Sanskrit and *bhāṣā* in this text. The division of linguistic labor is unequal: the first ten cantos are written exclusively in Sanskrit, and it is only in the last chapter that the other languages appear, all lumped together as though the very structure of the work were designed to cast the vernaculars in the role of dilettantish pretenders. The stated reason for keeping the single *deśabhāṣā* canto separate from the ten Sanskrit cantos attests to Jayarama's sense that linguistic hierarchies should be maintained. He repeatedly asserts that it would be inappropriate to include vernacular poems in the Sanskrit section of his *Campū*.[13] And yet while his insistence on keeping the Sanskrit and vernacular domains of expression absolutely separate appears to shore up traditional notions of Sanskrit purity and supremacy, we know that such a stance conflicts radically with the practices at the court of Shahaji Bhonsle, Jayarama's patron, where vernacular and Sanskrit poetry were part of a shared cultural arena.

Ultimately the actual execution of the work belies any notion of vernacular inferiority. The eleventh canto of Jayarama's *Campū* is almost as long as all the Sanskrit cantos put together, and it contains dozens of *deśabhāṣā* poems (including several in Brajbhasha) of great vibrancy.[14] If anything it is the vernacular poetry that shows real originality in the work, for half of the Sanskrit cantos consist of hyperconventionalized "head to toe" (*śikhanakha*) descriptions of Radha and Krishna, set pieces on the seasons (*ṣaṭṛtu-varṇanam*), and renditions of other, somewhat tired motifs such as the lovers' "water-play" (*jalakrīḍā*) and their flower-strewn bed (*puṣpaśayyā*). This mixed-language *Campū* serves as a metaphor for one of the most important developments in early modern courtly life: despite the earlier doctrines that denied its expressive validity, *bhāṣā* had begun to impinge dramatically upon the traditional dominance of Sanskrit.

If concerns about vernacular legitimacy loomed large in the consciousness of Sanskrit writers and occasionally engendered uncomfortable emotions, they were bound to be equally if not more pressing for Hindi writers, who were in a far weaker cultural position. The status of vernacular writing was certainly a central concern for Keshavdas, who, as founding author of the *rīti* style, made an indelible mark on literary history when he began to employ Brajbhasha as a medium for a range of new, classicizing genres. He wrote three *rītigranths*, the best known of which are the *Rasikpriyā* (Handbook for Poetry Connoisseurs, 1591) and *Kavipriyā* (Handbook for Poets, 1601), as well as imparting a new vernacular shape to Sanskrit styles such as the courtly epic (*mahākāvya*) and panegyric (*praśasti*). Keshavdas's personal profile, no less than his intellectual and literary one, points toward the major cultural shift that the early *rīti* tradition represents. He came from a lineage of Sanskrit pandits who had served the courts of Orchha and nearby Gwalior. By turning his attention exclusively to vernacular compositions he made a significant break with family tradition. He was profoundly aware of the literary frontier he was crossing, as is readily apparent from the opening to his *Rāmcandracandrikā* (Moonlight of Rāmcandra, 1601), a Braj *Rāmāyaṇa* that was styled in the tradition of Sanskrit *mahākāvya*:

> There was a *Sanāḍhya* Brahman by name of Krishnadatta Mishra. He had an exemplary character, garnering fame throughout the land. He held the title "king among pandits," and was endowed with every virtue. Krishnadatta had a son named Kashinatha, who had boundless wisdom—like Lord Ganesha. Kashinatha studied all the Sanskrit scholarly texts, and synthesized many different theories. To Kashinatha was born a slow-witted [*mandamati*] son, the poet Keshavdas. He wrote *Moonlight of Rāmcandra* in the vernacular [*bhāṣā*].[15]

The self-description *mandamati,* like Kavindra's *lāj* or Jayarama's peculiar manner of quarantining non-Sanskrit poetry in his *Campū*, appears to signal a feeling of apprehension about vernacularity and a deferential attitude toward classical authority. But aside from the obvious fact that the slow-witted do not know they are and do not declare it, other indications prompt us not to take this slow-witted poetic persona completely at face value.

Complicating Keshavdas's ostensible tone of vernacular humility in this passage is his paradoxical appropriation of Sanskrit literary prestige in the verses that follow. In a scene well suited to Keshavdas's own poetry of vernacular beginnings, Valmiki, venerated as the first poet of Sanskrit literature (*ādikavi*), appears to Keshavdas in a dream and inspires the fledgling Brajbhasha author to write his own version of the *Rāmāyaṇa* (vv. 1.7–21). Valmiki's presence at the very outset of the *Rāmcandracandrikā* evokes a complicated metatextual resonance about Keshavdas's own literary beginnings as Hindi's first *rīti* author, but it also has the effect of tapping into Sanskrit textual authority and rescripting it to shore up the claims of vernacular writing. For if seeking blessings from a hallowed Sanskrit predecessor appears to suggest humility, its opposite is also in evidence: the usurping of Sanskrit cultural space by the suggestion that a *bhāṣā Rāmāyaṇa* can take its place. The persona of the slow-witted vernacular poet seems to have been a placating gesture toward Sanskrit literary authority, but one deployed in the very act of transgressing it. Thus it would be a mistake to interpret it too literally as a reflection of true vernacular incompetence.[16]

The Paradox of Vernacular Newness

Unfortunately this point seems to have been lost on many modern scholars, who, perhaps taking Keshavdas too much at his word, have failed to read him or later *rīti* writers with the care they deserve. The near universal assessment of modern Hindi criticism is that the field of Braj *alaṅkāraśāstra* lacks the scholarly merits of its Sanskrit counterpart, a claim that warrants more careful exploration. Consider first the illogicality of two widely divergent constructions of what it meant to make the transition from Sanskrit to vernacular authorship, depending on whether the text under scrutiny is a *bhakti* or a *rīti* work.[17] In a *bhakti* context vernacularization is hailed as "liberation" from the classical language, where the homely dialects of everyday people fought for representation in the literary field.[18] When it comes to *rīti* poets' use of the vernacular, however, and their strong reliance on Sanskrit models and methods, modern critics have not emphasized the new, creative aspects of the transformation. *Rīti* works have often been dismissed as paltry imitations of more authoritative classical studies, as though the choice to use Braj instead of Sanskrit apparently suffices in itself to prove that *rīti* scholars are men of diminished intellectual powers.[19] The inadequacy of these assumptions becomes obvious if we look more closely at the theoretical works of *rīti* authors and try to make sense of their methodologies. What does innovation look like from the reference point of early modern Hindi intellectuals? How did scholars of this period contend with the weight of the past? To pinpoint the exact nature of vernacular newness can seem, in the case of many *rīti* authors, an elusive prospect. Newness, particularly its premodern manifestations, can exist in a range of subtle forms, in which case finely calibrated interpretive tools are needed to identify it. In trying to understand the idioms in which nonmodern pioneers speak, it is helpful to recall the words of Sudipta Kaviraj, who has usefully distinguished between modern and premodern modes of cultural change: "Modern rebellions announce themselves even before they are wholly successful; revolutions in traditional cultures tended to hide the facts of their being revolts."[20] If Kaviraj is correct, we will almost certainly fail to see alternative forms of newness if we adhere too closely to the paradigm of how change looks from the viewpoint of Western modernity. An anachronistic imposition of present ways of seeing on the past is one major failing of modern approaches to the intellectual life of the *rīti* period. In what follows I try to steer clear of such hermeneutic presentism, striving to listen instead to the voices of Brajbhasha writers and what they have to tell us about the composition and reception of their works.

Rīti authors are inconsistent when it comes to literary change, both in professing it and in executing it. In some cases they seem manifestly uninterested in developing bold new theories. It is often possible to identify one or more classical sources for the definition portion of any given *rīti* text, and some *lakṣaṇ*s of Brajbhasha *rītigranth*s are paraphrases of Sanskrit models, which would be considered plagiaries by the scholarly standards of today. Still, no matter what the aspirations of a *rīti* author in terms of theoretical innovation, the accompanying example verses—the poetry—are almost invariably original. This is in notable contrast to most works of Sanskrit *alaṅkāraśāstra*, in which literary principles were illustrated by excerpting existing Sanskrit poems from famous classics. Many *rīti* authors, however, did show considerable interest in *alaṅkāraśāstra* as an intellectual, and not just a poetical, enterprise. And yet there is a curious contradiction in their practice. For all their apparent radicalism in eschewing the time-honored language of Indian courtly letters and the trumpeting of their vernacular works as new theorizations, many early Brajbhasha scholars also insist that they have not departed from existing Sanskrit traditions. How can we reconcile both claims?

The paradoxical nature of vernacular newness is perhaps nowhere more apparent than in chapter 3 of Keshavdas's *Kavipriyā*. After preliminary chapters on his court, his king, and himself, the author embarks upon his treatment of vernacular literary theory with the classical subject of *doṣa*s, literary flaws that mar the beauty of poetry. In composing this constellation of introductory literary principles Keshavdas does not strictly follow Dandin's *Kāvyādarśa* (Mirror of Literature, seventh century), otherwise a major Sanskrit source book for the *Kavipriyā*.[21] Rather he begins with a discussion of several new categories of literary flaws, which at first makes the work appear refreshingly bold. Yet this boldness ultimately proves to be not only measured but seriously compromised. The "flaw of blindness" (*andhadoṣa*), for instance, may be an entirely new Keshavdasian category, but it is one intended precisely to proscribe poems that violate tradition.[22] On the one hand the poet is questioning the authority of Sanskrit, forging a new vernacular style, and engaged in writing one of the first treatises on Brajbhasha poetics; on the other he tells his readers that they should under no circumstances contravene literary tradition. Since a developed tradition of *alaṅkāraśāstra* did not yet exist for Braj, it is difficult to see how the inviolable poetic path (*pantha*) to which he refers could be anything other than a Sanskrit one.

As is the standard procedure in a *rītigranth*, Keshavdas reinforces his definition of the *andhadoṣa* with an example verse that develops his point. Here the example verse is presented in the form of a parody, which serves as a humorous warning about the potential aesthetic disaster that lies in wait for an inexperienced poet striking out on his own:

> Seeing her soft lotus-like breasts in bloom,
> the moon face of her lover beams in delight.
> Her eyes dart quickly like monkeys,
> the corners red like Sindur powder.
> Her lower lip is sweet like butter,
> seeking metaphors for her beauty Keshavdas despairs.
> There she stands, that desirable woman,
> like lightning or a roaming deer—
> she moves slowly like an elephant. (v. 3.8)

The mixed metaphors and infelicities in this verse are painfully obvious to any experienced reader of classical Indian poetry. Note how the most egregious errors concern the poet's flagrant disregard for tradition. First of all, a woman's breasts should be firm like lotus buds, not soft like blooming lotuses. The images in the next line are a precarious combination because according to Sanskrit poetic convention (*kavisamaya*) the moon causes certain lotuses to wither. In line 3 Keshavdas's imaginary clumsy poet gets one image right—the part about women's eyes darting quickly—but when it comes to the standard of comparison (*upamāna*) he makes a serious blunder in choosing the animal: in Sanskrit poetry beautiful women are doe-eyed (*mṛgākṣī*), not monkey-eyed! Furthermore when it is a question of the movement of eyes, fish (*mīna*) or wagtails (*khañjana*) are preferable images because they are consecrated by tradition as metaphors for fast-moving objects. In line 5 the hapless poet has bungled things again. Lower lips are indeed soft and sweet, but they should be compared to the red *bimba* fruit, not to pale yellow butter. The message any would-be poet takes away from this opening passage of the *Kavipriyā* is that vernacular composition must be rooted in classical imagery. For Keshavdas the foundational premise of vernacular poetics seems automatically to constrain its newness.

This ambivalence between innovation and adherence to tradition is not peculiar to Keshavdas; it would continue to define the scholarly behavior of many Brajbhasha intellectuals. Consider the words of Cintamani Tripathi, one of Keshavdas's most important successors, when he expresses a similarly contradictory logic about the nature of vernacular newness in the opening to his magnum opus, the *Kavikul-kalptaru* (Wish-Fulfilling Tree for the Brotherhood of Poets, ca. 1670): "I, Cintamani, have carefully considered the precepts of books written in the language of the gods [i.e., Sanskrit], and I am expounding a theory of vernacular literature. . . . I describe the system of vernacular literature according to my intellectual ability."[23] If his lexical choices have the significance I think they do, Cintamani viewed himself not so much as a translator of his Sanskrit source texts, but as someone engaged in a new theorization (*vicāra*) of vernacular literature (*bhāṣā kavita*). The statement "according to my intellectual ability" (*budha anusāra*) also seems to promise an original contribution to scholarship. But clearly the very enterprise of writing new literary theory in Brajbhasha was not only complicated but epistemologically fraught. That one can apparently develop such a theory only upon consulting Sanskrit precepts reveals a core dependency on the classical language.

According to My Own Understanding

Despite the frequently overpowering demand for compliance with Sanskrit literary norms, the corpus of Brajbhasha *rītigranths* does contain much that is unmistakably new. As one of the first works of the Hindi *rīti* tradition, Keshavdas's *Rasikpriyā* is a paradigm for the styles of newness that manifest themselves in early vernacular scholarship. At first glance the *Rasikpriyā* appears to be a very close adaptation of the *Śṛṅgāratilaka* (Ornament of Passion), composed by the Sanskrit rhetorician Rudrabhatta in perhaps the eleventh century. Keshavdas follows virtually the same order of treatment of the subject matter as his source, and significant lexical borrowings in the definition verses show his reliance on Rudrabhatta to be beyond doubt. Looking no further than these obvious similarities one would erroneously conclude, as has often been concluded about *rīti* writers across the board, that Keshavdas simply plagiarized from his Sanskrit predecessor. The reality is much more interesting. The *Rasikpriyā* is both new and not new in complex ways. The *Śṛṅgāratilaka* may well be Keshavdas's guide through the principles of *alaṅkāraśāstra*, but as often as not he veers off on his own detours.

One such detour is to invent variations on his predecessor's organizing categories, particularly in places where the original Sanskrit text provides only a cursory treatment of the subject. A good way to demonstrate this process is to compare Keshavdas's treatment of lovers' meeting places (*milana-sthāna*) in chapter 5 with the discussion in the Sanskrit source text. Rudrabhatta lists the possible occasions for lovers' rendezvous briskly, in a single verse, and then considers the subject closed, not bothering to furnish even one example. Keshavdas, seizing this opportunity for creative ramification, develops the idea into a major theme of an entirely new chapter on the various aspects of falling in love. He gives a complete example of nearly every occasion for the meeting of lovers mentioned in passing in the *Śṛṅgāratilaka*, and he also proposes new categories of his own.[24] As though to hold up a signboard marking out his vernacular innovations, Keshavdas closes this particular chapter with a statement that was to become the refrain of *rīti* poet intellectuals: "I have said this according to my own understanding" (*kahe apnī mati anusāra*, v. 5.41). However else he may think of his relationship to tradition, in the writer's own estimation he was often intending to create new knowledge.

Alongside Sanskrit *alaṅkāraśāstra* earlier Hindi poetry of the *bhakti* style also contributed in significant ways to the shaping of Keshavdas's scholarly profile. Among all his works the *Rasikpriyā* in particular is steeped in a *bhakti* worldview, which serves to differentiate the work markedly from its Sanskrit source text. Perhaps the most obvious point of departure is that the *nāyakas* and *nāyikās*, the heroes and heroines, who people Rudrabhatta's poems are generic, whereas the main actors in Keshavdas's verses are not just any handsome man or woman but objects of veneration to him: the deities Krishna and Radha.

Keshavdas's reverential stance toward Krishna and Radha translates into numerous points of theoretical divergence from his predecessor. For instance, neither scholar endorses literary representations of lovers who pine so much for their beloved as to reach the point of death (*maraṇāvasthā*), but whereas Rudrabhatta gives the reason that such poems lack beauty (*asaundaryāt*), for

Keshavdas the crucial point is that his poems are about a god, and he could not possibly describe the death of someone immortal and indestructible.[25] And when it comes to the traditional three types of *nāyikā in* Sanskrit, Keshavdas entirely omits the *sāmānyā nāyikā*, the "public woman" or courtesan: "As for the third type of woman, why should I describe it here? The best poets say not to depict tasteless [*birasa*] subjects. Here I have described all the *nāyikās* according to my own understanding of them" (vv. 5.39–40).[26] The omission of the *sāmānyā nāyikā* makes perfect sense in Keshavdas's more *bhakti*-oriented universe: in a text where Radha is the principal *nāyikā* (whether stated explicitly or left implicit), it would hardly have been possible to discuss the morally questionable figure of the courtesan.

A devotional orientation toward Krishna and Radha also colors Keshavdas's treatment of the theory of *rasa*, or emotion in literature. In the opening to *Rasikpriyā* he argues that Krishna is *navarasamaya*—that is, the deity embodies in his earthly avatar all the nine *rasas*—which the eighteenth-century commentator Surati Mishra considered one of the major postulates of the work. The sentiment of passion (*śṛṅgāra rasa*), given priority of place by all literary theorists both Sanskrit and Braj, is in Keshavdas's formulation further specified as the purview of Krishna.[27] When it comes to his treatment of the various affective responses and physical gestures (*bhāvas, hāvas*) that interact to contribute to the full complement of *śṛṅgāra rasa*, the love of Radha and Krishna is posited as the main substratum: "Passion [*śṛṅgāra*] arises from the love of Radha and Krishna. From the force of their emotion arises my theory [*bicāra*] about the physical gestures [*hāvas*] of love" (v. 6.15). In this case too Keshavdas's new formulations of his subject matter are nothing if not absolutely deliberate, as is evident from the way he concludes the discussion: "Keshavdas has described the various gestures of Radha and her lover according to his understanding of them. May master poets forgive his audacity" (v. 6.57). Keshavdas again foregrounds his new approach, although in this case (if we are to take him at his word) the poet's otherwise bold assertion of independence from the Sanskrit source material is tempered by a qualm about whether he is being too audacious. Whether or not the request for forgiveness is wholly ingenuous, perhaps it was obligatory, given the power of the vernacular's rival.

Keshavdas and his numerous successors in seventeenth- and eighteenth-century India contributed to a larger intellectual historical process by which Brajbhasha began to challenge the traditional dominance of Sanskrit. And yet if the confidence levels of Brajbhasha intellectuals increased over time, as the vernacular embodiment of *alaṅkāraśāstra* not only took hold but eventually supplanted that of the classical language, many *rīti* writers continued to express deference to their classical predecessors and voiced anxieties about their own abilities to contribute new theorizations. As late as 1746, after countless *rītigranths* had been written in Brajbhasha, Bhikharidas, one of the greatest Braj rhetoricians, is still compelled to say:

> I studied the Sanskrit texts
> *Candrāloka* and *Kāvyaprakāśa*.
> I understood them,
> and made their ideas beautiful in the vernacular.
> From other sources, too, I adopted the path of poets. . . .
> But even though I may express my own opinions,
> I still feel anxiety about that which
> I have created myself [*rahai svakalpita saṅka*].
> Therefore, I have mixed my own opinions
> with classical precepts—
> may poets forgive any faults.[28]

A century and a half after Keshavdas had shown scholars of systematic literary thought that such systematicity was not only possible but necessary in the vernacular, the very execution of the project apparently remained a source of anxiety. Or was it simply anxiety? Mixing older Sanskrit ideas with newer vernacular ones—innovation through renovation—was the modus operandi of *rīti* intellectuals, and this dual process of simultaneously reprising and reconfiguring the dominant tradition may need to be seen as far more than an act of deference. Affiliating with the dignity and power of a literary culture of the past also helped to ensure Brajbhasha's intellectual and aesthetic success in the present. The simultaneous advocacy of both newness and conformity to Sanskrit tradition—far from being the puzzle it first seemed when we were confronted with Keshavdas's theories about blindness to tradition from the *Kavipriyā*—may actually be emblematic of a more complex power play on the part of Brajbhasha poets.

As moderns it is not always easy to make out the significance of what may seem like mere micro refinements of preexisting theories, such as the ones presented here. This is not because newness isn't there, but because the revolutions of modernity have made our minds far less attuned to subtle gradations of newness. Moreover, there is little place for neoclassicism in the cultural landscape of today. Here it is relevant to point out that neoclassicism of the type found in the works of *rīti* authors is a well-attested literary stance in many parts of Eurasia during the early modern period. Intellectual historians of Europe, for instance, have widely recognized how Romance languages developed both out of opposition to and in imitation of Latin. Early modern French in particular is understood to be a historically versatile and at times contradictory idiom because it could be both a simple language of common speech and a *vulgaire illustre*, a highly refined literary language. By appropriating the very features that made Latin elevated, French writers associated with early modern courts imparted it with dignity, with majesty, with reason.[29] The point is not that *rīti* traditions and seventeenth-century French literary history should be unreflectively assumed to be analogous, although tracing cross-regional parallels can be an illuminating exercise. But it is surely a matter of some surprise that whereas neoclassicists like Corneille and Racine are celebrated greats of the French literary canon, Keshavdas and many of his fellow *rīti* poets are routinely denigrated in Indian literary criticism and historiography. What are the particular historical conditions that allow one culture to embrace its literary heritage and another to despise theirs?

Colonized Epistemological and Literary Spaces

Clouding our vision here may be a colonial-period legacy of ridiculing traditional Indian epistemological methods. Normative texts (*śāstras*) are one of the most critical of scholarly infrastructures in premodern India, but they have almost never been serious objects of research for outsiders. As has been insightfully discussed by Bernard Cohn, colonial administrators routinely dismissed Brahmanical intellectual practices as being focused on memory, repetition, and long taxonomical lists that appeared to befuddle rather than clarify matters through their sheer amplitude. Implicit in such a construction was the criticism that only unintelligent or intellectually depleted people could possibly confine their analysis to the minutiae of type and subtype rather than larger issues of "substance." British patterns of knowing were, in contrast, presented as based on reasoned argument and discriminating.[30] Such (mis)characterizations of Indian epistemology and unfavorable comparisons with Western modes of scholarship were but one arm of a larger body of colonial discourse that tended to characterize the cultural terrain of late medieval India as exhausted and therefore in need of the restorative influence of British rule.

If Indian knowledge practices in general were thus dismissed, a culturally generous approach to Indian literary styles was hardly likely to be forthcoming.[31] Even early Western scholars who did avow the excellence of Indian literature frequently complained that it was stilted and overly elaborate; following rigid literary systems was thought to stifle creative spirit, impeding access to the more "natural" forms of expression favored by Europeans since the heyday of Romanticism.[32] Although Indian *rasa* theory has attracted the attention of a few intellectuals, for the most part canonical Indian systems have never been taken seriously in Western academic writing.[33] Off-putting terms like *mannerist* or its Hindi equivalent, *rītibaddh* ("bound by convention"), foreclose rather than enable discussion of the creativity and power of traditional poetics theory. And since the late nineteenth century a nationalist preoccupation with newer themes of reform, social justice, and political independence, combined with the assimilation of modern Western genres such as the novel, has contributed to an almost total repudiation of earlier poetic modes.

Although nowadays the principal *rīti* literary systems such as *nāyikābheda* and manifold classifications of *alaṅkāras* are apt to be dismissed as tired relics from a feudal courtly culture, I believe it is incumbent on us, as modern students of this premodern literature, to be wary of simply rejecting the traditional categories out of hand. That is easy enough to do. Far more challenging is to try to understand what these categories meant to the people who used them, and why they mattered so much. In the final section of this essay I invite readers to step away from modern prejudices about *rīti* literature to consider some additional issues of importance in reassessing the tradition, such as how traditional methodologies served as an axis for the functioning of courtly literary communities, and what some of the actual uses of the popular *rītigranth* genre were.

The Rītigranth *in Practice*

A useful point of departure for better understanding the function of literary systems in the cultural circles of premodern India is a passage from the unpublished *Saras-sār* (Essence of the Aesthetically Endowed) of Ray Shivdas, which portrays with great liveliness a gathering of Brajbhasha poets that took place in Agra in 1737:

> In Agra there was once
> A meeting of the poets' community [*kavi-samāja*].
> Those who had a penchant for poetry came
> And met with glad hearts.
> All the well-known poets met.
> They decided to create a new book,

Having established new categories
And expressive modes [*rasa*].
Thus, the poets met and shared their ideas,
Each according to his ability,
With deference to literary systems [*rīti*].
With pleasure, all who were present listed
The possible categories.
According to the extent of their intellect
They set out the extensive range of categories,
With the idea that other poets
Would correct any shortcomings.
The poets were of differing opinions
But wise authorities were present,
In keeping with whose opinions
This new book was composed.[34]

This vignette of a premodern literary conference affords access to an intellectual vista centered on concerns certainly very different from those of today, but no less valid for being so. Several points merit attention. First, it is hard not to be struck by the dynamism of the literary environment that is described here, which directly contravenes the modern proposition that late precolonial literature was part of a waning cultural climate. Second, note how delving into the intricacies of specific components of the classical literary system is fundamental to the scholarly enterprise for this community. This corroborates much of what we have already observed in the works of Keshavdas: new knowledge was fashioned within the confines of the existing literary system by assessing the continuing viability of older *bhedas*, or classificatory distinctions, reconfiguring them as necessary and occasionally proposing new ones. And each poet brought "his own understanding" into play.[35] A final point to consider is what the *Saras-sār* suggests about how the Brajbhasha literary community was constituted. A detailed awareness of the plethora of literary types and subtypes formed the substratum of core knowledge that allowed a group of intellectuals to be in dialogue with one another and to participate in a network of meanings that were intelligible to all. This point requires further investigation, for it constitutes one of the most critical, if undertheorized, dimensions of *rīti* literary culture.

In the case of the Agra conference recorded with such enthusiasm in the *Saras-sār*, a group of Brajbhasha intellectuals was present at the same assembly, allowing us a glimpse of how the classical literary systems were, quite literally, a focal point around which scholars converged. But the actual physical copresence of scholars was not necessary for the constitution of a larger intellectual community. The community and national formations that, it has been argued, later became possible through the technology of print culture have been well documented in modern scholarship. Beginning in the modern period, according to the now classic image, two readers of the same newspaper, living in separate parts of a country, could find themselves participating in a shared cultural space across great distances without ever physically meeting.[36] But clearly the stimulus of print culture, albeit strong, is not a prerequisite for the development of such notional, or "imagined," communities, for a genre like the *rītigranth* enabled a strong sense of literary belonging from within the confines of a manuscript culture. A primary way of indicating shared participation in this community was to write a work of literary theory. In the case of the Agra conference the fashioning of the *rītigranth* was a collective enterprise, but in most cases individual poets contributed their "own understanding" to the larger literary and intellectual community in a single-author work.

Another explanation for the popularity of the *rītigranth* genre is its role in underwriting the courtly culture of performed poetry. It cannot be emphasized too strongly that the *rīti* works that come down to scholars today as inert, arcane entities had a more eclectic, "multimedia" literary life during their heyday. Brajbhasha poetry was not just read in private; it was recited, it was sung, it was danced. In courtly contexts it was the focal point of competitions such as *samasyāpūrti*, in which the patron or person overseeing the event would propose a point of departure (*samasyā*) for the assembled poets. This *samasyā* might be the last word or phrase or line of a poem, or perhaps a theme such as a particular type of *nāyikā*.[37] Poets would then be evaluated on the quality of the verse that they spontaneously composed or "completed" (*pūrti*). Success in this kind of competition required a solid background in the various domains of *alaṅkāraśāstra* encompassed by scholarly writings in the *rītigranth* style.

What of the audiences who read or listened to poetry being declaimed? In order to achieve the necessary interpretive skills they too had to be versed in the *rīti* system. This was particularly crucial because the most popular Brajbhasha verse form in courtly settings was the *muktak* (freestanding) poem. As its name suggests the *muktak* is not part of a larger narrative structure. The charm of this verse style is that the reader or listener (*rasika*) steps into the middle of a story. The full story is never told in the poem itself, especially in the case of a short couplet, where there is room for only the sparsest of narrative details. Consider the complex literary infrastructure that must be in place for even a two-line Brajbhasha *muktak*, like the following one from the *Bihārī-satsaī* (seven hundred poems by Bihari, 1650?), to generate meaning: "Why do you drive me crazy with all your lies? / You can't hide the truth. / Your eyes, dripping with redness, / tell the tale of last night's pleasures."[38] How is it that a short poem such as this—wherein the speaker, the addressee, and the subject of the conversation are never directly revealed—is readily comprehensible to its audience?

As far as the minimal narrative content of the poem goes, we are simply told that upon seeing somebody's red eyes another person gets angry. But the metadiscourse of *rīti* poetics allows us easily to fill in the rest of the story. In the case of this particular poem we need above all to ascertain the characters. A reader familiar with the basics of *nāyikābheda* is likely to surmise that Bihari has depicted an encounter between an angry female character (*khaṇḍitā nāyikā*) and an unfaithful lover (*śaṭha nāyaka*). According to the conventions of *rīti* literature red eyes in a man are a clue that he has been up all night making love to a rival. His eyes may be red either from lack of sleep or because during the heat of passion things got a little messy and betel juice (the proper location of which is, of course, the mouth) got into his eyes. Bihari's *dohā*s have frequently been celebrated for their quality of being "a small pot that contains the ocean" (*gāgar meṃ sāgar*). The reason this *rīti* poet can say so much in so few words, however, is because a structured literary system provides the context in which to interpret his poetry.[39]

How did an aspiring poet or poetry connoisseur learn this system—the price of entrance into the learned courtly circles of early modern India? One way was to study a *rītigranth*, perhaps with the help of a teacher or pandit. Court pandits—Sanskrit and, in later periods, Braj—were instrumental in the education of young princes and children of the nobility, and some *rīti* texts, like the *Kavipriyā*, served as companions to teaching.[40] In addition to instructing younger students, court pandits also served as mentors to kings, for whom literary connoisseurship was de rigueur and original literary composition strongly encouraged. Many *rītigranth*s were written explicitly at the request of royal patrons, and kings commissioned copies of the most authoritative works produced at other courts for their personal libraries. Reading and learning the principles of *alaṅkāraśāstra* alone did not transform one into a scholar of this subject. Perhaps yet another way to account for the proliferation of the *rītigranth* genre was that in some cases the writing of such texts itself was part of the education process (perhaps like a Ph.D. in Hindi literature in modern times?). Creating a new treatise on *alaṅkāra*s or *nāyikābheda* demonstrated that a pandit had sufficient mastery of the subject to carry out various tasks: performing in a courtly assembly, educating others, and delighting his patrons with a beautiful, well-structured anthology of Braj poetry.

Conclusion

Exploring in some detail the thought-world and cultural practices of seventeenth- and eighteenth-century *rīti* writers prompts us to reconsider current constructions of the intellectual life of late precolonial India. Failure to examine in sufficient depth the modalities of courtly writers has led to many unfortunate and inaccurate representations of *rīti* literary culture. In seeking to understand the logic and function of Brajbhasha literary science we not only deepen our awareness of the epistemological domains of premodern Indian life; we also enrich the field of Hindi studies by encouraging scholarly analysis of literary realms beyond the familiar domain of *bhakti* poetry.

Rīti authors have frequently been criticized for their narrow focus on the minute details of the various *bheda*s of classical literary science. Although in modern times this deep concern with precise categorization is generally viewed as both artistically and intellectually stilted, during the *rīti* period it constituted a vibrant, indeed indispensable compositional approach. The *rītigranth* genre should also be appreciated for its role in enabling the production and interpretation of courtly poetry. Intelligibility and success in courtly venues depended on the familiarity of both poets and audiences with classical literary conventions. The *rītigranth* was thus a primary tool for enabling the social and communicative processes of connoisseurship. The writing of *rītigranth*s also had a largely overlooked symbolic value insofar as it betokened membership in a widespread community of Brajbhasha poets and intellectuals. The knowledge system of vernacular *alaṅkāraśāstra* constituted a literary consensus that was continually being renegotiated by *rīti* authors through their participation in assemblies and their contributions to scholarship.

The assessment of the Brajbhasha *rītigranth* as largely derivative of Sanskrit sources, and therefore intellectually insignificant, is inaccurate. Many *rīti* works of *alaṅkāraśāstra* exhibit a complex weaving together of classical ideas with fascinating innovations. The newness that we see in *rīti* texts is not earth-shaking, at least not by contemporary measurements. But it is a newness we should take seriously by attempting to comprehend the logic and functioning of a fledgling branch of vernacular knowledge as it began to put forward increasingly strong claims to an existence separate from Sanskrit. Carving out a new domain of vernacular writing from a Sanskrit mold was not a process undertaken lightly; it engendered a range of anxieties about transgressing age-old language hierarchies. But alongside the uncertainties we hear an unmistakable voice of strength, an excitement about new literary and intellectual possibilities evident in the oft-repeated phrase of the *rīti* poet scholar: "I have composed this according to my own understanding."

Notes

This essay is revised from "The Anxiety of Innovation: The Practice of Literary Science in the Hindi *Riti* Tradition," *Comparative Studies of South Asia, Africa and the Middle East* 24, no. 2 (2004), 45–59. Parts of the argument have been presented in different scholarly venues, and I owe much to various listeners and readers over the years. An early seminar series on Indian knowledge systems hosted by Sheldon Pollock in the spring of 2003 was the inspiration for the first draft. Another version of the paper was tried out on patient colleagues at the Triangle South Asia Consortium colloquium series. Special thanks are due to Pika Ghosh and Shantanu Phukan for their detailed feedback. A conference on *kāvya* sponsored by Yigal Bronner and David Shulman at Hebrew University (Jerusalem, May 2004) helped to sharpen my thinking further.

1. Brajbhasha was the primary dialect of written Hindi for more than three centuries prior to about 1900, at which point Modern Standard Hindi (Khari Boli) began to achieve cultural dominance. Because I am dealing exclusively with premodern texts I use the terms *Hindi* and *Brajbhasha* synonymously.

2. The term *rītikāl* was coined by Ramchandra Shukla in 1929, and it has remained in wide circulation ever since. Shukla, *Hindī sāhitya kā itihās*, 1.

3. Keshavdas, *Rasikpriyā*, vv. 7.1–3. All translations here and elsewhere from both Brajbhasha and Sanskrit are my own.

4. A particularly unhelpful, if regrettably typical, analysis of Hindi neoclassicism as reflecting a decline from the simplicity of *bhakti* and a simultaneous fall from the intellectual grace of Sanskrit is the treatment of Keshavdas in Jindal, *A History of Hindi Literature*, 142–48.

5. A welcome attempt to counter modern biases against courtly literature in the case of Persian is Meisami, *Medieval Persian Court Poetry*, 40–76.

6. The perceived limitations on vernacular expression from a Sanskrit point of reference are discussed by Pollock, chapter 1, this volume.

7. For recent work on Jagannatha, see Pollock, "The Death of Sanskrit," 404–12. Some useful remarks on interchanges between Sanskrit and the regional languages of south India can be found in Rao and Shulman, *A Poem at the Right Moment*, 187.

8. See *Śṛṅgāramañjarī*, ed. Mishra. This Braj translation was by the *rīti* poet Cintamani Tripathi, discussed below.

9. See *Śṛṅgāramañjarī*, ed. Raghavan, 2, 37.

10. On the importance of Hindi literature at Shivaji's and other Maratha courts, see Divakar, *Bhonslā rājdarbār ke hindī kavi*.

11. On Kavindra's sense of shame, see Divakar, introduction to *Kavīndracandrikā* (quoting Kavindracarya's *Kavīndrakalpalatā*, v. 13).

12. Jayarama Pindye, *Rādhāmādhavavilāsacampū*, 227. Note Jayarama's apparent confusion about how to handle the multilingualism of his *Campū*: he vacillates about whether he has actually written a Sanskrit or a *bhāṣā* work. See 244–46.

13. See Pindye, *Rādhāmādhavavilāsacampū*, 233, 237, 243.

14. The first ten cantos occupy forty-three printed pages, whereas the last canto alone comprises thirty-three. For further remarks on this text, see the essay by Sumit Guha in this volume.

15. Keshavdas, *Rāmcandracandrikā*, vv. 1.4–5.

16. Cf. McGregor, "The Progress of Hindi," 928.

17. The very distinction posited by Hindi critics between *bhakti* and *rīti* texts rarely withstands close scrutiny. An excellent discussion of this issue is Snell, "*Bhakti* versus *Rīti*?" Also see Busch, "Questioning the Tropes about '*Bhakti*' and '*Rīti*' in Hindi Literary Historiography."

18. A typical formulation is Jindal, *History of Hindi Literature*, 64.

19. An example of this theoretical approach is Shukla, *Hindī sāhitya ka itihās*, 129–33. Since Shukla's day, several Indian scholars have undertaken to identify the contributions of *rīti* intellectuals, arguing for both the *ācāryatva* (intellectual achievement) and *mauliktā* (originality) of the *rīti* oeuvre. See, for instance, Chaudhari, *Hindī rītiparamparā ke pramukh ācārya*; Kishorilal, *Rīti kaviyoṃ kī maulik den*.

20. Kaviraj, "Writing, Speaking, Being," 35.

21. The other two are the *Kāvyakalpalatāvṛtti*, a thirteenth-century poet's manual by Amaracandra Yati, and the *Alaṅkāraśekhara* of Keshava Mishra, written in Delhi in the generation preceding Keshavdas.

22. The *andhadoṣa* is defined as "against the path" (*birodhī pantha ko*), v. 3.6.

23. Cintamani Tripathi, *Kavikul-kalptaru*, vv. 1.3, 1.6.

24. Compare *Rasikpriyā*, vv. 5.24–38, with Rudrabhatta's original discussion in *Śṛṅgāratilaka*, v. 2.38.

25. Compare the arguments in *Rasikpriyā*, v. 8.54, with those in *Śṛṅgāratilaka*, v. 2.28.

26. The three classical types of *nāyikā* are "one's own" (*svakīyā*), "the wife of another" (*parakīyā*), and the "public woman" (*sāmānyā*).

27. See the opening chapter of the *Rasikpriyā*, particularly vv. 1.2, 1.16. Surati Mishra's perspective is evident from his commentary, *Jorāvar prakāś*, 54–57. The concept of *navarasamaya* does have an analogue in Rudrabhatta's idea of Shiva as *sarvarasāśraya* (encompassing all the *rasas*) in *Śṛṅgāratilaka*, v. 1.1.

28. Excerpted from *Kāvyanirṇay*, vv. 1.5–7.

29. Fumaroli, "L'apologétique de la langue française classique," 157–58.

30. See Cohn, *Colonialism and Its Forms of Knowledge*, 51–53.

31. Macaulay's infamous characterization of the "native literature of India" is too well known to need quoting.

32. A. B. Keith, who apparently liked Sanskrit literature enough to write an entire book on the subject, decried its "obscurity of style," "taint of artificiality," and several other literary tendencies that he considered indicators of a "defect of the Indian mind" (*A History of Sanskrit Literature*, 9–10).

33. Exceptions are Pollock, "Theory and Method in Indian Intellectual History," "Working Papers on Sanskrit Knowledge-Systems on the Eve of Colonialism I," and "Working Papers on Sanskrit Knowledge-Systems on the Eve of Colonialism II."

34. This portion of Ray Shivdas's unpublished *Saras-sār* is excerpted in Gupta, *Surati Miśra aur unkā kāvya*, 21–22.

35. Note in particular the phrases "each according to his ability" (*jathā jog*) and "according to the extent of their intellect" (*apnī mati paramāna so*) from the *Saras-sār* passage.

36. Anderson, *Imagined Communities*, 32–36.

37. For instance, one poetic challenge presented to pandits in Shahaji Bhonsle's assembly concerned the elucidation of the difference between *nāyikā*s both "conscious" and "unconscious" of the arrival of puberty (*jñātayauvanā* and *ajñātayauvanā*) according to Bhanudatta's classical description of them. Jayarama, *Rādhāmādhavavilāsacampū*, 233.

38. *Bihārīsatsaī*, v. 11.

39. Bihari was one of the rare *rīti* poets who did not write a *rītigranth*. But the interpretation of his work is often dependent on the system. Such poets are known as "based on system" (*rītisiddh*) in Hindi criticism. See Mishra, *Bihārī*, 44–45.

40. Keshavdas states this unambiguously: "Keshav wrote the *Kavipriyā* so that boys and girls would understand the subtle ways of poetry. May scholars look leniently upon any mistakes," v. 3.1. A similar example is by Udaynath Kavindra, court poet to King Gurudatt Singh of Amethi, who wrote a *rītigranth* for the education of his son (cited in Sharma, *Rītikālīn kavi kālidās trivedī granthāvalī*, 20).

References

PRIMARY SOURCES

Bihārīsatsaī of Biharilal. Ed. Sudhakar Pandey. Varanasi: Nagari Pracarini Sabha, 1999.

Jorāvarprakāś of Surati Mishra. Ed. Yogendrapratap Singh. Allahabad: Hindi Sahitya Sammelan, 1992.

Kavikul-kalptaru of Cintamani Tripathi. Lithograph. Lucknow: Naval Kishore Press, 1875.

Kāvyanirṇay of Bhikharidas. *Bhikhārīdāsgranthāvalī,* vol. 2, ed. Vishvanath Prasad Mishra. Varanasi: Nagari Pracarini Sabha, 1957.

Rādhāmādhavavilāsacampū of Jayarama Pindye. 1922. Ed. V. K. Rajvade. Pune: Varda Books, 1989.

Rāmcandracandrikā of Keshavdas. *Keśavgranthāvalī,* vol. 2, ed. Vishvanath Prasad Mishra. Allahabad: Hindustani Academy, 1955.

Rasikpriyā of Keshavdas. *Keśavgranthāvalī,* vol. 1, ed. Vishvanath Prasad Mishra. Allahabad: Hindustani Academy, 1954.

Śṛṅgāramañjarī of Akbar Shah. Ed. V. Raghavan. Hyderabad: Hyderabad Archaeological Department, 1951.

Śṛṅgāramañjarī of Cintamani Tripathi. Ed. Bhagirath Mishra. Lucknow: Lucknow University Press, 1956.

Śṛṅgāratilaka of Rudrabhatta. Ed. R. Pischel, trans. Kapildev Pandey. Varanasi: Pracya Prakashan, 1968.

SECONDARY SOURCES

Anderson, Benedict. *Imagined Communities: Reflections on the Origin and Spread of Nationalism.* 2nd ed. London: Verso, 1991.

Busch, Allison. "Questioning the Tropes about '*Bhakti*' and '*Rīti*' in Hindi Literary Historiography." *Bhakti in Current Research, 2001–2003,* ed. Monika Horstmann. Delhi: Manohar, 2006.

Chaudhari, Satyadev. *Hindī rīti-paramparā ke pramukh ācārya.* Delhi: Hindi Madhyam Karyanvay Nideshalay [Delhi University], 1992.

Cohn, Bernard S. *Colonialism and Its Forms of Knowledge.* Princeton, N.J.: Princeton University Press, 1996.

Divakar, Krishna. *Bhonslā rājdarbār ke hindī kavi.* Varanasi: Nagari Pracarini Sabha, 1969.

———. Introduction to *Kavīndracandrikā* of Kavindracarya Sarasvati. Pune: Maharashtra Rashtrabhasha Sabha, 1966.

Fumaroli, Marc. "L'apologétique de la langue française classique." *Rhetorica* 2, no. 2 (1984), 139–61.

Gupta, Chotelal. *Surati Miśra aur unkā kāvya.* Allahabad: Smriti Prakasan, 1982.

Jindal, Kailash Bhushan. *A History of Hindi Literature.* 1955. New Delhi: Munshiram Manoharlal, 1993.

Kaviraj, Sudipta. "Writing, Speaking, Being: Language and the Historical Formation of Identities in India." *Nationalstaat und Sprachkonflikt in Süd- und Südostasien,* ed. Dagmar Hellmann-Rajanayagam and Dietmar Rothermund. Stuttgart: Steiner, 1992.

Keith, A. B. *A History of Sanskrit Literature.* 1900. New York: Haskell, 1968.

Kishorilal. *Rīti kaviyoṃ kī maulik den*. Allahabad: Sahitya Bhavan, 1971.

McGregor, R. S. "The Progress of Hindi, Part 1: The Development of a Transregional Idiom." *Literary Cultures in History*, ed. Sheldon Pollock. Berkeley: University of California Press, 2003.

Meisami, Julie Scott. *Medieval Persian Court Poetry*. Princeton, N.J.: Princeton University Press, 1987.

Mishra, Vishvanath Prasad. *Bihārī*. 1950. Varanasi: Sanjay Book Center, 1998.

Pollock, Sheldon. "The Death of Sanskrit." *Comparative Studies in Society and History* 43, no. 2 (2001), 392–426.

———, ed. "Theory and Method in Indian Intellectual History." Special issue of *Journal of Indian Philosophy* 36, nos. 5–6 (2008).

———, ed. "Working Papers on Sanskrit Knowledge-Systems on the Eve of Colonialism I." Special issue of *Journal of Indian Philosophy* 30, no. 5 (2002).

———, ed. "Working Papers on Sanskrit Knowledge-Systems on the Eve of Colonialism II." Special issue of *Journal of Indian Philosophy* 33, no. 1 (2005).

Rao, Velcheru Narayana, and David Shulman. *A Poem at the Right Moment*. Delhi: Oxford University Press, 1999.

Sharma, Ramanand. *Rītikālīn kavi kālidās trivedī granthāvalī*. Ghaziabad: K. L. Pacauri Prakashan, 2003.

Shukla, Ramchandra. *Hindī sāhitya kā itihās*. 1929. Varanasi: Nagari Pracarini Sabha, 1994.

Snell, Rupert. "*Bhakti* versus *Rīti*? The Satsaī of Bihārīlāl." *Journal of Vaishnava Studies* 3, no. 1 (1994), 153–70.

CHAPTER FIVE

Writing Devotion

The Dynamics of Textual Transmission in the *Kavitāvalī* of Tulsīdās

IMRE BANGHA

The literary culture of India in the middle of the second millennium CE was marked not simply by vernacularization and a dramatic decline in the production of Sanskrit literature, but also by the infusion of Sanskritic learning into the vernaculars. In many fields in north India the achievements of Sanskrit were being vernacularized. In the case of Hindi, for example, there is an extensive literature on astrology, aesthetics, and erotics following closely its Sanskrit models. Sanskrit epics and *purāṇa*s were being widely transcreated into the literary versions of the spoken languages. Apart from transcreations of whole works, there were subtler ways in which Sanskrit learning penetrated into vernacular domains. The aesthetics of new literary works were based on those described in Sanskrit treatises. Thus not only the Hindi *rīti* poetry trained on Sanskrit conventions but even the more spontaneous devotional works abound in *alaṅkāra*s, figures of sense and meaning, can be analyzed according to the *rasa* conventions.

One of the fields in which vernacular literatures followed Sanskrit was the vernacularization of manuscript production[1] that in the Hindi belt in north India for more than a millennium had been associated with Sanskrit (along with Prakrit and Apabhramsha) and since the thirteenth century with Persian and Arabic. The most influential Hindi devotional authors may have lived in the fifteenth and the sixteenth centuries, but *bhakti* texts begin to materialize in the late sixteenth century. While Sanskrit and Perso-Arabic manuscripts appear in relatively high numbers at earlier times, today there are only a few extant Hindi manuscripts dated prior to 1600, and most of them are manuscripts of Jain works and hardly any of them contain devotional texts. Although there have been several claims that certain manuscripts date to the sixteenth century or serve as an accurate record of sixteenth-century traditions, scholars have raised doubts about most of them.[2] Some devanāgarī manuscripts, however, date without a doubt from prior to 1600. The Jain *Pradyumna-carit*, for example, is extant in three manuscripts copied in 1548, 1577, and 1591.[3] Another twenty-seven manuscripts of Jain or Rajasthani works date from between 1493 and 1594.[4] The few non-Jain and non-Rajasthani manuscripts include a *Qutub-śatak* from 1576, and the famous "Fatehpur manuscript" with the songs of Sūrdās, Kabīr, and other devotees dates from 1582.[5] From the seventeenth century through the nineteenth, however, hundreds of thousands of Hindi manuscripts are extant.

It is a well-known fact that the most successful early *bhakti* poets, Kabīr, Mīrā Bāī, and Sūrdās, produced oral literature, chiefly songs with a refrain (*ṭeka*) set to a certain rhythm (*tāla*) and with a dominant mood (*rāga*). Their songs, called *pada*s, were not considered poetic literature (*kabitā* or *kabitt*) but *bhajan*s, that is, a means to express devotion. One of the poets, Kabīr, is credited with couplets that are openly inimical to books and writing.[6] Most of their poems were committed to writing after the death of the author and after an extended period of oral transmission that may not only have changed them through modifications called *geyavikāra*s,[7] but also expanded their literary corpora of perhaps a few hundred independent poems into thousands. Paradoxically the changes and expansion in oral tradition can be documented only through manuscripts, specifically manuscripts that bear a date of transcription. For example, the critical edition of Kabīr's *pada*s lists 593 songs found in manuscripts dated between 1570 and 1681, but according to one estimate over 2,000 different religious songs (*pada*s, *śabda*s, *bhajan*s, and *ramainī*s) attributed to him have so far been published.[8] The earliest available manuscript with the songs of Sūrdās, which is dated 1582, has only 239 different *pada*s. From 1624 we have a book with 492 and from another half-century later (1676) a collection with 1,472 songs,[9] and the modern published *Sūrsāgar* has over 5,000. The most dramatic amplification happened with Mīrā Bāī's songs. At present we have at hand only 6 poems of hers in seventeenth-century manuscripts, but at the end of the twentieth century Winand Callewaert found 5,197 *pada*s with her poetic signature.[10] From the inflated corpora that we have at our disposal we cannot determine what the early *bhakti* poets composed, and their personal identities tend to recede from us.

The fourth great devotional poet, Tulsīdās, differs from the previous poets because, as I will argue, he committed to writing many of his works at the moment of their birth. Although he may not have been the first devotional poet who *wrote*, he was the most successful in introducing *bhakti* into manuscript culture and thus appropriating a territory—right in the heartland of Sanskrit learning in north India, in Benares—that up to then had been dominated by Sanskrit.

In the title of his excellent study on performing the *Rāmcaritmānas*, *The Life of a Text*, Philip Lutgendorf implies that a text comes to life when performed. One can, however, argue that texts have life when they are *communicated*; this includes written transmission, in which texts may not change as much as in oral transmission, yet they are still prone to amplification, purification, and intentional or unconscious changes. What were the implications of infusing *bhakti* into the written world? How did written devotion work? We have only a few Hindi critical editions of *bhakti* poetry at our disposal, so we are not able to make definite statements about transmission, yet some common traits can be observed across several texts.

The observations regarding some relatively well-documented instances of early modern editing can shed light on some major processes that may have taken place during the two-thousand-year production of other manuscripts. During transmission the sequence of stanzas and lines can be altered, and lines, phrases, and words can acquire variant readings. There are cases when copies of the work are virtually identical, with the variation consisting of nothing more than the occasional spelling error, the insertion of paratextual material in the form of chapter or verse citations, or the appending of commentary, as is the case with the Bengali *Caitanya Caritāmṛta*. Because of the tight sectarian control in the reproduction of the text there is decidedly little variation in the manuscripts.[11] This was not the case, however, with most Hindi-language material, as the few critical editions at our disposal show. Most texts underwent significant changes after the death of their author.

In his study of the transmission of Kevalrām's (b. 1617) *Rāsa māna ke pada*, Alan Entwistle examines a composite text, which resulted from an editor's copying an exemplar of which the folios were in slight disorder and comparing that exemplar intermittently with another source text. The outcome was omission or conflation as well as correction of some omissions and insertion of *padas*.[12] Rupert Snell in his critical edition of the eighty-four *padas* of Hit Harivaṁś (early sixteenth century) suggests that a portion of eleven stanzas in the middle of the collection (songs 39–49) may represent an accretion to a preexisting collection of *padas* since textual phrases from this section very rarely show correspondences with the remainder of the text, while the rest frequently contains phrases that are repeated. Snell also suggests that the inclusion of these stanzas is the result of a conscious amplification because it is approximately in the same sequence that the predominance of Radha becomes established for the first time in the text, and thus this part confirms the more recently developed sectarian priorities of the Rādhāvallabh school.[13] In her edition of the poetry of another sixteenth-century devotional poet, Svāmī Haridās, Lucy Rosenstein distinguishes two phases in the development of the textual transmission, one before the canonization of Haridās's poetry and one after. An early manuscript of the canonized version from 1755 suggests that the canonization took place sometime before that date. Two or three of the sixteen manuscripts inspected by Rosenstein contain traces of the period before canonization and are closer to a period of oral transmission. Furthermore she suggests that two subrecensions of the canonized version exist.[14] In her edition of the *Rās-pañcādhyāyī* of Harirām Vyās (fl. 1550), Heidi Pauwels hints that the redaction of the *Vyās vāṇī* into a recension that she calls the Vrindaban vulgate took place sometime between 1667–68 and 1737. The earliest manuscript of the vulgate is from 1737. Pauwels had the good fortune to find a dated early manuscript (from 1667–68) that does not contain the vulgate version, but interestingly it shares peculiarities with the *Rās-pañcādhyāyī* today attributed to Sūrdās. All other manuscripts discovered by Pauwels fall into the Vrindaban vulgate recension, which in turn can be divided into two scribal subrecensions.[15]

Another genre in which the study of transmission can yield interesting results is hagiography. In his examination of the *Dādū janma līlā*, a late sixteenth-century biography of Dādū Dayāl by his disciple, Jan Gopāl, Winand Callewaert found that the number of verses in the four manuscripts representing the most ancient layer of the text is the smallest and that there are more in the manuscripts of the second layer (the oldest of which dates from 1654), and even more in a modern edition from 1947. The amplifications and omissions have resulted in a text with more miracles, a story explaining that Dādū was not born as a cotton carder but as a Brahman, and passages emphasizing his superiority over other great men, such as Akbar. The wording has been changed to suggest Dādū's divine origin and to identify his guru as Hari and not Bābā Būḍhā. Several times Jan Gopāl's incorrect rhyme is corrected in the manuscripts of the second layer.[16]

Surveying the sources of these texts one can observe that, apart from the *Rāsa māna ke pada,* where only two complete manuscripts were consulted, only edited versions (sometimes more than one version, as the case of the *Dādū janma līlā* suggests) tended to be transmitted, and only a few extravulgate handwritten books survived. This may result either from tight sectarian control or from the relative clarity of an edited text. Sectarian control similar to that of the *Caitanya Caritāmṛta* in the case of the three Vrindaban devotees, the *Haritraya* of Harivaṁś, Haridās, and Harirām Vyās, was introduced probably in the seventeenth century or early eighteenth, after the production of the edited vulgates. The initial lack of such uniformity may be due to the fact that Harivaṁś and Haridās produced songs to be transmitted orally. Clarity was reconstructed either by purging the text of its mistakes or by simplifying difficult original phrases when the editor was not able to interpret them.

These are devotional texts. Relatively little attention has been paid to the transmission of secular works. The interplay of the secular and the devotional can be observed in the case of the Nimbārkī renunciate Ānandghan (ca. 1700–1757), whose poetry was drastically changed in an editorial process. His quatrains can be interpreted either as expressions of Vaishnava devotion (as is done in the collection *Sujānhit*) or as poetry in a courtly style influenced by Persian literary ideas (as in the collection *Ghan-Ānand kabitt*). The text in the early incomplete versions of the *Sujānhit* has been altered in such a way that many occurrences of the word for the beloved, (*su*)*jāna*, loaded with both Hindi and Persianate meanings, have been altered into clearly religious or secular expressions, such as (*ju*) *syāma* (Krishna) and *su pyārī* (that [female] beloved). This was done in order to avoid the possibility of identifying Krishna with Ānandghan's worldly beloved, a Muslim dancer. In spite of the fact that Ānandghan's three earliest dated manuscripts (from 1727, 1729, and 1743) give these readings, they must be secondary since the multilayered meanings with the ambiguity of the word *sujāna* peculiar to a much larger corpus of quatrains is lost in them and the text becomes pedestrian. No later manuscript follows this practice; these early copies must represent an attempt to defend Ānandghan from sectarian accusations. It was around 1748 that a courtier and a Sanskrit and Hindi poet from Jaipur, Brajnāth Bhaṭṭ, edited another collection, known today as *Ghan-Ānand kabitt*, to emphasize the nonsectarian, all-encompassing aspect of love. Brajnāth changed the sequence of the poems of the *Sujānhit*, which had more worldly love-oriented poems in the beginning and more devotional ones toward the end, into portions dealing with more mundane aspects of *śṛṅgāra rasa* (the sentiment of love in poetry) such as *saṁyoga* (love in union), *viraha* (the pangs of separation), and *māna* (wounded pride). Brajnāth mentions at the end that in acquiring and editing these poems he has had to bear a lot of criticism and has lost his "honour and standing" thereby.[17]

In this essay I examine the textual transmission of another early modern corpus, one of the "minor" works of Tulsīdās (1532 or 1543?–1623?), the *Kavitāvalī* (Series of Poems/Quatrains), probably compiled in the second decade of the seventeenth century. This is a collection of some 350 loosely connected quatrains in strict meters and with a more individual approach than is found in most of the author's other works. Tulsī's favorite themes are collected here, and although arranged into seven cantos (*kāṇḍas*) according to the *Rāmāyaṇa* tradition, it does not have a linear narration.

The collection has enjoyed immense popularity. Initially it was transmitted in handwritten books, and although no autograph copy survives about seventy copied manuscripts have been traced in the past hundred years, hinting at the fact that several hundreds must have been prepared over the centuries. Since its first printed edition in 1815 the *Kavitāvalī* has been published about 120 times.

The Kabitt *Form*

The force that keeps the distinct parts of the collection together is not simply that of a linear narrative but rather the poetic form: the entire *Kavitāvalī* is written in *kabitt*s (quatrains), more specifically in syllabic *kavitt*s and moraic *savaiyā*s, with the addition of a few *chappay*s. People in modern India might interpret the title as "Series of Poems," taking the word *kavitā* in its modern sense. But this is somewhat misleading, and the work in most manuscripts and early editions is called *Kabitta(baddha) rāmāyaṇ* (*Rāmayaṇa* Composed in Quatrains) or *Kabittabaddha rāmcaritra* (The Deeds of Rāma Composed in Quatrains).[18] Excepting the designation *Tulsīdāsjīkṛt kabitt* found in one manuscript, all titles indicate that the scribes considered it a precomposed, unified work and not a random collection of Tulsī's *kabitt*s.

The early devotional poets conveyed their message most effectively in *pada*s, which normally have a loose moraic meter suitable for emotional expression through singing. The *Kavitāvalī* is a devotional work written not in *pada*s but in the *kabitt* form. The importance of the form can be judged by the fact that Tulsī's various works are organized on the basis of the meter involving a context of performance. The *Rāmlalā nahchū, Pārvatī-maṅgal,* and *Jānakī-maṅgal* are in the *maṅgal* and *haṁsagati* meters, used in songs sung at auspicious occasions. These works are composed in Avadhi, and their form also indicates a closer link to popular culture and to oral transmission. The *Kṛṣṇa-gītāvalī, Gītāvalī,* and *Vinay patrikā* are in Brajbhasha *padas*, suggesting a context of devotional singing. The *Dohāvalī* and *Rāmājñā praśna* are in couplets (*dohā*s and *sorațhā*s), the *Vairāgya sandīpanī,* just like the *Mānas* and other epics, in *dohā*s and *caupāī*s. Similarly to the *Kavitāvalī,* some works, such as the *Barvai Rāmāyaṇ* and the *Gītāvalī,* are divided into seven cantos according to the *Rāmāyaṇa* tradition, implying that Tulsīdās produced various *Rāmāyaṇas* in various meters.

While *padas* written in different dialects were the main form of devotional singing, *kabitts* were products of Brajbhasha specifically. Although they were also vehicles of devotional messages, along with the couplet *dohā*, from the end of the sixteenth century they became the major meters in court poetry.

The *kavitt* form, with its reliance on sequences of stressed and unstressed syllables, is especially suitable for conveying a sense of violence, heroism, or fear. The whole of the *Sundarakāṇḍa*, with its description of the burning of Laṅkā, and most of the *Laṅkākāṇḍa* are written in *kavitts*.

Themes and Structure

The arrangement of the first six *kāṇḍas* of the collection follows the Rama story, providing us with glimpses at some of its most enchanting points, while the *Uttarakāṇḍa* discards the narrative structure and comprises poems celebrating Rama's name, virtues, or grace, descriptions of the dark Kali age, of places of pilgrimage, of the *gopīs*' love for Krishna, or descriptions of Shiva, prayers for release from calamities such as the pestilence in Benares, and so forth. Several poems expound Rama's grace with reference to Tulsīdās himself. Even the first part of the *Kavitāvalī* is not strictly linear but like a series of miniature illustrations to an epic tale with which everyone is familiar. The poems sometimes have a style full of *alaṅkāras* of which any *rīti* poet could be proud.

The structure of the *Kavitāvalī* is unvarying in the modern editions, where it is a collection of 325 independent quatrains. I refer to this as the vulgate text, within which the only major variation is that the *Tulsī-granthāvalī* gives an extra *savaiyā* in a footnote but immediately rejects it as inauthentic. The collection is structured according to the seven *kāṇḍas* of the *Rāmāyaṇa*, but the distribution of the poems is rather uneven. There are altogether 142 quatrains in the first six *kāṇḍas*, with only one poem in the *Araṇyakāṇḍa* and one in the *Kiṣkindhākāṇḍa*; 183, more than half of the total, are in the *Uttarakāṇḍa*. Certain editions give the forty-four *kabitts* of the vulgate *Hanumānbāhuk*, the ailing Tulsī's prayers to Hanuman, as an appendix to the *Kavitāvalī*.

The oldest available manuscript, dating from 1691,[19] comprises only the *Uttarakāṇḍa* together with the vulgate *Hanumānbāhuk*. Tulsī's poems related to Hanuman, however, also acquired an existence independent of the *Kavitāvalī*. From the late eighteenth century onward several collections of *Hanu-*

mānbāhuk sarvāṅgarakṣā proliferated under the name of Tulsīdās and probably served as charms or prayers against diseases. The most widespread version included about half of the poems from what later became the vulgate *Bāhuk*, one from the *Kavitāvalī*, and some thirty-six spurious ones. It was these collections that were first called *Hanumānbāhuk*. The earliest documented *Hanumānbāhuk* (*sarvāṅgarakṣā*) manuscript is from 1778.[20] I have not been able to find an independent vulgate *Bāhuk* dated prior to 1860.[21] The emergence of the *Bāhuk* as an independent collection may be attributable to the growing cult of Hanuman.[22]

The Transmission

A peculiarity of manuscript transmission is that no two manuscripts present the same text. The difference between manuscript and print cultures is well illustrated by the story of two Gujarati clerks. On seeing a printed magazine the clerks sat down, one reading his version aloud and the other comparing it with the printed words in his copy. When they found that the two copies tallied word for word, their amazement was unlimited.[23] While manuscript transmission makes difficult, if not impossible, the reconstruction of what the author wrote, it can give information on the transmission and the reception of the text that may not be present in printed versions. For example, variations in the handwriting and changes of pen and paper may be evidence of the fluctuations in the degree of the scribe's concentration or his adoption of different postures while copying.[24] While the old manuscripts, composed with a varying degree of textual corruption, show a complex interrelationship, almost all modern redactions and commentaries of Tulsī's minor works, including the edited Gita Press volumes as well as the huge number of critical studies, directly or indirectly rely on the texts of their exemplar, the *Tulsī-granthāvalī* published in 1923, which in turn is based on a late nineteenth-century edition of the minor works, the so-called *Dūbejī* edition (1886) by Ramgulam Dvivedi and Lala Chakkanlal.[25]

The *granthāvalī* is not a critical edition: the editors do not indicate the original source for the text of an individual composition and give variants only occasionally. Only the *Rāmcaritmānas* and the *Barvai Rāmāyaṇ* have critical editions based on a consistent collation of the most important manuscripts, and since the still authoritative book on Tulsīdās by Mataprasad Gupta, first published in 1942, not many scholars have touched upon the textual problems in his corpus.[26] No history of the transmission of the *Kavitāvalī* has so far been undertaken; however, the material collected for a forthcoming critical edition by the Tulsīdās Textual Study Group, a group of students and academics in Budapest, Oxford, and Miercurea Ciuc in Romania, can serve as a basis for studying its spread.[27] Because a large number of manuscripts are lost and the available material is only partially processed, the reconstruction of the history of the text is only fragmentary. For the edition I have collected copies of thirty-seven complete or fragmentary manuscripts from Punjab, Rajasthan, Uttar Pradesh, Bihar, the United Kingdom, and the United States (listed in the appendix). References to about forty more have been found in various books, but some of them are kept locked away by their custodians and some have disappeared altogether.

Of these thirty-seven books, twenty have dated colophons either at the end of the *Kavitāvalī* or at the end of another work written by the same hand. The twenty given dates indicate a time frame between 1691 and 1887. Naturally it is possible that some of the undated books fall outside this period, but I do not have the tools to determine this. However, none of the forty-odd other manuscripts referred to in books but not inspected bear a date prior to 1691 or later than 1887.

The geographical spread of seventy manuscripts is indicated on the map.[28] Manuscripts that bear the name of the place of copying, the *lipisthān*, are marked with black dots and squares. Manuscripts with no *lipisthān* indicated were assigned to the place where they are preserved today (with the exception of the London and Harvard manuscripts); these are marked with empty dots and squares (squares represent manuscripts inspected by me). Clearly, handwritten books were circulated widely in north India. That is why, for example, the Dhaka manuscript is now found in Allahabad. Nevertheless most of the manuscripts with an indicated *lipisthān* are preserved in collections near the place of copying.

One can observe from the map that copies of the *Kavitāvalī* were produced and circulated over most of the modern Hindi belt, with the exception of present-day Madhya Pradesh and Chattisgarh, and, as the case of the Hoshiarpur and Dhaka manuscripts indicate, even outside it. The spread of the manuscripts is also indicative of a premodern Braj-Hindi cosmopolis, where people living in an area stretching from Rajasthan to Bihar and speaking various dialects shared a literature written in literary Brajbhasha. According to this map the spread of literary Brajbhasha is somewhat more limited than that of the then Indian lingua franca, Hindustani. Although frontiers were permeable, the limits of Brajbhasha manuscript production may have been determined by religio-cultural and literary factors. Braj literary culture did not penetrate

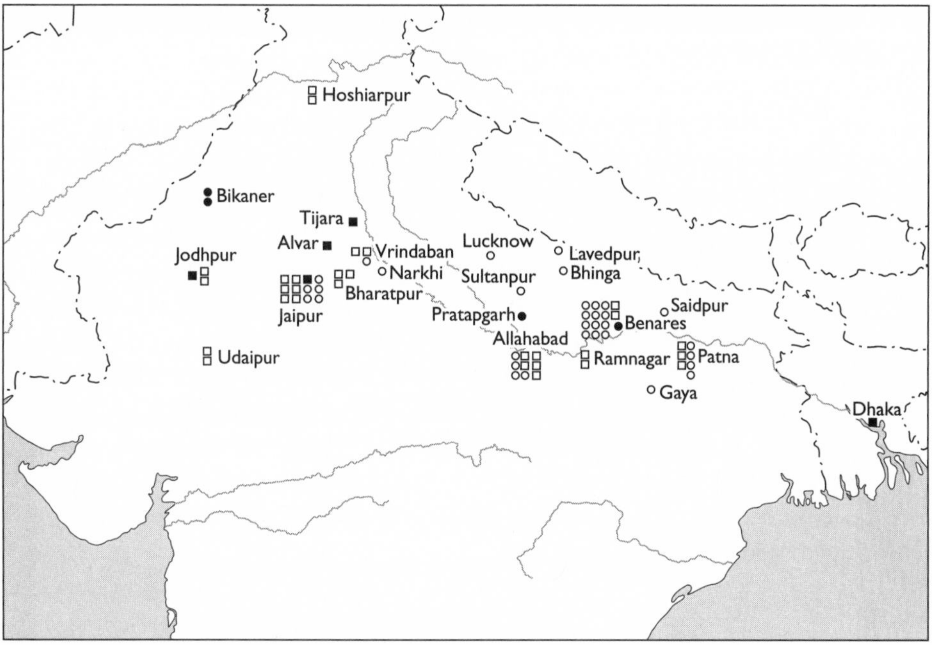

■ place of copying of manuscripts, accessible today
● place of copying of manuscripts, not inspected for the project
□ present location of manuscripts without indication of place of copying, accessible today
○ present location of manuscripts without indication of place of copying, not inspected for the project

Map of north India showing the provenance of the *Kavitāvalī* manuscripts. Designed by the author.

considerably into the Indo-Persian and Buddhist worlds of the northwest and north, while in the east it encountered another thriving literary culture dominated by Bengali. One knows from the history of Brajbhasha that it was more successful in penetrating into Gujarat and the Deccan. However, it never became the dominant literary idiom in those regions, and this may account for the lack of *Kavitāvalī* manuscripts from there.

One can also observe in the manuscripts regional variations in their script and orthography. Manuscripts from the east, for example, tend to mix *kaithī*-script characters into their *devanāgarī* and also tend to confuse the characters *sa* and *śa* very easily. The *kaithī* script was used widely in eastern Uttar Pradesh and in Bihar, and in the late nineteenth century it was even an official script in Bihar.[29] One major difference between Sanskrit and Braj manuscript transmission is the lack of standard grammar and orthography in Braj, manifest in the inconsistent use of word endings and conjuncts and in the different levels of Sanskritization, as opposed to an orthography that reflects pronunciation.

As far as the *Kavitāvalī* is concerned, the use of the *kabitt* form connected with the written tradition, the almost uniform sequence of poems, and the nature and the relatively small number of the variant readings show that the extant texts stem from written versions; no phase of oral transmission was involved, although oral tradition must have influenced it. (Even today many people know several of Tulsī's quatrains by heart, and the inversion of lines found in some manuscripts also hints at the influence of oral transmission.) In the words of Philip Lutgendorf, "Tulsīdās oral tradition exists as a complement to a distinct and well-attested literary corpus."[30] The transmitted text must be very close to that of the first edition(s) prepared in all probability by the poet himself. Alongside a moderate tendency on the part of editors toward expanding his literary corpus by interpolating other material, one can also observe a drive to purge his oeuvre of weak poems or of variations on the same theme conceived as redundancy.

The Two Recensions

An examination of other minor works of Tulsīdās shows that some existed in several forms in handwritten books before their text became standardized in the second half of the nineteenth century. The *Barvai Rāmāyaṇ* (Rāmāyaṇa in *barvais*), for example, has a short recension containing 69 *barvai* couplets and a longer recension with 405 *barvais*. The two share only fourteen couplets. The former shows a style full of *alaṅkāras* and may have been influenced by works of poetic virtuosity such as Rahīm's *Barvai nāyikā-bhed* or Keśavdās's *Rāmcandracandrikā*. The longer recension tells the complete Rama story.[31] The shortest composition attributed to Tulsīdās is the *Rāmlalā nahchū* (Rāma's Nail-paring Ceremony), an auspicious song describing Rama's nail paring apparently before his wedding. Its two forms contain either forty couplets (twenty *sohar* stanzas), as published in the *granthāvalī*, or twenty-six couplets (thirteen *sohars*), as found in a manuscript allegedly dating from 1608. The two versions have only twenty lines in common and, according to Danuta Stasik, the version published in the *granthāvalī* is not only more exhaustive but also renders a better image of a social occasion which involves Rama, the hero and god, as well as his family.[32] It is indeed this version that was copied by later scribes, since the later manuscripts whose description I was able to consult also give this version.[33] The difference between the two versions may be accounted for by the oral transmission also suggested by the genre. As far as the dramatic difference between the two *Barvai Rāmāyaṇs* is concerned, more research is required. Maybe Tulsīdās re-edited an earlier version late in his life, as the date 1622 inserted in one of the manuscripts of the longer tradition suggests.[34] It may also be that a tendency similar to the case of the growth of the biographies of Tulsīdās (longer and longer biographies were "discovered" when there was need for them among the readers)[35] worked behind the composition of a complete *Rāmāyaṇ* in the *barvai* meter sometime before the nineteenth century.

In some cases there is reasonable evidence to say that Tulsīdās edited his works several times, as Mataprasad Gupta has demonstrated in the case of the *Gītāvalī* and the *Vinay Patrikā* as well as that of the *Mānas*. The *Gītāvalī*, a retelling of the Rama story in *padas*, and the *Vinay Patrikā*, a collection of devotional songs, developed from two collections referred to in manuscript colophons as *Padāvalī Rāmāyaṇ* and *Rāmgītāvalī*, respectively, of which Gupta saw two related manuscripts apparently written by the same hand in 1609.[36] The interrelationship of the two is illustrated by the fact that five songs of the *Rāmgītāvalī* related to the Rama story but not present in the *Padāvalī Rāmāyaṇ* have found their way into the *Gītāvalī*. Gupta also observes that five *padas* relating to the same theme, that is, the dialogue between Trijatā and Sītā, were in different places in the *Padāvalī Rāmāyaṇ* but were grouped together in the *Gītāvalī*. From the absence of other *Padāvalī Rāmāyaṇ* and *Rāmgītāvalī* manuscripts Gupta inferred that Tulsīdās himself had edited the texts and that these edited versions became authentic and spread.

On examining the *Mānas* Gupta found that its first version might have been the second half of the *Bālakāṇḍa* (from v. 184, that is, without the initial frame of the story) and the *Ayodhyākāṇḍa*. This section shows unity in form (eight *ardhālīs* in each *caupāī* stanza) and theme: the speaker is the poet himself and the story is linear, starting with the causes of Rama's birth. In a verse from the *Uttarakāṇḍa* Tulsīdās claims that his work contains five hundred *caupāīs*. This may refer precisely to this original core of 506 *caupāī* stanzas. The second version contains *Bālakāṇḍa* 36–183 (the story of Shiva and the causes of Vishnu's incarnation as Rama) and the remaining *kāṇḍas*, with Yājñavalkya, Shiva, and Bhuṣuṇḍī as speakers. In a third and last phase Tulsīdās prepared the high-soaring introductory part (*Bālakāṇḍa* 1–35) and finished the poem.[37] While accepting Gupta's main propositions, Charlotte Vaudeville argues that Gupta's phases one and three were in fact one phase resulting in a proto-*Rāmcarit*, a pious account of Rāma's deeds without any pretension to becoming a holy book. In its avoidance of Braj forms and its rhythmic pattern, it was close to the early works of Tulsīdās, especially the *Jānakīmaṅgal*. The style of the poem, however, changes at the beginning of the Araṇyakāṇḍa. This, according to Vaudeville, must have coincided with a major disruption in Tulsī's life, his move from Ayodhya to Benares. The second phase of writing is marked by three major modifications in his approach. Here the poet presents the poem as a holy text, a *tantra*, revealed by Shiva to Parvati. In this part he also gives evidence of wider reading. Embedding the Rama story into a Shiva-Parvati dialogue reminds the reader of the *Adhyātma-Rāmāyaṇa*, while the introductory part is influenced by the *Bhāgavata Purāṇa*, and the later parts of the *Mānas* are under the influence of the *Bhuṣuṇḍī-Rāmāyaṇa*. Furthermore the poem becomes more didactic and less interested in the narration of events, and the regular occurrence of metrical forms gives way to a somewhat more variegated style.[38]

A close look at Tulsī's other works shows that he not only re-edited his earlier compositions but also kept developing his ideas. Preoccupations of an earlier work can recur in later poems in a more refined way, as is the case with some ideas present in the *Rāmcaritmānas*, which return in the *Gītāvalī*.[39]

The *Kavitāvalī* manuscripts available in complete form show a variation of about ninety poems. On the basis of this one can distribute the manuscripts into two groups of similar size. The first contains those that reach *Uttarakāṇḍa* v. 180 and normally include the *Hanumānbāhuk*, and the second group those that reach at most *Uttarakāṇḍa* v. 161 and do not include the *Hanumānbāhuk*. I call the former the Longer Recension and the latter the Shorter Recension.

The sequence of poems in the manuscripts of the two recensions shows further peculiarities that justify their being grouped together. For example, six poems in the middle of the *Uttarakāṇḍa* (vv. 7.91–96) are missing from all representatives of the Shorter Recension but are present in all specimens of the Longer. A calculation of all omissions and apocrypha shows that the archetype of the Shorter Recension contained 288 to 290 quatrains. Manuscripts of the Longer Recension include all *kabitt*s that found their way into the modern published versions, as well as eleven or thirteen apocryphal *kabitt*s missing from the vulgate text, out of which three always and another three occasionally correspond to the apocrypha of the Shorter Recension. Most manuscripts of the Longer Recension contain the *Hanumānbāhuk*. In this way the archetype of the Longer Recension contained 382 or 380 *kabitt*s.

Dividing the manuscripts into two recensions is further supported by an examination of the variant readings. Since our earliest manuscript dates to about seventy years after the poet's death, it is very likely that its text had already undergone several changes by the time it was copied. On the whole, however, the manuscripts show a relatively small number of variant readings. By far most of them arose from nonstandardized orthography and scribal errors, such as confusing similar-looking characters. The major variants, considerably fewer in number, include synonyms such as *prīti* (love) instead of *neha* (affection) and *tapa* (heat) instead of *dāha* (burning), confusion over difficult readings, correction of metrical licenses such as omitting the word *jaga* (world) from the beginning of a line in a dactylic *savaiyā*, and the replacement of some compromise words. For example, in a quatrain making fun of ascetics longing for women, in the phrase *bindhi ke bāsī udāsī tapī bratadhārī mahā* (executors of great vows, indifferent ascetics, dwelling in the Vindhya mountains) in our Bharatpur1 (B_L) manuscript the expression *bindhi ke bāsī* (dwellers of the Vindhya mountains) was changed to *audha ke bāsī*, saving face for the ascetics but creating the muddled meaning "executors of great vows, indifferent inhabitants of Ayodhya."

Manuscripts belonging to the same recension are also likely to share the same major variants. On the basis of the major variants from fifteen poems in different parts of the twenty-four substantially long manuscripts, a cluster analysis distributes the manuscripts into two groups with two subgroups each. On the two-dimensional distance model in figure 1, manuscripts that share more variants are closer to each other. (The manuscripts are labeled according to the place of their copying or, when that is not known, according to the place where they were found. The eccentric Patna2 manuscript is omitted.)

The distribution of the manuscripts is far from random; rather they tend to converge into two major galaxies, which correspond to the two recensions. The group on the left represents the Longer Recension, the one on the right the Shorter Recension. There are also variants within each recension, but their weight is usually less than that of the variants that define the two recensions. This shows that the authority of either recension was not questioned apart from some stray eccentric manuscripts. However, the manuscripts of the Shorter Recension form a more compact group than those of the Longer. This means that they have fewer variants within the recension in comparison with manuscripts belonging to the other recension. In the 137 quatrains already collated in places where readings are not shared by the two recensions, the Shorter Recension accepts a common reading 243 times, while the Longer Recension accepts one only 120 times. The smaller number of variant readings within the Shorter Recension suggests later editing and shorter time of transmission after editing. One may argue that it could also be due to stricter sectarian control. However, since there were two recensions in circulation (at least from the second half of the eighteenth century onward, as the dating of

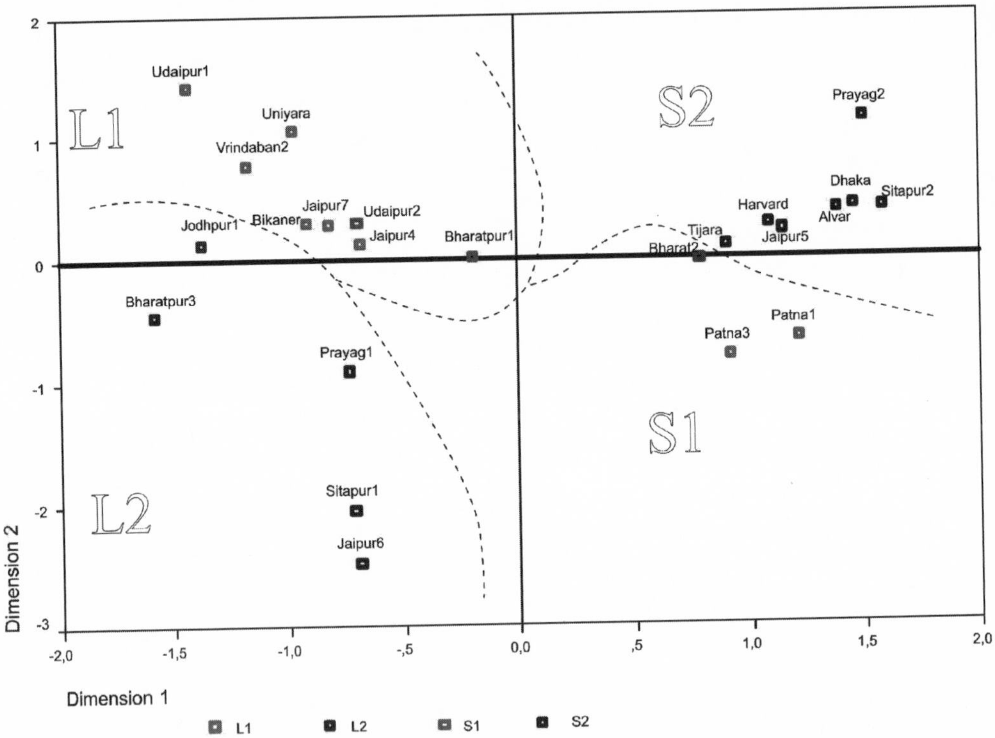

Figure 1. Euclidean distance model. Prepared by Daniel Balogh.

the manuscripts shows), it would be illogical to control the diffusion of only one recension.

According to the quantity of manuscripts giving the same major variants, one can distribute the readings into three categories: (1) an eccentric reading given by only one or two scribes, (2) a reading shared within a subrecension but not within a recension, and (3) a reading shared within the bulk of (at least four-fifths) of a recension. A high number of variants consist of meaningful words, but many do not make sense at the level of the line. By far the largest number of the meaningful variants comes from our third category, dividing the manuscripts into two recensions. This suggests a conscious intervention to the text when the archetype of the Shorter Recension was prepared.

On the basis of the 137 quatrains already collated, it can be observed that either the meaningful variants not shared in the two recensions show insignificant differences (e.g., *soca* [concern, sorrow] versus *soka* [sorrow]) or those of the Longer Recension are better but sometimes more difficult. Variants within the Shorter Recension tend to be more simplistic or nonsensical. For example, v. 89 in the *Uttarakāṇḍa* is about Rama's name, which is more powerful than Rama himself. In the first line it is illustrated by the case of the poet Vālmīki:

> *rāma bihāya marā japate bigarī sudharī kabi-kokila hū kī*
>
> Having abandoned Rama, simply by repeating the word *marā*, "it is dead," even the corrupted fate of the poet-cuckoo came right.[40]

The third line refers to Draupadī's calling on God's name when Duryodhana tried to strip her naked:

> *nāma pratāpa baḍe kusamāja bajāi rahī pati pāṁdubadhū kī*
>
> Through the great power of the name, the honor of the Pāṇḍavas' wife was saved openly in a wicked assembly.

This is, however, the reading only of the Longer Recension. The Shorter, probably because of confusing the character *na* with the *kaithi*-script *ra* (⊣), reads *rāma* instead of *nāma* (name), not only producing *punarukti doṣa*, the error of repeating the same word in the same sense, but also creating a contradiction with the first line, which emphasizes that the god's name is greater than the god himself.

It is a generally accepted philological rule that a more difficult but still meaningful reading, the *lectio difficilior*, tends to be more authentic than a simpler one, which is normally the result of the scribe's not understanding the complexity of the text. The more sensible nature of the more difficult reading is illustrated by the third line in v. 100 in the *Uttarakāṇḍa*, in a reference to Rama's taking the side of a dog against a Brahman mendicant. Since Brahmans cannot be punished physically, Rama cleverly made him the abbot of an extremely corrupt monastery, a position the dog had held in his previous life.

> *sāhiba sujāna jāna svāna hū ko pakṣa kiyo*
>
> The clever lord knowingly took even the side of the dog.

Again this is the reading only of the Longer Recension. The Shorter Recension reads *jinha* (who) instead of *jāna* (knowingly), destroying the internal rhyme and taking away one shade of the meaning, the emphasis on Rama's cleverness.

Another case of a better *lectio difficilior* can be observed in the first line of the apocryphal quatrain that follows *Ayodhyākāṇḍa* v. 11, which is a description of Sita's languor:

sūkhi gaye ratanādhara mañjula kañja se locana cāru cucvai

The jewels of her lips went dry and her lovely eyes, which are like charming lotuses, are dropping tears.

Here the Longer Recension has more difficult readings than the somewhat clumsily rhyming versions of the Shorter. The readings of the Shorter Recension also do not show as much variation among different manuscripts as those of the Longer, where the scribes were at odds. The scribes of the Longer Recension seem to have felt uncomfortable with the word *cucvai*, a present singular third-person form of the verb *cucānā* (to drip, to ooze), slightly distorted for the sake of rhyme. The fourteen manuscripts of this recension that I have consulted have nine different variants (*vitai, ciccai, cuvaĩ, citaicyai, cucavai, cala cvai, cucce, citai, cusvai*) either breaking the rhyme or making further distortions. The Shorter Recension simplifies the case with *citai* (she looks up), resulting in the flaw of repeating the same word in the same sense, since *citai* also figures in the third line. Nevertheless nine out of eleven manuscripts give this reading. The two rhymes that follow in the same verse are similarly problematic and present several variants.

If one is looking for a pattern of change within the Shorter Recension one can observe a drift of reverence toward Rama and toward the poet by not accepting readings that seem to contradict Rama's noble image or phrases that are grammatically imperfect. A few examples of consistently different variants from the first two cantos show this clearly. The editor changed the expression "that Rama" (*so rāma*) into "Lord Rama" (*śrī rāma*) in *Bālakāṇḍa* 10; spoke deferentially of Lakshmana in the plural (*bhāi bhale*) in *Ayodhyākāṇḍa* 2 instead of the singular of the Longer Recension (*bhāi bhalo*); put the relative pronoun *jo* (who) into the correct Braj plural *je* in *Ayodhyākāṇḍa* 5, thinking that the villagers were not supposed to know that Rama was a princely youth (*rājakiśora*) on his way to the forest; and corrected the word to "young man" (*kiśora*) in *Ayodhyākāṇḍa* 24, compensating for the metrical loss later in the same line. The editor also refused the idea in *Ayodhyākāṇḍa* 18 that the village women had goose bumps (*pulakī*) on seeing Rama; he got rid of the word and inserted a redundant word later in the line. The reading of the Longer Recension is as follows:

tulasī suni grāmabadhū vithakī pulakī tana au cale locana cvai

Tulsī says, hearing this the village women were delighted, had goose bumps on their bodies and their eyes were filled (with tears).

The Shorter Recension, however, consistently reads:

tulasī suni grāmabadhū vithakī tana aura cale jala locana cvai

Tulsī says, hearing this the village women were delighted in their bodies and their eyes were filled with water.

It is not only the textual variants shared within a recension that point to the authority of the editor. No *kabitt* of Tulsīdās is found in the other collections of his works, although the syllabic line pattern of the *kavitt* was frequently used in the *padas* of his *Gītāvalī* and *Vinay Patrikā*. In these collections, however, there are no quatrains, and the songs with the *kavitt*-type lines have a refrain and five or more lines. This fact suggests that toward the end of his life Tulsīdās himself, or maybe someone else with an authority that was respected by later generations, collected all the *kabitts* not present in earlier collections and edited them into the archetypal Longer Recension.

This inference is further supported in the case of the Longer Recension by the content of the poems themselves. The last twenty-two poems of the recension refer to astrological events and an epidemic (*mahāmārī*) in Benares, indicating a date for this section of around 1615.[41] Many of these twenty-two quatrains as well as the *Hanumānbāhuk* note that their poet is highly respected, and others refer to old age and suffering from diseases. A celebration of recovery would have been a good opportunity to show the working of God's grace, as the poet did in the case of his childhood deprivations. The lack of any reference to recovery in any work suggests that the *Kavitāvalī*, together with the *Hanumānbāhuk*, is among the poet's last works.[42] Indeed tradition holds that the *Kavitāvalī* contains Tulsī's last poem, a quatrain (v. 7.180) about glimpsing a kite (*kṣemkarī*), an auspicious bird when setting out for a journey.[43] The style of these last twenty-two poems is so consistent with that of the previous quatrains that no one has ever questioned their authenticity. The most obvious explanation for the presence of the two recensions is therefore that the archetype of the Shorter Recension was a composite work edited on the basis of a somewhat corrupt text but under the influence of the structure of an earlier, shorter version of the *Kavitāvalī*, which had been compiled in all probability by the poet himself before his illness.

Tulsīdās edited his works again and again. One can therefore assume that at least the structures of the two recensions of the *Kavitāvalī* are his or a close disciple's two editions. The study of transmission, however, is made more complicated by the process of contamination, resulting in the fact that some manuscripts are not copies of one single source but rather composite versions.

This is clearly shown in some manuscripts, where one can observe corrections executed by a second hand on the basis of a third manuscript.

Problematic Poems and Scribal Argumentation

As discussed above, the Longer Recension in all probability came into being by adding twenty-two more poems together with the *Hanumānbāhuk*, all written in Tulsī's last years, to an original *Uttarakāṇḍa* that ended at v. 7.161. The authenticity of these twenty-two poems and of the *Bāhuk* cannot be questioned, and they are rightly included in modern editions. Another twenty-two, namely the nine poems omitted from the Shorter Recension and the thirteen apocrypha omitted from the vulgate and at least two manuscripts, may or may not have been written by Tulsī. These poems and various explanations for their omission have been examined elsewhere.[44]

A poem could become suspect for various reasons. Arguments relating to syntax, metrics, and structure figure more often than the contents of the stanza. No theological reasoning was involved in the argumentation, since the majority of the scribes did not discard the one about a *nirañjan* (attributeless god), even though in *Uttarakāṇḍa* v. 128 Tulsīdās propagated image worship against the worship of an attributeless deity, referring to the fact that to save Prahlāda, God appeared from a stone pillar and not from the heart. The inclusion of these mutually contradictory poems in the same collection can be linked to the tendency toward synthesis in Tulsīdās, treated in other studies.[45]

One can further observe that many of the suspicious poems show some kind of metrical, stylistic, or structural weakness in the eyes of the copying scribes. It is difficult to imagine how these poems found their way into a popular collection if they were not there originally, and in most of the cases one can presume omission rather than insertion. The concept of weakness, however, is relative, since it seems that Tulsī's instances of poetic license were regarded as flaws only by later generations. A major poetic license Tulsīdās took was the freedom to add or omit one or two syllables in the beginning of a line in a *savaiyā*. Five out of the twenty-two apocryphal poems have such unmetrical lines, which occur in several places in the vulgate as well.[46]

Three suspect cases are due to a supposed *samasyāpūrti*, the widespread poetic practice of writing a new poem on a given phrase or line. (In other words, if two poems contained the same line or phrase, one of them became suspicious.) However, it is not necessary to assume that another poet wrote a quatrain on the same line. It may have been the original poet himself; a similar process was not questioned in several other cases within the vulgate *Kavitāvalī*.[47]

Yet another argument for omission is structural looseness or obscurity. A later editor, or perhaps Tulsīdās himself, may have discarded the poems that he found aesthetically weak. The case of a *kavitt* from the *Hanumānbāhuk* strongly suggests that it was a later editor who discarded the lame poems. The third line of the vulgate of *Hanumānbāhuk* v. 40 is present only in the manuscripts of the L1 subrecension (see figure 1); an entirely different reading is given in the L2 manuscripts, and the whole poem is omitted from the extremely "purist" Sitapur1. What picture does the oldest (Prayag1) manuscript present? It has only three lines, and the suspicious third line is missing. There is, however, a calculated lacuna at the end to indicate a missing line. We may suspect that incomplete poems were originally also part of the collection, especially when we take into consideration that we are dealing with Tulsī's last poems and the dying poet may not have been able to revise and complete them all. Later copyists, however, did not accept the fact that the poet-saint may have written unfinished quatrains and either completed *Hanumānbāhuk* v. 40 or simply omitted it.[48] Here the idea that the poet-saint must have produced only perfect poems was at work. The case of the second apocryphal quatrain following *Uttarakāṇḍa* v. 106 is similar. The second line is missing from Prayag1, and there is some empty space at the end. The quatrain is omitted from Sitapur1 and from many other manuscripts, including those of the Shorter Recension.[49]

Some of the apocrypha can be weak but authentic, such as the stanza in which Rama scolds Sita for her languor as they make their way to the forest (the verse following *Ayodhyākāṇḍa* v. 11), verses with poetic license (such as in *Uttarakāṇḍa* vv. 91–96), or verses that share a phrase or a line and were suspected as *samasyāpūrti*. Some of the apocrypha can indeed be products of other poets, such as the ones at the end of the *Araṇyakāṇḍa* and *Kiṣkindhākāṇḍa* in the Shorter Recension to rectify the disproportion created by the small number of poems in these cantos, or maybe the one about the attributeless, *nirañjan* (stainless) god (the verse following *Uttarakāṇḍa* v. 148), written perhaps by the Nirañjanī poet Tursīdās.

Conclusion

In his study about performing the *Rāmcaritmānas* Philip Lutgendorf found traces of enmity toward the *Mānas* among Tulsī's contemporaries. It is not simply the hagiographical tradition that mentions it; even as late as 1887 F. S. Growse observed that many pandits "still affect to despise his work." Moreover several passages at the beginning of the *Mānas* suggest a concern to anticipate criticism, while in *Vinay Patrikā* v. 8.3 the poet explicitly complains to Shiva that the god's "servants" in Benares have been tormenting him. The sources suggest that the opposition came from local Brahmanical circles. One can also observe, continues Lutgendorf, that Tulsī's poetry garnered an enthusiastic reception among other groups. Toward the end of his life the poet is mentioned as the Vālmīki of his age by Nābhādās in Galta, almost a thousand kilometers away. The use of *kaithī* script for the transmission of his texts, a writing system favored in economic and political contexts, according to Lutgendorf, suggests *kāyastha* and merchant links. The fact that the Persian translation of the epic is second only to the Sanskrit translation also reveals a link with Persianate circles, including not only Muslims but also Hindu *kāyasthas*.[50]

The relatively small number of nonorthographic variants and the more or less unquestioned structure of the collection suggest that the *Kavitāvalī* goes back to a written source. One of the most important motives for the Brahmanical opposition must have been that Tulsīdās had vindicated for *bhakti* the written culture belonging so far to Sanskrit (and to Prakrit and Apabhramsha).

On the basis of our previous observations we can attempt to reconstruct some phases of the textual history of the *Kavitāvalī*, although this may hardly be more than speculation. As has been mentioned, the Shorter Recension includes neither Tulsī's last poems nor several apocrypha, while most manuscripts of the Longer Recension contain both. This is a very strong indication that some time around 1610 Tulsīdās collected his previously written quatrains and, probably discarding some weaker and incomplete ones, arranged them into an anthology imitating the *kāṇḍa* structure of the *Mānas*. This must be the collection that contained poems up to *Uttarakāṇḍa* 148 (or later, up to 161); this compilation was circulated and later served as a basis for the Shorter Recension.

After the death of Tulsīdās an associate of his probably edited the text again. The weaker and incomplete poems that found their way into this recension hint at the fact that it was a disciple who reverentially collected every *kabitt*, including those to Hanuman, from the written and maybe in some cases oral legacy of his master and prepared what I call the Longer Recension. The relative proliferation of variants in this recension suggests an early date for its compilation.

During the transmission some suspicious poems were incorporated into the collection but discarded later by the editor of the Shorter Recension. Making use of a probably already corrupted text and of the structure of the early version up to *Uttarakāṇḍa* 148 or 161, the editor purged the text of many poems and possibly also inserted some quatrains, such as the ones at the end of the *Araṇyakāṇḍa* and *Kiṣkindhākāṇḍa,* to rectify the imbalance created by the small number of poems in this canto. The scarceness of the variants in the Shorter Recension as compared to the Longer must be due to the fact that the editing took place at a relatively late date, perhaps in the eighteenth century, since the earliest available manuscript of this recension dates from as late as 1772. Apart from its clearer though often weaker readings it may be an awareness of the already independent existence of the *Hanumānbāhuk* that contributed to the popularity of this recension that considered the two parts of this work as two independent compositions. I can make only tentative statements about the emergence of the two recensions since there are still some questions to answer. For example, why do we have a version with poems only up to *Uttarakāṇḍa* 148? Why did the editor take a version only up to *Uttarakāṇḍa* 161? Why do all manuscripts of these two versions show similarities that indicate a later date, though the lack of Tulsī's last poems suggests an earlier date of composition?

The emergence of the Shorter Recension did not suppress the Longer Recension; both had their sometimes independent, sometimes intertwining histories and came to include a few other apocryphal quatrains. Naturally there was contamination between the two recensions, resulting in occasional purging of the Longer Recension, as may be the case with the Sitapurı and Hoshiarpurı manuscripts, or inclusion of omitted poems into the Shorter Recension, as in the Tijara manuscript.

The two recensions also determined the publication history of the work. The first published edition of the *Kavitāvalī* from Calcutta in 1815 was based on the Shorter Recension, the later Benares and Lucknow versions, such as the one edited by Durga Mishra in 1858, on the Longer Recension.[51] The modern vulgate text is a composite version of the two that gradually developed during the nineteenth century and became standardized in the *Dūbejī* edition. It is this vulgate that was taken over by the Granthāvalī and the Gita Press in the twentieth century.

We have seen that the two standardized recensions were copied for more than two centuries. Several poems were added to them, and some others, even though in all probability written by Tulsī, were omitted. While in the case of Mīrā and Sūrdās collections the collector tendency was overwhelming and resulted in a dramatic increase in the number of poems, in the *Kavitāvalī* the written tradition did not permit many extraneous quatrains to be included, and the purging tendency of a purist editor kept the number of poems down. Most of the poems excluded from about half of the manuscripts show poetic failings rather than ideological digression, indicating that the elimination of aesthetic or structural deficiencies was a more important editorial preoccupation in the eighteenth and nineteenth centuries than faithfulness to the received text. In contrast to the more pedestrian purist selection, the weaknesses shown both in the earliest extant manuscript and in the suspicious poems present a brighter and more human poet with flaws, imperfections, and humility.

Appendix: The Kavitāvalī *Manuscripts*

Manuscript	Location	Preserved in	Number	Date (VS)	Written in
A_S	Alvar.	RORI	4872	1877	Alvar
Ti	Alvar.	RORI	4873	1891	Tijārā (Bahādurpur?)
Bh1	Bharatpur 1	RORI	72	1888	
Bh2	Bharatpur 2	RORI	404/151		
B_L	Bharatpur 3	RORI	22		
Ha	Cambridge, Mass.	Houghton Library, Harvard University	In1446		
H_L	Hoshiarpur 1	VVRI	916	1879	Maṇḍche
H_S	Hoshiarpur 2	VVRI	432		
Ja1	Jaipur 1	RORI	2423		
Ja2	Jaipur 2	RORI	9683		
D_L	Jaipur 3	City Palace	2987 (3)	(1773)	
J_L	Jaipur 4	City Palace	1779	1858	Jaipur
J_S	Jaipur 5	City Palace	3153	1899	
Ja6	Jaipur 6	City Palace	1930		
Ja7	Jaipur 7	City Palace	3437 (3)		
Bi	Jodhpur	RORI	12354	1919	Bikaner
M_L	Jodhpur 1	RORI	12357	1847	
Jo2	Jodhpur 2	RORI	22732 (15)		
Lo	London	Wellcome Inst.	MS Hindi 361		

Manuscript	Location	Preserved in	Number	Date (VS)	Written in
P_S	Patna 1	Caitanya, Library		1894	
Pa2	Patna 2	BRP	101 kha	1291 Hijri (1876/77)	
Pa3	Patna3	BRP	2895	1985?	
I_L	Prayag 1	HSS	3-9/5264	(1748-49)	
I_S	Prayag 2	HSS	3-6/1896	1829	Ayodhya?
Pr3	Prayag 3	HSS			
Dh	Prayag	HSS	3-7/2686	1944	Dhaka
S_L	Prayag	HSS	8003/4508	1890	Sitapur?
S_S	Prayag	HSS	7963/4484	1893	Sitapur?
Su	Agra	KMMI	411	1875	Śubhaṭṭpur—Jodhpur
U_L	Udaipur 1	RORI	2103		
Ud2	Udaipur 2	RORI	4137		
Un	Udaipur	RORI	5295	1912	Uniyārā
V_S	Varanasi	NPS	615	1848	
Vr1	Vrindaban 1	VRI	4725		
Vr2	Vrindaban 2	VRI	8401	1897	

Note: KMMI: K. M. Munshi Institute, BRP: Bihar Rashtrabhasha Parishad, HSS: Hindi Sahitya Sammelan, NPS: Nagaripracharini Sabha, RORI: Rajasthan Oriental Research Institute, VVRI: Vishveshvaranand Vedic Research Institute.

Notes

This essay was written within the framework of OTKA (Hungarian Scientific Research Fund) program no. T 038047. Several research trips to India were supported by the Max Müller Fund (Oxford), and the statistical analysis was sponsored by the Society for South Asian Studies (British Academy).

1. For a study of manuscript culture in precolonial India, see Pollock, "Literary Culture and Manuscript Culture in Precolonial India."

2. For example, a now apparently lost *Mānas* manuscript (Miśra, *Rāmcaritmānas,* 460) and the hardly accessible *gurmukhi Mohan* or *Goindval pothis* (claimed to be from 1570–72; see Callewaert et al., *Millennium Kabīr Vāṇī,* 10; Mann, *Goindval Pothis,* 16–25). Some other editors mention now unavailable sixteenth-century manuscripts, such as one of Tulsīdās's *Jānakī-maṅgal* dating from 1575 (Chakkanlal). The manuscript serving as the main source for the *Kabīr granthāvalī,* edited by Śyāmsundar Dās, is also claimed to date from 1504 (*Pratham sanskaran kī bhūmikā,* 1, 5–6). The colophon with this date, however, is written by a different hand and cannot be accepted as

genuine. Vaudeville, *A Weaver Called Kabir*, 29–30; see also Callewaert et al., *Millennium Kabīr Vāṇī*, 13–14.

3. One at Baddhicandjī kā Digambar Jain Mandir, one at Scindia Oriental Research Institute, Ujjain, and one at Abhay Jain Granthalay, Bikaner. See Nyāytīrtha and Kāslīvāl, *Pradyumna-carit*, 9–10. Premodern Hindi manuscripts are normally dated in the Vikrama Era. However, I give all dates converted into our Common Era by deducting fifty-seven years from the Vikrama year when there is no precise indication of month and day.

4. Six Jain manuscripts are in the Vishveshvaranand Vedic Research Institute in Hoshiarpur: *Mānpat kī caüpāī* no. 119 (1493), *Draupadī-carit caüpāī* no. 2 (1527), *Jambū-svāmī-caüpāī* no. 91 (1544), *Purandar-kumār kathā* no. 12 (1551), *Purandar-caüpāī* no. 46 (1565), and *Jambū-kumār-carit* no. 28 (1566). The Anup Sanskrit Library in Bikaner preserves four copies of Rāysinghjī's vernacular commentary on Śrīpati's *Ratnamālā* from 1569, 1584, 1588, and 1590 (Rajasthani Catalogue, 334, 330, 332, 333). There are four other works in Rajasthani on heroic or romantic themes: *Rāo Jatsī rā chand* by Vīṭhū Sūjā Nagarājot (Rajasthani Cat., 99), *Acaldās khacī rī vacanikā* by Gādan Sivdās (Rajasthani Cat., 99gha), *Tripurasundarī rī velī* by Jasvant (Rajasthani Cat., 272), and *Ḍholā mārū rī caupāī* by Vācak Kuśallābh (Rajasthani Cat., 45). There is also a *Phuṭkar Siṁhāsan battīsī caupāī* by Vinaysamudra (Rajasthani Cat., 169). There are two *Visāldev-rās* manuscripts from 1576 (Anup Sanskrit Library in Bikaner, Rajasthani Cat., 99) and 1589 (no. 56 in the Vishveshvaranand Vedic Research Institute, Hoshiarpur). I was able to locate further ten manuscripts of Jain and Rajasthani works in the Abhay Jain Granthalay in Bikaner. The fact that most early manuscripts are found in only three archives warns us that the distribution of extant early manuscripts may depend more on the success of preservation over the centuries than on the actual proportions of early Hindi manuscript production. If a collection has always been preserved well, like the ones that are now in Hoshiarpur and Bikaner, then we can find many early manuscripts at one place. The lack of other similar collections, however, indicates that there were indeed very few Nagari Hindi manuscripts in circulation in the sixteenth century.

5. The *Qutubśatak* manuscript is preserved in the Anup Sanskrit Library in Bikaner and was used in M. P. Gupta's 1967 edition of the *Qutubśatak*. See Orsini, "Qutubśatak, Delhi and Early Hindavi." The Fatehpur manuscript is preserved at the Sawai Mansingh II. Museum in Jaipur.

6. For example, *kabīr paṛhibā dūri kari, pustaka dei bahāi* (19.2; Kabīr says, get rid of learning and float away in a river your books). Other similar *dohā*s appear in the *Kathaṇīṁ binā karaṇīṁ kau aṅga* (The Chapter on Saying without Doing). *Kabīr granthāvalī*, 30.

7. A detailed study of *geyavikāras*, manifest in the inversion of *antarā*s and lines, in adjustments to the rhythm, in the use of fillers, in varying the beat, and in the modifications of the refrain, can be found in Callewaert and Lath, *Hindi Padāvalī of Nāmdev*, 63–82.

8. Callewaert et al., *Millennium Kabīr Vāṇī*, 2; Lorenzen, *Praises to a Formless God*, 205.

9. Hawley, *Three Bhakti Voices*, 199.

10. Hawley, *Three Bhakti Voices*, 98–115; Callewaert et al., *Millennium Kabīr Vāṇī*, 2. For more details, see Callewaert. "The 'Earliest' Songs of Mira (1503–1546)."

11. Pollock, "Literary Culture and Manuscript Culture in Precolonial India," 14.

12. Entwistle, *Rāsa Māna ke Pada of Kevalarāma*, 86–87.

13. Snell, *Eighty-four Hymns of Hita Harivaṁśa*, 326–32.

14. Rosenstein, *The Devotional Poetry of Svāmī Haridās*, 71.

15. Pauwels, *Kṛṣṇa's Round Dance Reconsidered*, 30–31.

16. Callewaert, *The Hindī Biography of Dādū Dayāl*, 17–20.

17. Bangha, *Saneh ko mārag*, 49–58, 139–53; Bangha, "Mundane and Divine Love in Ānandghan's Poetry."

18. Other, sporadically occurring titles include *Tulsīdāsjīkṛt kabitt, Rāmkabitāvalī, Kavitāvalī*, and *Kavitāvalī rāmāyaṇ*.

19. Manuscript no. 3–9/5264 at Hindī Sāhitya Sammelan, Allahabad.

20. Khoj 23–432u. See Gauṛ et al., *Hastalikhit hindī granthõ kā saṁkṣipt vivaraṇ II*, 612.

21. See Śyāmbihārī Miśra, *Hastalikhit hindī granthõ kī khoj kā vivaraṇ IV*, 433 (Khoj 09–323d).

22. About the growth of the cult, see Lutgendorf, *Hanumān's Tale*.

23. Narsiṁhrao Divatia, *Smaraṇamukur* (Bombay: Sahitya Prakashak, 1926), 18–19, quoted in Sitanshu Yashaschandra, "From Hemachandra to Hind Svaraj: Region and Power in Gujarati Literary Culture," in Pollock, *Literary Cultures in History*, 594.

24. Entwistle, *Rāsa Māna ke Pada of Kevalarāma*, 93.

25. *Tulsī granthāvalī II*; Chakkanlāl and Dvivedi, *Gosvāmī Tulsīdāskṛt dvādaś granth*, is allegedly based on some very old manuscripts, including the *Jānakī maṅgal* dating from 1575(!), the *Pārvatī maṅgal* from 1754, the *Gītāvalī*, the *Kṛṣṇa Gītāvalī*, and the *Vinay Patrikā* from 1660. I was not able to inspect this book, but as judged from the *Granthāvalī* that drew on its text of the *Kavitāvalī*, it is very good, and most of its readings can be found in our old manuscripts.

26. Miśra, *Rāmcaritmānas*, is based on the collation of twenty-four manuscripts that date from between 1615 and 1726. (The original of one manuscript is claimed to represent a 1584 tradition.) The text in the *Mānasāṅk* of the magazine *Kalyāṇ*, published by the Gita Press in 1938, was also prepared on the basis of the oldest available manuscripts (Lutgendorf, *Life of a Text*, 62), but given the lack of a proper apparatus listing the variant readings it cannot be considered a critical edition. Varmā's *Barvai Rāmāyaṇ* edition is the result of the collation of four manuscripts after the inspection of eleven handwritten books. Gupta, *Tulsīdās*.

27. In 2005 the Tulsīdās Textual Study Group became a research group of the Society for South Asian Studies (British Academy) under the name South Asian Language and Literature. The publication of the critical edition of the *Kavitāvalī* is ex-

pected by 2012. For more information, see the website of the Society for South Asian Studies within the British Academy.

28. It was after the submission of the essay that I came across a *Kavitāvalī* manuscript in the Gurmukhi script, which is today preserved at the Sikh History Department, Khalsa College, Amrtisar (no. 2189). This handwritten book, along with the large amount of Gurmukhi manuscripts of Braj poetry, further testify to the extent of penetration of Braj literary culture into the Panjab.

29. King, *One Language, Two Scripts*, 65–69.

30. Lutgendorf, *Life of a Text*, 11.

31. *Barvai Rāmāyaṇ*, 18–29.

32. Stasik, "Text and Context," 388.

33. Wellcome Institute ms. Hindi.45, Hindī Sāhitya Sammelan ms. 3-172/5757, Nāgarīpracāriṇī Sabhā ms. 632kh, Bihār Rāṣṭrabhāṣā Pariṣad ms. 280.

34. *Barvai Rāmāyaṇ*, 29.

35. The oldest documented *Barvai Rāmāyaṇ* manuscript dating from 1740 contains only forty-one couplets (*Barvai Rāmāyaṇ*, 24–26). The two oldest manuscripts with 405 *barvai*s consulted for the critical edition date only from 1816. The twenty-four folios of a manuscript from 1800 and preserved in Patna at the Bihār Rāṣṭrabhāṣā Pariṣad (no. 287), which I have not yet been able to consult, also suggest that it contains the Longer Recension. No earlier dated manuscripts are available for either of the recensions.

36. The *Padāvalī Rāmāyaṇ* manuscript was incomplete, containing only thirty-five *padas* of the *Sundarakāṇḍa* and nineteen of the *Uttarakāṇḍa*. The order of *padas* is different in the beginning of each *kāṇḍa*. The *Rāmgītāvalī* manuscript was complete, though with some serious lacunae, containing 158 *padas* in an order largely different from that of the *Vinay Patrikā*. Gupta, *Tulsīdās*, 212–19.

37. Gupta, *Tulsīdās*, 254–65.

38. Vaudeville, *Étude sur les sources et la composition du Rāmāyaṇa de Tulsī-Dās*, 319–23.

39. Gupta, *Tulsīdās*, 237–41.

40. Translations from the Braj, if not indicated otherwise, are mine.

41. The *Rudrabīsī*, or "Twenty Years of Rudra" (7.170 and poem no. 240 in the *Dohāvalī*), refers either to 1566–85 (according to Kannu Pillai, *Indian Ephemeries*) or to 1598–1618 (according to Sudhākar Dvivedī). (For both references see Gupta, *Tulsīdās*, 183.) The *Mīn kī sanīcarī*, or "Saturn in Pisces" (7.177), took place from March 1583 (Caitra Śukla 5 VS 1640) through May–June 1585 (Jyeṣṭha VS 1642) and again from March 1612 (Caitra Śukla 2 VS 1669) through May–June 1614 (Jyeṣṭha VS 1671), according to Sudhākar Dvivedī. See Gupta, *Tulsīdās*, 186, 504–8. There were three major epidemics during Tulsī's lifetime. The famines of 1555–56 and of 1595–98 are supposed to have been followed by pestilence, and a new disease, bubonic plague, appeared in 1616. See Smith, *Akbar the Great Mogul*, 397–98; *Tuzuk-i-Jahangiri*, 330, 442.

42. Tulsī's most influential biographer, Mātāprasad Gupta, believed that the *Kavi-*

tāvalī, together with the *Hanumānbāhuk*, dated from the period 1610–23 (Gupta, *Tulsīdās*, 251–54). Several scholars, including the English translator F. R. Allchin, followed his ideas (introduction to *Kavitāvalī*, 63–66). Śyāmsundar Dās and Pitambardatt Barthwal, however, relying on the *Mūl gosāiṁ-carit* (claimed to have been preserved in a manuscript from 1791; for a study of the different Tulsī biographies, see Lutgendorf, "Quest for the Legendary Tulsīdās"), argue that the *Barvai Rāmāyaṇ*, the *Hanumānbāhuk*, the *Vairāgya-sandīpanī*, and the *Rāmājñā-praśna* are Tulsī's last works (see Gupta, *Tulsīdās*, 69), while Madanlāl Śarmā and Gītārāṇī Śarmā consider the *Vinay Patrikā* to be his swan song, although they do not provide any argument for this supposition (Śarmā and Śarmā, *Kavitāvalī*, 56).

43. Tulsīdās, *Kavitāvalī*, Allchin translation, 66.

44. Bangha, "The *Kavitāvalī* of Tulsīdās."

45. See Lutgendorf, *Life of a Text*, 10.

46. Vv. 1.20, 2.5, 2.7, 6.5, 6.13, 6.33, 7.1, 7.12, 7.34, 7.43, 7.47, 7.49, 7.51, 7.52, 7.88, 7.103, 7.106, 7.132, 7.147, 7.153, 7.154.

47. Vv. 1.2 and 1.6, 1.3 and 1.4, 2.1 and 2.2, 7.40 and 7.41, 7.43 and 7.44, 7.112–14 share the same last line. Partial similarity can be observed in the cases of 5.11 and 5.12 (only the last word is shared), 6.1 and 6.2 (parallel construction). The last word is similar in 6.44 and 45, in 7.75 and 83, and in 7.88–90.

48. Similar uncertainty about an incomplete poem can be observed in the *Guru Granth Sahib*. See Deol, "Sūrdās."

49. This is not an overall model, however, and the cases of omission are more complex. Prayag1 omits one or more lines in three more cases in the *Uttarakāṇḍa* (107, 111, 123), when all later manuscripts give the vulgate reading. The apocryphal *savaiyā* 7.141+ (the + sign refers to an apocryphal quatrain that follows a numbered vulgate stanza) is missing from the handwritten books of the Shorter Recension. The fourth line is absent in the Jodhpur1 and in Udaipur1 manuscripts, while in Jaipur4 the last line of the poem is erased and a new first line is added. The same structure with the same four lines is followed in three more manuscripts (Jaipur7, Tijara [marginalia], Uniyara). Lines can also be omitted on purpose, as was done with the embarrassing last line of v. 2.11+ in two manuscripts (Vrindaban1 and Patna2).

50. Lutgendorf, *Life of a Text*, 8–10. He gives the reference to Growse as *The Rāmāyaṇa of Tulasīdāsa* (Cawnpore: E. Samuel, 1891; reprint edition New Delhi: Motilal Banarsidass, 1978), lv. It should be mentioned that the *kaithī* script does not necessarily mean *kāyastha* affiliation, as its name suggests, since it was more widely used in the eastern part of the Hindi belt.

51. Tulsīdās, *Kavitta Rāmāyaṇa*, edited by Sārasvat; Tulsīdās, *Kavitta Rāmāyaṇa*, edited by Miśra. Miśra mentions that the text is based on the manuscript of Paṇḍit Raghunāthdās Gosvāmī.

References

Items are quoted according to the latest edition mentioned. In the Hindi books where the year of publication is given in the Vikrama Era, fifty-seven years have been deducted to get the CE date.

Bangha, Imre. "The *Kavitāvalī* of Tulsīdās: Dynamics of Textual Transmission in Premodern India." *Comparative Studies of South Asia, Africa, and the Middle East* 24, no. 2 (2004), 33–44.

———. "Mundane and Divine Love in Ānandghan's Poetry: An Eighteenth-century Literary Debate." *The Banyan Tree*, ed. Mariola Offredi. New Delhi: Manohar, 2000.

———. *Saneh ko mārag*. New Delhi: Vāṇī, 1999.

Chakkanlāl, Lālā, and Rāmgulām Dvivedī. *Gosvāmī Tulsīdāskṛt dvādaś granth*. (*Jānakī maṅgal* from VS 1632, *Pārvatī maṅgal* from VS 1811, *Gītāvalī, Kṛṣṇa Gītāvalī, Vinay Patrikā* from 1717. sab purānī pustakõ se śodhkar.) Kāśī: Sarasvatī Yantrālay, 1886.

Callewaert, Winand M. "The 'Earliest' Songs of Mira (1503–1546)." *Annali* (Istituto Universitario Orientale, Napoli) 50, no. 4 (1990), 363–78.

———. *The Hindī Biography of Dādū Dayāl*. Delhi: Motilal Banarsidass, 1988.

Callewaert, Winand M., and Mukund Lath. *The Hindi Padāvalī of Nāmdev: A Critical Edition of Nāmdev's Hindi Songs with Translation and Annotation*. Delhi: Motilal Banarsidass, 1989.

Callewaert, Winand M., et al. *The Millennium Kabīr Vāṇī*. Delhi: Manohar, 2000.

Deol, Jeevan Singh. "Sūrdās: Poet and Text in the Sikh Tradition." *Bulletin of the School of Oriental and African Studies* 63, no. 2 (2000), 169–93.

Entwistle, Alan W. *The Rāsa Māna ke Pada of Kevalarāma: A Medieval Hindi Text of the Eighth Gaddī of the Vallabha Sect*. Groningen: Egbert Forsten, 1993.

Gauṛ, Kṛṣṇdevprasād, et al., eds. *Hastalikhit hindī granthõ kā saṁkṣipt vivaraṇ II*. Benares: Nāgarīpracāriṇī Sabhā, 1964.

Gupta, Mātāprasād. *Kavitāvalī*. 1941. 6th ed. Allahabad: Hindī Sāhitya Sammelan, 1949.

———. *Tulsīdās: ek samālocnātmak adhyayan*. 6th ed. Allahabad: Lokbhārtī, 2002.

Hawley, John Stratton. *Three Bhakti Voices: Mirabai, Surdas and Kabir in Their Times and Ours*. New Delhi: Oxford University Press, 2005.

Kabīr granthāvalī. 1928. Ed. Śyāmsundar Dās. 22nd ed. Benares: Nāgarīpracāriṇī Sabhā, 2001.

King, Christopher R. *One Language, Two Scripts: The Hindi Movement in Nineteenth Century North India*. 1994. New Delhi: Oxford University Press, 1999.

Lorenzen, David N. *Praises to a Formless God*. Albany: State University of New York Press, 1996.

Lutgendorf, Philip. *Hanuman's Tale: Messages of a Divine Monkey*. New York: Oxford University Press, 2007.

———. *The Life of a Text: Performing the* Rāmcaritmānas *of Tulsīdās*. Berkeley: University of California Press, 1991.

———. "The Quest for the Legendary Tulsīdās." *Journal of Vaishnava Studies* 1, no. 2 (1993), 79–101.

Mann, Gurinder Singh. *The Goindval Pothis*. Cambridge, Mass.: Harvard University Press, 1996.

McGregor, Ronald Stuart. *Hindi Literature from Its Beginnings to the Nineteenth Century*. Wiesbaden: Otto Harrassowitz, 1984.

Miśra, Satyanārāyaṇ. *Santkavi tursīdās nirañjanī sāhitya aur siddhānta*. Kanpur: Sāhitya Niketan, 1974.

Miśra, Śyāmbihārī. *Hastalikhit hindī granthõ kī khoj kā vivaraṇ IV*. 2nd ed. Benares: Nāgarīpracāriṇī Sabhā, 1998.

Nyāytīrtha, Cainsukhdās, and Kastūrcand Kāslīvāl, eds. *Pradyumna-carit*. Jaipur: Keśralāl Bakhśī, 1960.

Orsini, Francesca. "Qutubśatak, Delhi and Early Hindavi." Unpublished paper.

Pauwels, Heidi. *Kṛṣṇa's Round Dance Reconsidered: Harirām Vyās's Hindi Rās-pañcādhyāyī*. London Studies on South Asia no. 12. Richmond, U.K.: Curzon, 1996.

Pollock, Sheldon. "Literary Culture and Manuscript Culture in Precolonial India." *History of the Book and Literary Cultures*, ed. Simon Eliot, Andrew Nash, and Ian Willison. London: British Library, forthcoming.

———, ed. *Literary Cultures in History: Reconstructions from South Asia*. Berkeley: University of California Press, 2003.

Rājarkar, Bhagatsiṁh Hanamantrāv. *Kavitāvalī: sandarbh aur sandarbh*. Kanpur: Grantham, 1976.

Rāmcaritmānas. Edited and commentary by Hanumān Prasād Poddar. *Kalyāṇ (Mānasāṅk)*. Gorakhpur: Gita Press, 1938.

Rāmcaritmānas. Ed. Viśvanāth Prasād Miśra. Benares: Kāśīrāj, [1962].

Rosenstein, Lucy L. *The Devotional Poetry of Svāmī Haridās: A Study of Early Brajbhāṣā Verse*. Groningen: Egbert Forsten, 1997.

Śarmā, Madanlāl, and Gitārāṇī Śarmā. *Kavitāvalī: bhakti darśan aur kāvya*. Delhi: Rajesh, 1990.

Smith, Vincent A. *Akbar the Great Mogul, 1542–1605*. Oxford: Clarendon Press, 1917.

Snell, Rupert. *The Eighty-four Hymns of Hita Harivaṃśa: An Edition of the Caurāsī Pada*. Delhi: Motilal Banarsidass, 1991.

Stasik, Danuta. "Text and Context: Two Versions of Tulsīdās's Rām-lalā-nahachū." *Studies in Early Modern Indo-Aryan Languages, Literature and Culture*, ed. Alan W. Entwistle and Carol Solomon. Delhi: Manohar, 1999.

Topsfield, Andrew. *Court Painting at Udaipur: Art under the Patronage of the Maharanas of Mewar*. Zürich: Artibus Asiae, [2001].

Tulsī granthāvalī II: mānasetar ekādas granth. Ed. Rāmcandra Śukla, Bhagavāndīn, and Brajratnadās. 2nd ed. Benares: Nāgarīpracāriṇī Sabhā, 1974.

Tulsīdās. *Barvai Rāmāyaṇ*. Ed. Rāmkumār Varmā. Allahabad: Hindī Sāhitya Sammelan, [1967]. (The front page of the book erroneously gives Shaka 1889.)

———. *Kavitta Rāmāyaṇa*. Ed. Durgā Miśra. Benares: Divākar Chapekhānā, 1858.

———. *Kavitta Rāmāyaṇa*. Ed. Baburām Sārasvat. Khidirpur: Sanskrit Press, 1815.

———. *Kavitāvalī*. Trans. F. R. Allchin. London: George Allen and Unwin, 1964.
———. *Kavitāvalī*. Trans. S. P. Bahadur. New ed. Delhi: Munshiram Manoharlal, 1997.
———. *Kavitavali*. Trans. Indradevnārāyaṇ. Gorakhpur: Gita Press, 2001.
The Tuzuk-i-Jahangiri or Memoirs of Jahangir. Vol. 1. Trans. Alexander Rogers and Henry Beveridge. London: Royal Asiatic Society, 1909.
Vaudeville, Charlotte, *Étude sur les sources et la composition du Rāmāyaṇa de Tulsī-Dās*. Paris: Librairie d'Amérique et d'Orient, 1955.
———. *A Weaver Called Kabir*. 1993. New Delhi: Oxford University Press, 1997.

CHAPTER SIX

The Teaching of Braj, Gujarati, and Bardic Poetry at the Court of Kutch

The Bhuj Brajbhāṣā Pāṭhśālā (1749–1948)

FRANÇOISE MALLISON

Kutch certainly is the "place apart," as it was called by B. N. Goswamy and A. L. Dallapiccola in their book on the court's eighteenth-century painting.[1] This region of Gujarat, one of the districts of Saurashtra, forms a link between Sind, with which it has in common its language (Kacchi being a variety of Sindhi) and many cultural features, and Gujarat, to which it has always belonged politically and culturally. In spite of these powerful ties Kutch remained isolated geographically because of the two swampy wastelands surrounding it, the Big and the Little Rann. The feudal organization of the area into specific Bhayyads (*bhāyāta*), where the local lords, all united with each other through family links, shared power, lasted until the final establishment of the Jādejā dynasty. The Jādejā kings kept remarkably good relations with the local Muslim population, due, among other reasons, to their earlier links with a branch of the Sammas (*Samā*), a Rajput lineage converted to Islam, in Sind.[2]

Khengār I (r. 1548–85) united Kutch under his control—brilliant ruler that he was—founded the city of Bhuj, had the first palace erected, and also founded the Moṭī Pośāl for the training of the bards, the keepers of the royal history.[3] A royal poet (*rāj kavi*) belonging to the caste of the Cāraṇ bards was always to be present at the court. He was not only the official poet, a function which he might share with persons belonging to a different caste or of another status; he was also of premier importance at the time of the rituals establishing, transmitting, and consecrating kingship.[4] As a matter of fact in Kutch every secular power at whatever level in the hierarchy of the Bhayyads or even of mere Jāgīrdārs (holders of small freeholds or villages) needed the presence of a bard more than that of the usual Brahman priest (*rāj guru*) to represent power, a feature apparently characteristic of Kutch. Lakhpati Siṃha (r. 1741–61), the founder of the Bhuj Brajbhāṣā Pāṭhśālā (the Braj Language School in Bhuj), was the tenth successor of Khengār I.

Isolated within their natural borders, these rulers knew how to maneuver cleverly between the Gujarati suzerain and the Mughal emperor while achieving prosperity for their kingdom at the same time. Thus the promise to reduce the incidence of piracy along the coast in order to promote the security of the pilgrims on their way to Mecca made in 1617 by Rāo Bhārmal I (r. 1585–1632) to the Mughal emperor Jahangir brought him the privilege of minting coins, which remained in use in Kutch until Indian independence in 1947.[5] Lakhpati himself obtained the favor of the weak Mughal emperor Alamgir II; in 1757 he received from Alamgir II the royal title of *mirzā mahārāv* and the coveted, prestigious insignia of the Māhī Marātib (golden fish).[6]

While still the heir to the throne, Lakhpati seems to have visited the Mughal court at Delhi and remained fascinated by the refinement of art, music, and literature displayed there.[7] When Lakhpati rose to power in 1741 by removing his father, the very popular Rao Desaljī (r. 1718–41), he could afford the luxury of a refined courtly style with support for the arts. In order to hold his collections he had the so-called Mirror Palace, the Āīnā Mahal, built. For its decoration and furniture he availed himself of the services of a gifted craftsman, Rāmasiṃha Mālam. He was called "the Navigator," as after being rescued at sea off the coast of East Africa by a Dutch merchantman he had spent eighteen years in Holland instructing himself in the arts and techniques of architecture, tile and enamelwork, glassblowing, clock making, gun casting, and foundry. Two more times Lakhpati sent him to Europe to purchase material. A long time before any British influence in the region, the chatoyant baroque decoration of the Āīnā Mahal became a testament to this ruler's attraction to other cultures and other aesthetic norms.[8]

Lakhpati was also a patron of music, dance, and poetry.[9] He is credited with five major Brajbhāṣā works, among them some in a technical style (such as the *Suratataraṅgiṇī* and the *Rasaśṛṅgāra*) and some in a narrative vein (such as the *Hari bhakti vilāsa*), in addition to short poems in the Kacchi language. He owed this talent to his master and friend, the Rājakavi Hamīrdān Ratnu of Jodhpur, with whom he established in 1749 the Bhuj Brajbhāṣā Pāṭhśālā, also called the Kāvyaśālā (the Poetry School).[10] For two centuries the new school occupied three subsequent buildings, two of which were within the palace compound. The third and last of these was probably built during the reign of Prāgmaljī II (r. 1860–75), on the occasion of the construction of the Prāgmaljī Mahal. About fifty students, originating from Kutch, Saurashtra, Rajasthan, and even Punjab and Maharashtra, were admitted every year; of these about ten were usually able to complete the five-year curriculum.[11]

The manuscripts (whether original works or compiled textbooks) must have occupied a lot of space. K. C. Simha found a list from the time of the last of the Jain *ācāryas*, Jīvan Kuśal (ca. 1876), that records about eleven hundred manuscripts.[12] After the proclamation of Indian independence, when the mahārāja was no longer able to assure the upkeep of the institution, which at that time attracted only four students, the school was liquidated. The dispersion of its archives was unavoidable. Some manuscripts went to the *ācārya,* who at that time was a Gaḍhvī from the bardic community, and some to the attendants of the ruler of the kingdom. Some survive in the archives of the Āīnā Mahal museum at Bhuj, at the library of the Department of Gujarati at the University of Rajkot,[13] and in the Agarcand Nahata Collection in Rajasthan. There are about three hundred manuscripts available at present,[14] most of which date from the school's last hundred years.

The institution was entirely financed by the government, as the ruler had allotted to it the revenue produced by the village of Rehā.[15] The amount yielded was 3,000 rupees a year, enough to cover the outlays for the running of the school and the expenses of the students,[16] but neighboring rulers from Rajasthan sometimes agreed to pay the fees of the bardic students sent by them.

The curriculum is a matter of debate among the authors who have done research on the school, but they almost all endorse the important testimony of the poet Dalpatrām (1820–98).[17] Belonging to a family of Śrīmālī Brahmans from Wadhwan (Vaḍhavāṇa in the north of Saurashtra), he had been given shelter as a boy by the Svāmīnārāyaṇīs (a Vaiṣṇav movement in Gujarat), who had ascertained his talent and sent him to Bhuj. As a court poet, like many of those who had completed the course at the Bhuj Brajbhāṣā Pāṭhśālā he had a hard time finding a patron to support him. He made appearances at the courts of Wadhwan, Dhrangadhra, Wankaner, Idar, and Bhavnagar until the decisive meeting with Alexander Kinloch Forbes, the English judge who had discovered the bardic literature of Gujarat and who became his friend and patron in 1848. Due to this friendship Dalpat was to become the first "modern" writer of the Gujarati language, an exemplary link between traditional Gujarat and the colonial era. Although considered to be a rhymester according to Western standards, he nevertheless amuses his readers with surprising verses on strange topics. If the Bhuj Brajbhāṣā Pāṭhśālā escaped oblivion, this is largely due to his reporting about its teaching and status in an article titled "Bhūjamāṃ [*sic*] kavitānī śālā viṣe" published in *Buddhiprakāśa,* the journal he edited on behalf of the Gujarati Vernacular Society founded by Forbes in 1848.[18] The account he gives consists essentially of a list of about thirty texts supposed to be thoroughly known, even by heart, by the students. This list is impressive. It starts with *Bhāṣā vyākaraṇa*, a grammar of Braj attributed to Kavi Ratnajit and dated 1717, according to H. Bhayani,[19] followed by textbooks of prosody and poetic composition (see the list in the appendix), model texts (e.g., *Bihārī Satsāī*), didactic works, and texts in the *purāṇic* or epic vein. Dalpatrām's article ends on a curious note: "The textbooks on poetics are composed after Sanskrit models but it is impossible to convey in Braj the integral nature of their Sanskrit content."[20]

Whatever Dalpatrām's misgivings, the syllabus being taught in the vernacular was meant to inculcate in the minds of the future local bards (Cāraṇī, Gaḍhvī, Bāroṭ) and court poets a learned and refined poetic art, that of the *rīti kāvya* fashionable in all the princely states of northern India, including at the imperial court in Delhi. Several authors wonder about the prominent position given to the teaching of classical poetry in Braj, in a remote Gujarati court where Kacchi idiom predominated in daily affairs. According to Nirmala Asnani, the Braj language developed into a medium for poetry in Kutch because of the spread of Vaishnavism in the region.[21] This seems questionable as neither the court of Kutch nor the population of the kingdom was attracted by Vaishnavism. If the composition of Vaiṣṇav lyrical poetry was included in the curriculum, this was done in Gujarati, as the Gujarati Vaiṣṇavas sing in Braj the songs of the saint-poets from the north, especially in their liturgical celebrations, but sing Gujarati devotional songs at home. Govardhan Sharma wonders whether Braj might stand for *piṅgal*, a composite of Braj and *diṅgal* (the particular literary idiom the bards employed for their poems of praise and war, from Rajasthan to Saurashtra and Kutch).[22] He is right, as sometimes the term *piṅgal* stands for Braj poetry. But it is nevertheless obvious that the term *Braj* as used in the very name of the Kāvya Pāṭhśālā refers to the learned court poetry fashionable in northern India at Lakhpati's time, that is, the *rīti* poetic art in which Lakhpati himself was supposed to have excelled. Lakhpati instituted courtly rituals in Kutch capable of impressing all his vassals and of amplifying through the luster of his patronage his royal image, which, as in the case of other local sovereigns, benefited from Mughal decline. The important local presence of the bardic castes on the one side and the princes' aspirations to promote their court poets on the other may finally explain the success of the strange institution the Bhuj Brajbhāṣā Pāṭhśālā proved to be.

Besides the acquisition of composition techniques for the different poetic forms (*rīti kāvya, diṅgal, piṅgal*), as shown by Dalpatrām, the curriculum included the arts of elocution and singing, the preparation of manuscripts, knowledge of the political and military arts, medicine, horsemanship, and more.[23]

Nirmala Asnani lists the different languages employed—to different degrees—in the manuscripts that have survived or that are mentioned in the rare testimonies concerning the *Kāvyaśālā*: transregional languages (Sanskrit and Persian and the literary idioms Braj and *diṅgal*), regional languages (Gujarati, Punjabi, Marathi, and the more modern Urdu, Khari Boli Hindi, and Rajasthani), and the local language (Kacchi).[24] This array delineates the limits of the cultural area in which the Bhuj Brajbhāṣā Pāṭhśālā could exercise some influence, from which its masters and students hailed.

A student's progress was assessed by means of yearly written and oral exams during which the student submitted his own compositions to an assembly of poets. Among the standard exercises figure the *pūrti*, in which the student had to complete a verse line, and above all the *bāvanī*, a poem of fifty-two lines, each starting with a different letter of the alphabet, and of which a few manuscripts are still available.[25] Gifted students were allowed to double their exams in one year and thus were able to leave school earlier. At a later period a certificate of the state of Kutch was delivered to each student after the completion of the course.[26]

Learning was supposed to be rapid, the timetable of this secular sort of *gurukul* being meant for quick progress. After the morning devotional rites there were classes for two hours, from nine to eleven, then individual cooking because of the different social statuses of the students. Classes resumed from two to five, and again there was kitchen work until eight o'clock, followed by group discussions.[27] This atmosphere of discipline reminds one of an artist's training in music, dance, or drama. The comparison seems especially appropriate in the case of performing artists cum writers, whose work was meant to be experienced orally rather than to be read only.

The successive heads of the school belonged to different groups. At first there was a long succession of Jain Munis. The first of them, Kanak Kuśal of Kishangarh of the Tapāgaccha branch, was called to this position by Hamīrdān Ratnu because of his gift for Sanskrit composition as well as Ardha-maghadi, Rajasthani, Braj, and more. His disciple Kuṃar Kuśal succeeded him. Both were made first court poets and granted the title of *bhaṭṭāraka*, just like the great Hemacandra ācārya in fact. As renunciates they did not take part in the *darbār* but were present at the rites performed in honor of the tutelary Devī Āśāpūrā. They were followed by six or seven Jain Munis, from guru to disciple, for more than a century, until the replacement of Jīvan Kuśal by a Brahman,

Prāṇ Jīvan Tripāṭhī, around 1875.[28] He was followed by his disciple Gopāl Jagdev Brahmabhaṭṭa (Gopa Kavi) at the end of the reign of Prāgmal II (r. 1860–76) and the beginning of that of Khengār III (r. 1876–1942). They were followed by a lineage of Cāraṇ bards and Bhāṭs: Keśavjī, Hamīrjī, Devīdān, and finally Śambhudān Gaḍhvī. The long presence of Jain Munis at the head of the institution is not surprising since the wealthy Jain community was an essential element of the Kutch economy. Also, in Gujarat as well as in Rajasthan the Jain renunciate is the symbol of the learned scholar who does not hesitate to compose in the vernacular and has the necessary technical knowledge for making and preserving manuscripts. Nor is it surprising that other classes of local scholars should have competed for the post. It should be noted, however, that there were only a few Brahmans.

The students of the Bhuj Brajbhāṣā Pāṭhśālā belonged to a number of groups from different origins and religions. Whether from Sind, Kathiawad, Gujarat, or Rajasthan or from even farther away, they could be bards, Rajputs, Lohanas, Brahmans, or renunciates, and of varying creeds such as Jain, Śaiv, Vaiṣṇav, Sant, or Śākta. Whatever their origin, they all received the same training. Secularism seems to have been the rule, which is not surprising in the context of the province of Kutch. Govardhan Sharma lists 331 names of students, while Nirmala Asnani lists 325.[29] Neither provides any dates, although Nirmala Asnani has painstakingly accessed the *vahīvaṅcā* of the bardic genealogies. The works produced by the students—the part that is still available (384 works listed by G. Sharma and 325 by N. Asnani)[30]—belong mainly to the later period of the school, starting, it seems, with the reign of Prāgmaljī II (r. 1860–75) and the construction of the new building. They are now classified according to the following categories: the poetic styles, such as *rīti kāvya, diṅgal-piṅgal*; devotional poetry of all shades (i.e., *saguṇ*, Śaiv, Vaiṣṇav, Śākta, *premākhyān*); textbooks of prosody, music, dance, and astrology; *ākhyāna*, and *purāṇic* tales. Some students became famous and had their works edited. Among the better known are the *ācārya*s Kanak Kuśal and Kuṃar Kuśal and several Svāmīnārāyaṇī poets, of whom the best known are Brahmānand (1772–1849, a converted bard who first wrote under his caste name, Lāḍū Bāroṭ) and Gop Kavi (1760–?).

How is one to understand the oblivion and neglect to which the unique institution that was the Bhuj Brajbhāṣā Pāṭhśālā was fated? One reason may be the decline of the Rajputs in Saurashtra as in Kutch, concomitant with the rise of merchant classes on whom the British colonialist efforts mainly relied and who were in favor of a Sanskritization of social patterns at the expense of the princes and their allies, such as the rural classes, and their clients, such as the bards.[31] Considering the part taken by the Paṭhśālā in teaching Gujarati language and literary composition of all sorts, it is interesting to note that the histories of Gujarati literature composed on Western lines neglect the bardic vein—despite the fact that it was a learned discourse—and that it surfaces again at the time of Gandhi's movement under the veil of folklore.[32] It is strange that the Native School Book and School Society established at Bombay in 1822 asked a Maharashtrian pandit to write the Gujarati grammar, as, according to K. B. Vyas, "Gujarati scholars were not easily available for the task."[33] It is also strange that Narmadāśaṅkar (1833–86), the first modern writer of Gujarat after Dalpat, educated at the Elphinstone Institution at Bombay, felt the necessity to compose for himself a manual of prosody and a dictionary which, as he said, were nowhere available.[34] It is hard to explain this rupture in spite of the fact that Dalpatrām had stored within him all the potentialities of modernity within the framework of a traditional system he was able to put at the disposal of his English friend A. K. Forbes.[35]

A critical study of the Bhuj Brajbhāṣā Pāṭhśālā is indeed still awaited, as it is becoming possible to scrutinize more closely and minutely the sources available. This could do justice to the real importance of bardic and courtly literature and, taking into account the ambitions of the institution founded by Lakhpati in 1749, would help to reconstitute, if only to a limited extent, the intellectual history of the premodern period.

Appendix

The textbooks quoted by Dalpatrām:

1. *Bhāṣāvyākaraṇa* (Amadāvādamāṃ paṇa che)
2. *Mānamañjarī nāmano kośa*
3. *Anekārtha mañjarī*
4. Piṅgaḷa judāṃ judāṃ nīce lakhelāṃ che, paṇa tenī matalaba ghaṇīkharī eka che, tenāṃ nāma
   Ā māheluṃ eka piṅgaḷa jovānī jarūra che. Ghaṇāṃ jovānī jhājī jarūra nathī:
   — *Chandaśṛṅgāra piṅgaḷa* (sahuthī sāruṃ che)
   — *Cintāmaṇī piṅgaḷa*
   — *Chandabhāskara piṅgaḷa*
   — *Hamīra piṅgaḷa* (cāraṇī bhāṣāmāṃ che)

— *Lakhapata piṅgaḷa* (mārāvāḍī bhāṣāmāṃ che)
— *Nāgarāja piṅgaḷa* (māgadhī bhāṣāmāṃ che)

5. *Bhāṣābhūṣaṇa* (aḷaṅkārano saṅkṣepa grantha che)
6. *Vaṃśīdhara* (aḷaṅkārano vistārathī grantha che)
7. *Kavipriyā* (saṭīka moṭā grantha kavitānā guṇadoṣanā khulāsano che)
8. *Rasarahasya* (saṭīka, temāṃ vyaṅjanā, dhvanī, lakṣaṇā vagere kavitānī jhīṇī bābato che)
9. *Kāvya kutohala* (rasarahasyanī saṅkṣepa vārtāno che)
10. *Rasikapriyā* (kavitānā nava rasa che, te mahelā mukhya śaṇagāra rasanuṃ varṇana che)
11. *Sundara Śaṇagāra* [*sic*] (te paṇa eja matalabano che)
12. *Bihārīśataśāī* (kavitānī ghaṇīka taharenī yukitaono che. Saṭīka ā grantha Kalakatāmāṅ chapāyelo che)
13. *Vrandaśataśāī* (dṛṣṭānta—siddhāntanā sāhityano grantha che)
14. *Devīdāsa kṛta Rājanīti*
15. *Nathurāma kṛta Rājanīti*
16. *Sundaravilāsa* (jñāna varṇanano che)
17. *Jñānasamudra* (jñāna viṣe che, temāṃ cṃhanda ghaṇī jātanā che)
18. *Pṛthvīrājarāso* (Candabāroṭano banāvelo, paṇa jāṇyāmāṃ Bhūjamāṃ je grantha sampūrṇa nahī hoya. Ane Bundina rājāne tyāṃhāṃ che evuṃ sāṃbhaḷyuṃ che)
19. *Avatāracaritra*
20. *Kṛṣṇa Bāvanī*
21. *Kavitābandhī Rāmāyaṇa*
22. *Rāmacandra Candrīkā* (*piṅgala viṣeśo che*)
23. *Rāgamālā*
24. *Sabhāvilāsa*

Two additions not included in the list:

— E vagere grantho, ane te sivāya Naḍiyādamāṃ *Sāhityasindhu* nāmano grantha tyāṃhāṃnā Deśāīye karāvelo ghano sarasa che
— *Praviṇa* [sic] *Sāgara* Rājakoṭamāṃ banelo che

Source: Dalpatrām, "Bhūjamāṃ (*sic*) kavitānī śālā viṣe," *Buddhiprakāśa* 8 July 1858, 105–7.

Notes

This is a revised version of the paper "Braj, Gujarati and Bardic Poetry Patronized by the Rulers of Kutch: The Bhuj Braj-bhāṣā Pāṭhśālā (1749–1948)" read at the conference *Journées Gujarat* Instances of Patronage: Arts, Literature and Religion, Ecole Pratique des Hautes Etudes, Paris, 16 and 17 June 2003.

1. Goswamy and Dallapiccola, *A Place Apart*, chapter 2, "History."

2. Rushbrook Williams, *The Black Hills*, 71.

3. Rushbrook Williams, *The Black Hills*, 113. Khengār I appointed as its head a Jain Muni, Māṇek Vīrjī, who might have assisted him during the reconquest of his kingdom. Later the place became a temple of Ambājī Devī, where the descendants of Māṇek as Goḍjī (*gorjī*) were the keepers of the genealogic rolls.

4. Basu, *Von Barden und Königen*; Basu, "Caste through the Prism of Praise."

5. Rushbrook Williams, *The Black Hills*, 120–21.

6. Rushbrook Williams, *The Black Hills*, 137. These insignia are on display at the Āīnā Mahal museum of Bhuj.

7. Shah, *Mahārāva Lakhapatisiṃha*, 5. The court at Delhi seems to have always fascinated the rulers of Kutch and Sind, who, more or less directly, were its dependents. Thus a hagiographic legend tells how Māmaidev, the saint of the Bārmatī-panth (an untouchable community that was half-Hindu and half-Ismaili), was put to death by Bāmbhaṇiyā, the Samma king of Sind (fourteenth century?) for having shown him at his demand, miraculously on a rice plate, the court of the Sultan at Delhi, a feat judged dangerous once it had been accomplished. Bhagavant, *Mataṅga Purāṇa ane Meghavāla samājanī utpatti*, 344–60.

8. Jaffer, "The Aina Mahal."

9. One room of the Āīnā Mahal palace, named Phuvārā Mahal, was meant for the artistic activities of the ruler. There he taught or composed his poetry. Although situated on the second floor, it contained a pleasure pool with water sprayers, in the midst of which sat the royal throne, topped by a high ventilated roof. It was considerably damaged by the earthquake on 26 January 2001. See Jethi and London, "A Glorious Heritage," 57–58.

10. For a discussion concerning the exact date of the foundation, see the arguments summarized in Sharma, *Bhuja kī kāvyaśālā*, 14–16; Ajani, "Mahārāvaśrī Lakhapatajī ane vrajabhāṣā pāṭhaśālā," 115–16.

11. It could actually be lengthened to last seven or ten years, or shortened to three. See Sharma, *Bhuja kī kāvyaśālā*, 33; Asnani, *Kaccha kī brajabhāṣā pāṭhaśālā*, 102.

12. Simha, *Kavirāja Gopakṛta Kāvya Prabhākara*, 52.

13. They are part of the bardic manuscripts collected by the department, until recently under the direction of a Cāraṇ bard, the late Shri Ratudan Rohadiya. Two volumes of the catalogue are published: Dave and Rohadiya, *Charani Sahitya Pradeep*; Jani and Rohadiya, *Charani Sahitya Pradeep*, vol. 2.

14. Nirmala Asnani, in *Kaccha kī brajabhāṣā pāṭhaśālā*, lists 325 works composed by students or teachers of the institution; Govardhan Sharma, in *Bhuja kī kāvyaśālā*, lists 381 works, but neither he nor Asnani provides the origin and the exact references for quotations. On the dispersion of the manuscripts, see Sharma, *Bhuja kī kāvyaśālā*, 37; Simha, *Kavirāja Gopakṛta Kāvya Prabhākara*, 52.

15. There now exist two villages by the name of Rehā, Rehā nānā and Rehā moṭā, both situated toward the southeast of the *talukā* of Bhuj.

16. Shah, *Mahārāva Lakhapatisiṃha*, 209.

17. Simha, *Kavirāja Gopakṛta Kāvya Prabhākara*, 47–48; Shah, *Mahārāva Lakhapatisiṃha*, 210; Sharma, *Bhuja kī kāvyaśālā*, 30–31.

18. *Buddhiprakāśa*, 8 July 1858, 105–7. I owe thanks to Dr. R. T. Savaliya, who painstakingly copied by hand Dalpatrām's article from a unique, brittle copy of *Buddhiprakāśa* kept at the library of the B. J. Institute of Learning and Research, Ahmedabad (Gujarat).

19. Bhayani and Patel, *Kavi Ratnajit-viracita Bhāṣā-vyākaraṇa*, 3.

20. *Kavitā viṣenā je graṃtho upara lakhyāte saṃskṛtanā ādhārathī thayelā che. ane saṃkṛtamāṃ jeṭaluṃ varṇana che, te saghaluṃ vrajabhāśāmāṃ āvī śakhyuṃ nathī* (translation is mine). Dalpatrām does not comment further on the topic. Did he feel a certain intellectual discomfort because of a vernacular, Braj, being put to service in a field previously reserved for Sanskrit expression? On this subject, see the contribution of Allison Busch in this volume. There are no other extant mentions of the school by Dalpat, but his son, the well-known poet Nhānālāl Kavi (1877–1946), in a biography of his father, alluded to the Pāṭhśālā in six praise-filled but vague lines (Kavi, *Kavīśvara Dalapatarāma*, 159).

21. Asnani, *Kaccha kī brajabhāṣā pāṭhaśālā*, 42–61.

22. Sharma, *Bhuja kī kāvyaśālā*, 16–17. The study of Govardhan Sharma is oriented toward the *piṅgal* style of poetry as taught in the Kāvya Pāṭhśālā.

23. Sharma, *Bhuja kī kāvyaśālā*, 36.

24. Asnani, *Kaccha kī brajabhāṣā pāṭhaśālā*, 498–514. The author chooses to classify the languages according to their importance within the school. Concerning the problem of languages and the respective literary spaces they occupy, see Yashaschandra. "From Hemacandra to *Hind Svarāj*."

25. Ajani, "Mahārāvaśrī Lakhapatajī ane vrajabhāṣā pāṭhaśālā," 125–26.

26. Simha, *Kavirāja Gopakṛta Kāvya Prabhākara*, 49–51. The sample of certificate reproduced on 50 dates from 1919.

27. Sharma, *Bhuja kī kāvyaśālā*, 35.

28. The list of school heads is still a matter of discussion. See Sharma, *Bhuja kī kāvyaśālā*, 26–30; Simha, *Kavirāja Gopakṛta Kāya Prabhākara*, 40; Ajani, "Mahārāvaśrī Lakhapatajī ane vrajabhāṣā pāṭhaśālā," 128; Asnani, *Kaccha kī brajabhāṣā pāṭhaśālā*, 153–72, who counts fifteen Jain names. The difficulties arising at the moment of succession offer Jīvan Kuśal the occasion to address to the sovereign a versified letter containing details on the functioning of the school. For this interesting source, the *Araja*

Patrika, see Simha, *Kavirāja Gopakṛta Kāvya Prabhākara*, 42–44; Asnani, *Kaccha kī brajabhāṣā pāṭhaśālā*, 528–30.

29. Sharma, *Bhuja kī kāvyaśālā*, 69–74; Asnani, *Kaccha kī brajabhāṣā pāṭhaśālā*, 531–34.

30. See note 14.

31. On the evolution of traditional society in Saurashtra, see Tambs-Lyche, *Power, Profit and Poetry*.

32. Thanks to the work of Jhavercand Meghani (1897–1947). See his *Cāraṇo ane cāraṇī sāhitya*.

33. Vyas, "The First Grammar in Modern Gujarati," 300.

34. Mallison, "Contradictions and Misunderstandings"; Mallison, "Bombay as the Intellectual Capital of the Gujaratis," 80.

35. One century later one finds at least some echo of A. K. Forbes's great work, *Râs Mâlâ, or, Hindoo Annals of the Province of Goozerat, in Western India* (published in two volumes at London: Richardson Brothers, in 1856) in the method L. F. Rushbrook Williams used to get his information on the history of Kutch from royal and local bards.

References

Ajani, Umiyashankar. "Mahārāvaśrī Lakhapatajī ane vrajabhāṣā pāṭhśālā." *Kaccha: parisaṅvāda na prāṅgaṇamāṅ*. Bhuj: Ajani Prakashan, 2001.

Asnani, Nirmala. *Kaccha kī brajabhāṣā pāṭhaśālā evaṅ usase sambaddha kaviyoṅ kī kṛtitva*. Prayag (Allahabad): Hindi Sahitya Sammelan, 1996.

Basu, Helene. "Caste through the Prism of Praise: Charan Bards of Kacch." Unpublished manuscript, paper delivered at the conference *Journées Gujarat* Instances of Patronage: Arts, Literature and Religion, Ecole Pratique des Hautes Etudes, Paris, 16 and 17 June 2003.

———. *Von Barden und Königen: Ethnologische Studien zur Göttin und zum Gedächtnis in Kacch (Indien)*. Frankfurt: Peter Lang, 2004.

Bhagavant, Matang Malshi Ladha. *Mataṅga Purāṇa ane Meghavāla samājanī utpatti*. Gandhidham: Matang Jatashankar Malsi Bhagavant, 1992.

Bhayani, H., and Bholabhai Patel, eds. *Kavi Ratnajit-viracita Bhāṣā-vyākaraṇa*. Bombay: Forbes Gujarati Sabha, 1992.

Dalpatrām. "Bhūjamāṅ (*sic*) kavitānī śālā viṣe." *Buddhiprakāśa* 8 (July 1858), 105–7.

Dave, I. R., and Ratudan Rohadiya. *Charani Sahitya Pradeep*. Rajkot: Saurashtra University, 1981.

Goswamy, B. N., and A. L. Dallapiccola. *A Place Apart: Painting in Kutch, 1720–1820*. Delhi: Oxford University Press, 1983.

Jaffer, Amin. "The Aina Mahal: An Early Example of 'Europeanerie.'" 2000. Special issue of *The Art of Kutch. Marg*, ed. Christopher W. London, 51, no. 4 (2004), 62–75.

Jani, Balvant, and Ratudan Rohadiya. *Cāraṇī Sāhitya pradīpa*. Vol. 2. Rajkot: Saurashtra University, 1998.

Jethi, Pramod J., and Christopher W. London. "A Glorious Heritage: Maharao Lakhpatji and the Aina Mahal." 2000. Special issue of *The Arts of Kutch. Marg*, ed. Christopher W. London, 51, no. 4 (2004), 57–58.

Kavi, Nhānālāl D. *Kavīśvara Dalapatarāma*. Vol. 1. Ahmedabad: A. N. Kavi, 1933.

Mallison, Françoise. "Bombay as the Intellectual Capital of the Gujaratis in the Nineteenth Century." *Bombay: Mosaic of Modern Culture*, ed. Sujata Patel and Alice Thorner. Bombay: Oxford University Press, 1995.

———. "Contradictions and Misunderstandings in the Literary Response to Colonial Culture in Nineteenth Century Gujarat." *Literary Initiatives: Literature and Colonialism in South Asia*, ed. Stuart Blackburn. Special Issue of *South Asia Research* 20, no. 2 (2000), 119–32.

Meghani, Jhavercand. *Cārano ane cāraṇī sāhitya*. Ahmedabad: Gujarat Vernacular Society, 1943.

Rathod, R. K. *Kacchanuṃ saṃskṛtidarśana*. 1959. Ahmedabad: Navabharat Sahitya Mandir, 1990.

Rohadiya, Ratudan. *Gujarātanā cāraṇī sāhityano itihāsa*. Gandhinagar: Gujarat Sahitya Academy, 1985.

Rushbrook Williams, L. F. *The Black Hills: Kutch in History and Legend. A Study in Indian Local Loyalties*. London: Weidenfeld and Nicolson, 1958.

Shah, Kantilal M. *Mahārāva Lakhapatisiṃha—vyaktitva aura sāhityika kṛtitva*. Baroda: Hindi Vibhāg, Maharao Sayajirao University, 1978.

Sharma, Govardhan. *Bhuja (Kaccha) kī kāvyaśālā*. *Paramparā* 89, ed. N. Bhati. Jodhpur: Rajasthan Shodh Sansthan, Jodhpur University, 1989.

Shukla, Dayashankar, ed. *Kuṃvarakuśala kṛta Lakhapati—jasasindhu*. Baroda: Maharao Sayajirao University, 1992.

Simha, Kunvar Chandraprakash, ed. *Kavirāja Gopakṛta Kāvya Prabhākara kiṅvā Rukmiṇīharaṇa, tathā anya grantha*. Baroda: Maharao Sayajirao University (Hindi Anusandhan Granthamala 3), 1964.

Tambs-Lyche, Harald. *Power, Profit and Poetry: Traditional Society in Kathiawar, Western India*. New Delhi: Manohar, 1997.

Vyas, K. B. "The First Grammar in Modern Gujarati: An Unpublished Gujarati Grammatical Treatise of 1825 A.D." *Journal of the Gujarat Research Society* 17, no. 4 (1955), 287–300.

Yashaschandra, Sitamshu. "From Hemacandra to *Hind Svarāj*: Region and Power in Gujarati Literary Culture." *Literary Cultures in History: Reconstructions from South Asia*, ed. Sheldon Pollock. New Delhi: Oxford University Press, 2004.

Part III

Inside the World of Indo-Persian Thought

CHAPTER SEVEN

The Making of a Munshī

MUZAFFAR ALAM AND SANJAY SUBRAHMANYAM

The difficult transition between the information and knowledge regimes of the precolonial and colonial political systems of South Asia was largely, though not exclusively, mediated by scribes, writers, statesmen, and accountants possessing a grasp of the chief language of power in that time, namely Persian. More than any vernacular language or Sanskrit, it was in Persian that the officials of the English East India Company conducted its early rule, administration, and even diplomacy in the years leading up to and after the seizure of the revenues of Bengal in the mid-eighteenth century, and it was hence natural enough that they had to come to terms with the social group that was considered most proficient in this regard.[1] To be sure the Mughal aristocracy and its regional offshoots provided them with certain models of etiquette and statecraft, and various sorts of "Mirror of Princes" texts hence attracted the attention of Company officials. But the pragmatic realities of political economy that had to be dealt with could not be comprehended within the *ādāb* (norms of comportment) of the aristocrat, and the representatives of Company Bahadur were, in any event, scarcely qualified themselves to claim such an unambiguous status. The real interlocutor for the Company official thus was the *munshī*, who was a mediator and spokesman (*wakīl*) but also a key personage who could both read and draft materials in Persian and who had a grasp of the realities of politics that men such as Warren Hastings, Antoine Polier, and Claude Martin found altogether indispensable.[2]

Though the term *munshī* is recognizable even today, it has shifted semantically over the years. Aficionados of Hindi films since the 1960s will recognize the character of the *munshī* as the accountant and henchman of the cruel and grasping *zamīndār*, greasily rubbing his hands and usually unable to stand up to the immoral demands of his master.[3] Specialists on colonial surveying operations in the Himalayas and Central Asia will recall that some of those sent out on such ventures were already called "pundits" and "moonshees" in the mid-nineteenth century.[4] But it is not the latter set of meanings that is our concern in this brief essay. Rather we look at how, in the high Mughal period, one became a *munshī*, what attributes were principally called for, and what the chief educational demands were. The sources with which we approach such a problem fall broadly into two categories. Relatively rare are the first-person accounts or autobiographical narratives that are our principal concern here. The more common are normative texts, corresponding to the "Mirror of Princes" type, but falling in this instance into what we may term the "Mirror for Scribes." Thus, just as in the reign of Aurangzeb, an author such as Mirza Khan could write a text like the *Tuhfat al-Hind*, in which he set out the key elements in the education of a well brought-up Mughal prince, others in the same period wrote works such as the *Nigārnāma-i Munshī*, which were primarily concerned with how a *munshī* was to be properly trained and which technical branches of knowledge he ought rightfully to claim a mastery of.[5] Earlier still, from the reign of Jahangir (r. 1605–27), we have a classic text titled *Inshā'-i Harkaran*, the author of which, Harkaran Das Kambuh of Multan, claimed to have served as a scribe in the high Mughal administration. The significance of this text was such that the East India Company produced an edition and translation of it in the late eighteenth century so that it could serve as a model text for its own early administrators when they dealt with the knotty problems of inherited Mughal administrative practice and terminology.[6] The *munshī* was thus the equivalent in the Mughal domains of the south Indian *karaṇams*, whose careers and worldview have recently been the subject of an extensive treatment.[7]

Since such materials fell into a branch of knowledge that was regarded as secular, we are not entirely surprised to find that many of its authors, including Harkaran himself, were Hindus, usually Khatris, Kayasthas, or Brahmans. It has long been recognized that over the centuries of Muslim rule in northern India the frontiers of Persian came to extend far beyond the narrow circle of the emperor, the princes, and high nobles.[8] Akbar was the first among the Indo-Islamic kings of northern India formally to declare Persian the language of administration at all levels, which had not been the case under the Afghan sultans. The proclamation to this effect was apparently issued by his famous Khatri revenue minister, Todar Mal. It was accompanied by a reorganization of the revenue department as well as the other administrative departments by the equally famous Iranian noble Mir Fath-Allah Shirazi. How an eighteenth-century historian, Ghulam Husain Taba'taba'i, remembered and recorded this change is worth quoting: "Earlier in India the government accounts were written in Hindi according to the Hindu rule. Raja Todar Mal acquired new regulations [*zawābit*] from the scribes [*nawīsindagān*] of Iran, and the government offices then were reorganized as they were there in *wilāyat* [Iran]."[9] Persian was now on the ascendant, and it was not simply the royal household and the court which came to bear the Iranian impress. As *mutasaddī*s and minor functionaries, Iranians could be seen everywhere in government offices, even though they were not in exclusive control of these positions. A substantial part of the administration was still carried out by members of the indigenous Hindu communities which had hitherto worked in Hindawi; this was of great consequence, for our purpose, for these communities soon learned Persian and joined the Iranians as clerks, scribes, and secretaries (*muharrir*s and *munshī*s). Their achievements in the new language were soon recognized as extraordinary. To this development Akbar's reform in the prevailing *madrasa* education—again planned and executed by the Iranian Mir Fath-Allah Shirazi—contributed considerably. Hindus had already begun to learn Persian in Sikandar Lodi's time, and 'Abdul Qadir Badayuni even mentions a specific Brahman as an Arabic and Persian teacher of this period.[10] Akbar's enlightened policy and the introduction of secular themes in the syllabi at middle levels "stimulated a wide application to Persian studies." Hindus—Kayasthas and Khatris in particular—joined *madrasa*s in large numbers to acquire mastery in Persian language and literature, which now promised a good career in the imperial service.[11]

Beginning in the middle of the seventeenth century the departments of accountancy (*siyāq*) and draftsmanship (*inshā'*) and the office of revenue minister (*dīwān*) were mostly filled by these Kayastha and Khatri *munshī*s and *muharrirs*. As noted above, Harkaran Das is the first known Hindu *munshī* whose writings were taken as models by later members of the fraternity.[12] The celebrated Chandrabhan "Brahman" was another influential member of this group, rated second only to the *mīr munshī* himself, Shaikh Abu'l Fazl ibn Mubarak (1551–1602). Chandrabhan was a man of versatile skills and also wrote poetry of high merit.[13] There then followed a large number of other Kayastha and Khatri *munshī*s, including the well-known Madho Ram, Sujan Rai, Malikzadah, Bhupat Rai, Khushhal Chand, Anand Ram "Mukhlis," Bindraban Das "Khwushgu," and a number of others who made substantive contributions to Indo-Persian language and literature.[14] Selections and specimens of their writings formed part of the syllabi of Persian studies at *madrasas*. Certain areas hitherto unexplored or neglected now found skilled investigators, chiefly among Kayasthas and Khatris. They produced excellent works in the eighteenth century in the philological sciences. The *Mir'āt al-Istilāh* of Anand Ram "Mukhlis," the *Bahār-i 'Ajam* of Tek Chand "Bahar," and the *Mustalahāt* *al-Shu'arā'* of Siyalkoti Mal "Warasta" are among the most exhaustive lexicons compiled in India. These Persian grammars and commentaries on idioms, phrases, and poetical proverbs show their authors' keen interest, extensive research, and unprecedented engagement in the development of Persian in India.[15]

Underpinning these developments was undoubtedly the figure of the "ideal" or "perfect" *munshī*, which many of these men aspired to be. What did this mean in concrete terms? A passage from a celebrated letter written by Chandrabhan "Brahman" to his son Khwaja Tej Bhan is worth quoting in this context:

> Initially, it is necessary for one to acquire training in the [Mughal] system of norms [*akhlāq*]. It is appropriate to listen always to the advice of elders and act accordingly. By studying the *Akhlāq-i Nāsirī, Akhlāq-i Jalālī, Gulistān,* and *Bostān,* one should accumulate one's own capital and gain the virtue of knowledge. When you practise what you have learnt, your code of conduct too will become firm. The main thing is to be able to draft in a coherent manner, but at the same time good calligraphy possesses its own virtues and it earns you a place in the assembly of those of high stature. O dear son! Try to excel in these skills. And together with this, if you manage to learn accountancy [*siyāq*], and scribal skill [*nawīsindagī*], that would be even better. For scribes who know accountancy as well are rare. A man who knows how to write good prose as well as accountancy is a bright light even among lights. Besides, a *munshī* should be discreet and virtuous. I, who am among the *munshī*s of the court that is the symbol of the Caliphate, even though I am subject to the usual errors, am still as an unopened bud though possessing hundreds of tongues.

Chandrabhan then goes on to set out the details of a rather full cultural curriculum, in a letter that was clearly destined for a larger readership than his son alone. He writes:

> Although the science of Persian is vast, and almost beyond human grasp, in order to open the gates of language one should read the *Gulistān, Bostān,* and the letters of Mulla Jami, to start with. When one has advanced somewhat, one should read key books on norms and ethics, as well as history books such as the *Habīb al-Siyar, Rauzat al-Safā, Rauzat al-Salātīn, Tārīkh-i Guzīda, Tārīkh-i Tabarī, Zafar-nāma, Akbar-nāma,* and some books like these that are absolutely necessary. The benefits of these will be to render your language elegant, also to provide you knowledge of the world and

> its inhabitants. These will be of use when you are in the assemblies of the learned. Of the master-poets, here are some whose collections I read in my youth, and the names of which I am writing down. When you have some leisure, read them, and they will give you both pleasure and relief, increase your abilities, and improve your language. They are Hakim Sana'i, Mulla Rum, Shams-i Tabriz, Shaikh Farid al-Din 'Attar, Shaikh Sa'di, Khwaja Hafiz, Shaikh Kirmani, Mulla Jami, and Unsuri, Firdausi, Jamal al-Din 'Abd al-Razzaq, Kamal Isma'il, Khaqani, Anwari, Amir Khusrau, Hasan Dehlavi, Zahir Faryabi, Kamal Khujandi, 'Amiq Bukhari, Nizami 'Aruzi Samarqandi, 'Abd al-Wasi Jabali, Rukn Sa'in, Muhyi al-Din, Mas'ud Bek, Farid al-Din. 'Usman Mukhtari, Nasir Bukhari, Ibn Yamin, Hakim Suzani, Farid Katib, Abu'l 'Ala Ganjawi, Azraqi, Falaki, Sauda'i, Baba Fighani, Khwaja Kirmani, Asafi, Mulla Bana'i, Mulla 'Imad Khwaja, 'Ubaid Zakani, Bisati, Lutf-Allah Halawi, Rashid Watwat, Asir Akhshikati, and Asir Umami. May my good and virtuous son understand that, when I had finished reading these earlier works, I then desired to turn my attention to the later poets and writers and started collecting their poems and *masnawīs*. I acquired several copies of their works, and when I had finished them I gave some of them to some of my disciples. Some of these are as follows: Ahli, Hilali, Muhtasham, Wahshi, Qazi Nur, Nargis, Makhfi Ummidi, Mirza Qasim Gunai, Partawi, Jabrani, Hisabi, Sabri, Zamiri Rasikhi, Hasani, Halaki, Naziri, Nau'i, Nazim Yaghma, Mir Haidar, Mir Ma'sum, Nazir, Mashhadi, Wali Dasht Bayazi, and many others who had their collections [*dīwāns*] and *masnawīs*, and whose names are too numerous to be listed in this succinct letter.[16]

The extensive list cited here is remarkable both for its diversity and its programmatic coherence. The masters of the Iranian classics obviously found an appreciative audience even among the middle-order literati in big and small towns, as well as among village-based revenue officials and other hereditary functionaries and intermediaries. All Mughal government papers, from imperial orders (*farmāns*) to bonds and acceptance letters (*muchalka, tamassuk, qabūliyat*) that a village intermediary (*chaudhurī*) wrote were in Persian.[17] Likewise there was no bookseller in the bazaars and streets of Agra, Delhi, and Lahore who did not sell manuscript anthologies of Persian poetry. *Madrasa* pupils were in general familiar with the Persian classics, and Persian had practically become the first language of north India.[18] Those steeped in Persian appropriated and used Perso-Islamic expressions such as *Bismillāh* (in the name of Allah), *lab-bagur* (at the door of the grave), and *ba jahannum rasīd* (damned in hell) just as much as their Iranian and non-Iranian Muslim counterparts did. They would also now look for and appreciate Persian renderings of local texts and traditions. Lest it be forgotten, the religious scriptures were rendered in full into Persian by various Hindu translators, and these too came to be a part of the cultural baggage of the typical Kayastha or Khatri.[19] While we cannot present a detailed analysis of each of these texts, at least some of these translations clearly enjoyed circulation outside the relatively rarefied milieu of the court.[20]

Yet the core of the technical curriculum for a *munshī* must be seen as lying elsewhere, notably in epistolography, accountancy, and methods of fiscal management. The *Nigārnāma-i munshī* shows this clearly enough. It was written by an anonymous author who used the pennames "Munshi" and "Malikzada" and who had been a member of the entourage of Lashkar Khan, *mīr bakhshī* in 1670–71. The author then seems to have entered the service of the prince Shah 'Alam, and then gone on to hold a series of other posts into the mid-1680s. It was around the age of seventy, having accumulated considerable experience, that he thought to compose this didactic text. The *Nigārnāma* itself is made up of two sections (*daftars*), which follow an introduction largely devoted to the subject of *inshā'* and the work of prominent *munshīs* of the past. The first part is subdivided into four chapters, dealing with the drafting of different kinds of letters: those for princes of the royal blood, those written for nobles, those for *dīwāns*, and orders and letters of appointments. The examples chosen seem to be authored by the writer "Munshi" himself. The second *daftar* surveys examples of the work of other prominent *munshīs*, including royal orders, orders written on behalf of Prince Shah 'Alam, and other letters and reports, and includes a particular section devoted to one prominent *munshī*, a certain Uday Raj Rustamkhani. Clearly a pantheon of *munshīs* existed, and the great exemplars of style were never arbitrarily chosen. Thus this particular work includes, besides Uday Raj, letters and orders drafted by Shaikh 'Abdus Samad Jaunpuri, Mir Muhammad Raza, and Sa'adullah Khan. The last personage, a prominent *dīwān* of Shahjahan's reign, was clearly viewed as one of the heroes of the *munshī* tradition, for one version of the manuscript also reproduces the "Manual of the Diwan" in its first *daftar*.[21]

In a vein complementary to the *Nigārnāma* is the *Khulāsat us-Siyāq*, written by a certain Indar Sen in 1115 H (1703–4), which is made up of a preface, an introduction, three chapters, and a conclusion that includes some examples of arithmetic formulae that would be of use for the *munshī* in his accounting (*siyāq*) practice.[22] This work, written late in Aurangzeb's reign probably by a Kayastha author, is mostly concerned with fiscal management, and its three central chapters concern key institutions: the Diwan-i A'la, the Khan-i Saman, and the Bakhshi. The introduction sets out the transition from Hindawi accountancy to that in Persian in the time of Akbar and emphasizes the need for the *munshī* class to move with the times. Yet even more than the *Nigārnāma* text this is a rather narrow conception of what the role of the *munshī* in fact was. A more comprehensive view is provided by looking to the autobiographical materials from the same broad period, insisting as they do on the formation of the moral universe of the *munshī*.

Nek Rai's Premature Autobiography

The fate of the *munshī* was to wander, since his type of employment required him to travel with a peripatetic patron of the elite class. It is thus no coincidence that the chief text that we discuss here, though largely autobiographical in nature, explicitly uses the word *travel* (*safar*) in its title. It comes from the pen of a seventeenth-century member of a scribal group (probably a Kayastha, though we cannot entirely rule out the possibility that he was a Khatri), Nek Rai by name, and seems to emerge from a context in which Persian scribal skills were being ever more widely disseminated and available in increasing numbers to Khatris, Kayasthas, and even some Brahmans. We have noted that as early as the reign of Akbar, Khatris such as Todar Mal featured in a prominent place in the revenue administration, but the seventeenth century saw their numbers growing apace, and after 1700 there was a veritable explosion in their ranks. Earlier historians have noted this fact while surveying the writings of authors such as Bhimsen (author of the *Tārīkh-i Dilkusha*), who accompanied the Mughal armies into the Deccan in the latter decades of the seventeenth century.[23] However, the author of our account, Nek Rai—whose text is really a Bildungsroman of sorts with a thread of travel running through it—has thus far escaped the attention of historians of the Mughal period. Our discussion is based on a single manuscript of his work;[24] the text is titled *Tazkirat al-Safar wa Tuhfat al-Zafar*, and it was copied by a certain Ram Singh, on the patronage of Lala Hazari Mal, who may have been from the author's own family, on 10 Zi-Qa'da 1146 H (14 April 1734) in Hyderabad. Our discussion follows the thread of the narrative very closely, paraphrasing and commenting on it.

First-person prose narratives, while less common perhaps in Mughal India than in the Ottoman domains, still had a respectable place in Mughal belles-lettres.[25] The Mughal emperors themselves had shown the way, for Babur authored one such text (in Chaghatay Turkish), and Jahangir too distinguished himself as an author in this genre. In the course of the seventeenth century some other examples may be found, by authors such as 'Abdul Latif Gujarati, though a real explosion must await the eighteenth century and the phase of "Mughal decline." At the same time, the autobiographical account was not entirely known in the north Indian vernacular tradition, as the case of the celebrated *Ardhakathānak* of Banarasi Das demonstrates.[26] In this panorama the text by Nek Rai must count as an early example of an Indian first-person account in Persian, unusual for its time perhaps, but not quite unique. It shares a feature with Banarasi Das's text, namely its concern with the author's childhood and youth rather than with his mature years. In fact Nek Rai's account is even more "half a tale" than the *Ardhakathānak*, for it ends when its author has barely reached his early to mid-twenties.

We should note at the outset that Nek Rai's text is written in a deliberately difficult and flowery Persian and begins with the praise of God and of the "pen." The initial theme that is treated is not travel, as might be suggested in the title, but rather speech (*sukhan*) itself. The first page and a half of the manuscript is devoted to an exegesis on the invocatory phrase *Bismillāh al-rahmān al-rahīm*, addressed through the construction of its letters, the idea of justice that it embodies, and more. It would seem that Nek Rai's model here is the prose of Abu'l Fazl, and there follow some verses with allusions to the ancients and other prestigious figures. We then move from speech (*sukhan*) to the pen (*qalam*) in the space of some lines, as well as to a development on the subject of the craftsmanship of God. It is worth noting that as with Abu'l Fazl, the use of Arabic phrases here is quite limited. It is only once the initial framing in terms of the wonders of God's creation have been dealt with, that the proper text of his narrative begins. This prefatory *hamd* section is thus extremely artful and clever, and even manages to incorporate the name of the reigning monarch, 'Alamgir. A sample of it runs as follows:

> The account of the disturbed conditions of this sinful *faqīr*, who with the help of fortune and the support of thoughtfulness has entered the alley of the pen, and the field of paper, to venture a description, is on account of the grace of God. May this account be able to apply the kohl of experience to the eye. It is like a light-giving lamp in the night of thought. Just as the movement of the pen brings light onto the blank page, may this account bring light to the night in the city of transitory being. This is a pious account [*zikr-i khair*]:
>
> *Agarche nek niyam*
> *Khāk-i pā-i nikānam*

Even if I am not pious,
I am the dust of the feet of the pious.

The play here is obviously on the author's own name, Nek Rai. This is followed by another verse, perhaps more indicative of his own (non-Muslim) identity: "No wonder I am not thirsty / I am an earthen pot of basil." Nek Rai then explains the title of his text, *Tazkirat al-Safar wa Tuhfat al-Zafar*, since there is on the one hand some mention of travel in it, and on the other hand Dar al-Zafar (or Bijapur) finds a place in it. We then move at last to the beginning of his account proper, or the *āghāz-i dāstān*.[27]

In the thirteenth regnal year of Aurangzeb, or 14 Zi-Hijja 1080 (May 1670), on a Thursday (here we find some astrological details), Nek Rai was born, so he tells us, in the city of Amanabad-Allahabad; this corresponds, he states, to the year 1726 Samvat of Raja Bikramajit, the calendar that is preferred by the Indian Brahmans (*ba nazdīk-i barahmanān-i Hind*). This city in which he was born, he takes care to note, is also called Prayag; the town has excellent buildings and is on the banks of the River Ganges. A Brahman astrologer named Debi Dutt was called in at his birth by his father and grandfather, and on his advice the child was called Nek Rai. The name had some effect, in the sense of allowing him access to science and culture (*'ilm-o-adab*) as well as honor, distinction, and a good rank (*mansab*) already in his youth. He mentions the *zaicha*, or astrological chart, made at his birth, which he reproduces faithfully in the text; it is in the "Indian style," though the version he reproduces has Persian names and terms in it. He then proceeds at some length to explain the chart and the extent to which it has in fact influenced his life, as well as things that might have happened but which in fact did not.

We return then to a description of Allahabad itself (*tausīf-i sawād-i balda-i Ilahābād*). Nek Rai tells us that he will provide a view of the town that will show his command over the very art of description. It is not just a town located on a river but one that brings salvation to all of Hindustan. Its lanes and bazaars are wonderful, and in each instance in the description the metaphors relate to water: the lanes are like rivers, the walls like waves, and so on. The town has a fort made of stone, both powerful and beautiful, built by the monarch Jalal al-Din Akbar. It reaches up to the sky, but its reflections also plumb the water. Its walls are as strong as the *sadd-i Sikandarī*, Alexander's wall against Gog and Magog. Inside it is a building called the Chihil Sutun, with buildings of marble, which seem to emerge from the water itself. There follows a long description of storms on the Ganges, which happen every hundred years or so; they are apparently as powerful as the storm of the time of Noah and bring destruction, uprooting trees and flooding everywhere. The town has many gardens, such as the Jahanara Bagh. His own description of the town, writes the immodest Nek Rai, is as if he were weaving silk, even as his pen moves on silken paper.

We become quickly aware in the course of these initial pages that the entire family of Nek Rai is made up of *munshī*s. When Ilahwardi Khan Ja'far became governor of Ilahabad the author's grandfather and father were employed by him, the former as *dīwān* and the latter as *bak͟hshī*.[28] Unfortunately the khan died a month before Nek Rai's birth, and this caused numerous problems for the father and grandfather, suggesting the ultimate dependence of these service gentry on an elite class of patrons, to the extent that they sometimes took the name of the patron as a sort of surname. Amanullah Khan, the son of the deceased, did provide employment to them for a bit over two years, but then Nek Rai's grandfather himself died at the end of three years, "like a fruit-giving tree that had outlived its time." This grandfather is portrayed in the account as a formidable and rather wealthy man. When Ilahwardi Khan was governor of Shahjahanabad, the grandfather had built a fine house (*imārat-i 'ālī*) in that city. Then Ilahwardi was transferred to Akbarabad (Agra), and the grandfather followed him there and built another house (*kāk͟h-i aiwān*), with fine decorations. The grandfather also had a house in Mathura, described as a *hawelī-i dil-kusha*. He even built a house in the city of Benares, where he had a number of friends and where people in India go on pilgrimage (*matāf*); besides this, he also had a garden in Gorakhpur. This leads Nek Rai to cite a verse from Sa'di: "Whoever came [to the world], built a new house. / He then left, and another took care of it." Therefore the grandfather's death was a major blow to the family and an occasion of great mourning.

Soon after, Husain 'Ali Khan, brother of Ilahwardi Khan, gave employment to Nek Rai's father; this was in the mid-1670s, in the eighteenth regnal year of 'Alamgir, and Nek Rai was five years old. This implies incidentally that the father, Lal Bihari, had no employment for some two or three years, but that attachments to a particular patron's extended family remained strong. Husain 'Ali had at this time been sent as governor to Ilahabad. As for the young Nek Rai, he began his formal education at the beginning of his sixth year, in keeping with tradition (*az rū-i rasm-o-'ādat*), with the first Persian letters on a tablet (*lauh-i abjadk͟hwānī*). But soon enough Husain 'Ali was called back to the Mughal capital, and Nek Rai's father accompanied him to Delhi. Since he had some relatives in Agra, Nek Rai was sent off to stay with them for a time. His teacher there was a certain Durvesh Muhammad Jaunpuri, and he tells us that the light of understanding began to dawn within him under the tutelage of this first master. Two years were spent gaining initial training in reading, writing, and the rudiments of Persian, and Nek Rai was now about seven years old. He began to enter into Persian literature and had his first readings of Shaikh Sa'di; he also moved back to Delhi. It was at this stage that he was married off to the daughter of a certain Daya Ram, son of Bhagwan Das Shuja'i; a brief and rather conventional description of the marriage follows, and we are told that we are now in the twentieth regnal year.

At this time the writer's father decided at last to change patrons and came to be attached to a certain Tahir Khan, who was given the *faujdārī* of *sarkār* Mu'azzamabad or Gorakhpur (a powerful post, in view of the economic expansion in the area at this time). The father was given the post of *peshdast* and *mushrif* (overseer) in charge of the lands and commons (*kharāba*). Nek Rai moved there shortly thereafter with the rest of his family. He mentions Gorakhpur as an open and spacious place and pleasant to live in. But soon after he moved there, Tahir Khan was transferred from the spot and moved back to Delhi via Jaunpur and Ilahabad, since the alternative route through Awadh and Lakhnau was considered more difficult. On the way was the holy site of Kachhauchha, where the tomb of Shah Ashraf Jahangir was, where they went in pilgrimage to the Chishti saint.

The central story about Kachhauchha that is recounted in the text is as follows. When Shah Ashraf arrived there, only a few Muslims were in residence. Those who lived there were in fear of a certain pandit-yogi. When Shah Ashraf came through, people complained to him, and he made inquiries about the yogi. It was learned that he was a great practitioner of magic (*sihr-o-fusūn*) and like the very master of Harut and Marut (two fallen angels who were great magicians). But the shah with a glance began to burn the yogi, who was obliged to beg for mercy and admit defeat.

There is no doubt that the shrine has great power, writes Nek Rai. It is a place where diseases that are reputed to be incurable can be taken care of. Shah Ashraf during his lifetime had told the merchants (*baqqālān*) of the town that the expenses of those who came to the shrine would be defrayed from his own family's resources. The merchants would thus advance the grain in the knowledge that they were secure. Nek Rai's own paternal uncle, Pratap Mal, had an alcohol problem and had become dry like a stick. No doctor could cure him, and finally he was brought to Kachhauchha. Ceremonies were performed for three months (involving offerings of milk, etc.), and he recovered his health. Near the shrine was a garden inhabited by djinns and spirits (*āsebzada*) whose conversations could be heard by mortals. They would climb on trees and generally create a ruckus. But those who were possessed by spirits, especially women, could be cured by coming to the garden. The wonder of the place was that even if women were hung upside down, their clothes remained more or less in place; no indecent exposure occurred. This Nek Rai assures us is a true story and not some fantasy on his part; he also makes it clear that his personal identity encompasses a devotion to certain Sufi shrines, especially those of the Chishti order.

The return to Delhi occurs in the text after this digression in Kachhauccha, and it would seem that the absence from the Mughal capital was of no more than seven months. Delhi itself now merits brief mention, as a wonderful place with excellent buildings and beautiful women. The party arrived there together: Nek Rai, his father, and the rest of his family. It seems that there is a coming of age, for Nek Rai begins to speak now of the sensual pleasures of the town. Shahjahanabad is a place, he writes, where hundreds of handsome Yusufs pursue their Zulaikhas. It is a place where the air is like the breath of Jesus, bringing the dead back to life. This is another set of passages with allusions and comparisons, another moment when Nek Rai shows his mastery of, among other things, Old Testament metaphors. Thus the *dabīr*s of the town wield their pens like the staff of Moses and the trees on the bank of the river are like pearls in the beard of Pharaoh. He mentions the canal made in the time of Shahjahan by the great Iranian noble 'Ali Mardan Khan, whose waters are so sweet (*shīrīn*) as to be the envy of Farhad himself.[29] The great fort had been built by the second Sahib-Qiran (that is, Shahjahan) and was hence called Shahjahanabad. There are also old forts, such as Tughlaqabad, which touch the very sky. If Amir Khusrau, emperor of the land of speech, were alive today, writes Nek Rai, he would have taken his mastery of the word to the sky (the idea being that the objects of this time are so much better than those of Khusrau's time). Delhi is thus called "Little Mecca" (Kh̲wurd Makka) by people in this time. Every year people go to the shrine of Hazrat Khwaja Qutb al-Din Bakhtiyar Kaki (another Chishti saint) to prostrate themselves and attain what they want. Verses follow in praise of Qutb al-Din, taken from Amir Khusrau and other authors.

Nek Rai also recounts an incident involving the great Chishti Sufi Nizam al-Din Awliya and a verse of his disciple Amir Khusrau, the recitation of which had occasioned the death of a certain Mulla Ahmad Mimar. But despite these digressions, the central thread continues to be the education of the author. Nek Rai now resides in the town and studies with Shaikh Khairullah, nephew of Durvesh Muhammad, his earlier teacher. He studies the *Gulistān* and *Bostān* of Sa'di, the *Tūtīnāma,* and the *Sikandar Nāma* of Nizami, but soon the shaikh himself has to leave town, as he is given a post in the Lakhnau area, in *hawelī*

Selak. So Nek Rai studies instead with Sayyid 'Abdul Qadir Lahauri, to whom he is introduced by his father. He praises this new teacher, who is in his view one of the best-educated men of his time. At this time Mas'um Khan, the son of Shahnawaz Khan, is made *faujdār* of Gorakhpur, and Nek Rai's father (who already knew him) goes with him as *diwān* and *bak͟hshī*, advancing in his career "as Yusuf had in his time." Instead of going back to Gorakhpur, Nek Rai and his family are left behind in Mathura, still another praiseworthy town which steals the hearts of people, as noted by the poet Mulla 'Ali. The metaphor he uses now compares the town and its people to a text written in *nasta'līq* or in *nask͟h*. He even praises the style of speech there, which is more beautiful than elsewhere. The river plays a prominent role in this town, with extensive steps (the *ghāts*) of stone. Brahmans wearing sacred threads (*zunnārdārān-i Hind*) come from afar to reside there, some two thousand in number. The sweets of this town are so famous that people carry them away as gifts; they are made of milk and sugar and can be preserved for a number of days without going bad. This town is apparently particularly beautiful during the rainy season.

Nek Rai now enters into less savory aspects of the town of Mathura. It is said that when Bir Singh Dev Bundela, at the time of Akbar, killed 'Allami Shaikh Abu'l Fazl near Gwaliyar at the behest of Prince Salim, in appreciation for this Salim (when he came to the throne) gave all the goods of Shaikh Abu'l Fazl to the raja. The raja asked the sultan for permission to use this money, which was a large sum, to make an impressive place of worship (*ma'bad*) reaching up to the sky. This, he declared, was in the interests of the pursuit of the spiritual. When the *ma'bad* was built thousands of people came in pilgrimage, and festivals and fairs were held there. This went on until the emperor 'Alamgir, "in consideration of matters external to spirituality," decided to tear it down. The act was carried out by Husain 'Ali Khan, who was the *faujdār* of Mathura; he "made a mosque from the temple" (*az ma'bad masjid tartīb yāft*). Nek Rai is ironically disapproving of this act and cites a verse in this context attributed to the seventeenth-century poet Chandrabhan "Brahman," mentioning him by name:

Babīn karāmat-i butk͟hāna-i marā ay Shaikh
Ki chūn k͟harāb shawad k͟hāna-i K͟hudā gardad.

Look at the miracle of my idol-house, o Shaikh.
That when it was ruined, it became the House of God.

So although devoted to Chishti saints and a member of a family with an extensive tradition of service to the Mughals, Nek Rai sees the times in which he lives with a certain irony. This digression on the temple-turned-mosque leads him into a rather extended discussion of *wahdat al-wujūd* and *wahdat-i adyān*, the Unity of Being and the Unity of All Religion, which must be read as an implicit criticism of these acts during the time of Mughal rule. Remove the dust of bigotry from the cheek of the Beloved, he remonstrates; don't trust what you see, which is mere appearance (*zāhir-bīnī*). What is the difference after all between the stone and glass, though one may break the other? The religion of 'Isa and the religion of Musa seem to be different, but when you really look into it, they are the same. The appearance of each letter may be different, but when you put them together in a word, they gain a different sense. The wave, the drop, and the bubble seem different of course, but are they really so?[30] He suggests the possibilities of reconciliation between apparent opposites in a verse from Rumi: those who are prisoners of color will make even Moses fight with himself, while those who have gone beyond color (*bī-rangī*) can reconcile even Moses and the pharaoh.

This further digression being completed, Nek Rai now returns to the matter of his own education. It is in Mathura that he completes his study of the *Tūtī-nāma* and the *Sikandar-nāma*; he now begins to read Abu'l Fazl's letters, the writings of Jami, and the *Mu'ammiyāt-i Husainī*. He has been in Mathura barely a year when his father calls him and the rest of the family to Gorakhpur. This leads him to describe Qasba Gorakhpur; this place and Mathura do not qualify in his view for the more dignified term *Balda*. It is a place of no great population; from a distance it appears large, but when one approaches one realizes that it is spread out, like the inflated hearts of lovers. There is a shrine of Sayyid Ghalib Shahid which is so miraculous that lions frequent it without harming the humans, as in the proverb where the lion and the lamb coexist. The Ruhin River goes by the town. This brings him to a description of the residence made there by his grandfather, less than half a *kos* from the river, and even closer in the rainy season. These are clearly pleasurable memories for Nek Rai, of boat rides on the river and other leisure activities in the rainy season. The bananas and pineapples of the town come in for special praise, as do other fruits. This would seem to be linked to the garden the family has there, with its special vegetables and fruits, listed in some detail. In the bazaar one finds excellent fish, and the rice is available at two *man-i shāhjahānī* for one rupee. The lemons and mangoes too are truly delicious, juicy, and (in the case of the latter) extraordinarily sweet.

By now Nek Rai is ten, and he reads the *Qirān al-Sa'dain* of Amir Khusrau and other texts of greater complexity. These one or two years were truly happy ones, he states, but like all good things they had to come to an end.

People grew jealous of his father and complained about him to the *faujdār*, so that the family was obliged to return to Delhi, this time via Awadh, crossing the Ghaggar and Saryu, a difficult journey. A description follows of the banks of the Saryu, with its beautiful trees and scenic qualities. He takes a boat ride on the bow-like bend in the river, and its undulating waves bring out more poetry in him. Nek Rai, as usual, loves to compare everything to the Persian letters; if something is like a *be*, something else is like a *qāf*, and so on. They thus arrive in the town of Awadh, which he takes time to describe, linking it once more to obvious religious themes. For this is the place where Ram and Lachhman were born, he notes, and "we people" (*mā mardūm*) are attached to the faith of these gods. He feels obliged to give a brief version of the story of Ram Chandra, noting that at the age of ten this prince had learned the sacred sciences from Bishwamitra and began to bring out a hundred meanings from his pen. Then he went to the court of Raja Janak and won the hand of Sita at her *svyamwar* by bending the bow. Throughout this passage the characteristic obsession with the letters of the Persian alphabet pursues Nek Rai. The marriage of Ram and Sita takes place, and they return to Awadh, meeting Parashuram on the way. But once in Awadh problems begin with his stepmother. Ram is obliged to leave the town and go into the forest in exile (*bīshāgirdī*) with Lachhman and Sita, heading toward the Deccan. There he recounts the incident of the golden deer, leading to the futile chase by first Ram and then Lachhman, and Sita's kidnapping by Ravan to Lanka. The figure of Hanuman suddenly appears; he leaps over the sea (*daryā-i shor*) to go and find her. Hanuman sets fire to the town and returns. With his army of monkeys Ram then builds a bridge over the sea and reaches the island of Lanka. The battle begins, and Ravan's son Indrajit wounds Lachhman. Hanuman then flies off to find the Sanjivani and brings back the mountain, crossing a thousand leagues in the wink of an eye.

Eventually Ram kills Ravan and sends him to Hell (*wāsil-i jahannum*) with an arrow, and Ram, Lachhman, and Sita return to Awadh. Ram begins to rule, as his father has meanwhile died. It is written in the history books (*ki chunānche dar tawārīkh-i Hind*) that when his end approached he gathered together some persons and left the town, in the direction of the Sarais of Eternity. Even today, writes Nek Rai, when one comes to Awadh one feels an unseen presence here, and the people of his own party too felt it.

Four days later the party covered the forty leagues to the town of Lakhnau. The stone fort and bazaars of this town strike our author favorably, to say nothing of the excellent bridge with high arches on the Gomti River, which passes below the town. The same river flows toward Jaunpur. The town itself is highly populated, and the matchlockmen (*bandūqchīs*) of the area are well known in all of India. But this competence also causes problems for the *faujdārs* of the area, as there is much potential rebelliousness here. From Lakhnau the party moves on to Qannauj, and Nek Rai makes some snide remarks on the miserly nature of the people there. Nearby is Makanpur, where one finds the shrine of the mystic Badi' al-Din Madar, known as Shah Madar, of the *silsila* of 'Abd al-Qadir Jilani. This is a relatively brief mention with some praise, but no stories of his prowess are added. In fact it is not even clear whether Nek Rai actually visits Makanpur, since it seems to be a bit out of the way. The party is then quickly on its way to Agra.

The experiment with new patrons has clearly failed, ending in jealousy and unhappiness. The father of Nek Rai now goes to Gwaliyar, to enter once more into the service of an old employer, Amanullah Khan, the son of Ilahwardi Khan. The khan sends him as *amīn* of his own *jāgīr* at Jalesar. Jalesar is described as a place with a mud fort, which is however as strong as one made of stone. Here too is a *dargāh* of one Sayyid Ibrahim, which is a place that is frequented by pilgrims, especially on Thursday evenings.[31] The town is noted for its enamel work (*mīnāgarān*) and for embedding enamel pieces into pots and dishes, including pieces with calligraphy. There are several furnaces in which special stones that are found in the region are treated. It seems there is some link between the enamel workers and the *dargāh* of Ibrahim, who may have been some sort of founder (*pīr*) of the settlement. Nek Rai describes the process of enamel work as like wax melting in the hands of the biblical David. He even describes his visit to one of the workshops (*kārkhāna*) to inspect the work there.

He remains in Jalesar for about a year. In this period he continues his education with the letters of Abu'l Fazl and completes the other texts that he has mentioned earlier. It is here that his knowledge of Abu'l Fazl becomes deeper, and he even cites some crucial passages and aphorisms from his letters (*az qalam-i 'Allāmī Shaikh Abu'l Fazl īn nikāt-i chand*), including reflections on the question of religion (*mazhab*). There is clearly a continuity between this and the earlier passage on *wahdat al-wujūd*. The continuing influence of Abu'l Fazl on the *munshī* class is evident here, in terms not only of his political philosophy but also of his understanding of the working of a bureaucracy, the seven key principles for the functioning of a state, and so on. In a similar vein Nek Rai quotes from the *Mu'ammiyāt-i Husainī* and from Jami, but these are less important than his quotations from Abu'l Fazl.

Among other new texts that he picks up at this time are the *Dīwān-i 'Urfī Shīrāzī*, *Kulliyāt-i Hakīm Auhad al-Dīn Anwarī*, the *Tuhfat al-'Irāqain*, and

Dīwān-i Afzal al-Dīn Khāqānī, all of which are briefly mentioned and commented upon. For example, Khaqani struck him for his profound use of words. He also cites some verses from some of these books. This is the time when he begins to read contemporary Mughal poets (*tāzā gūyān*), which, as their name suggests, give him a sense of freshness. Among these are Sa'ib Tabrizi and Mirza Jalal "Asir," the latter of whom comes in for particular praise; verses from both poets are cited in the text. He also cites some of his own verses, which follow the style of Tabrizi. Similarly he reads the poems of Ghani Kashmiri, with his use of double-entendre (*san'at-i īhām*), and those of Wahid Tahir. The list goes on with Haji Muhammad Jan Qudsi, the *malik al-shu'arā'* of the time of Shahjahan, and Abu'l Barakat Munir, Talib Amuli, Kalim, and Muhammad Quli Salim. In each case Nek Rai gives us a few examples and his own appreciation of their particular skills. He also reads the *Majālis al-'Ushshāq* of Sultan Husain Baiqara and the section on poets in Khwandmir's *Habīb al-Sīyar*. The list extends on and on, to Maulana 'Arif al-Din 'Ali Yazdi's *Zafar Nāma*, all of which seem to be parts of the complete education of the *munshī*. This is also the occasion for him to point to the crucial differences between Iranian and Indian poets. These include Munir Lahauri's critique of the style of the *tāza gūyān*, as well as the comments of Mulla Shaida, with a brief mention of who these authors were. Nek Rai mentions several Indian poets who are not so well known, such as Mulla Anwar Lahauri, Mulla 'Ata'i Jaunpuri, and Mulla Tufaili Fatehpuri, and offers his own praise of them, thus locating his own position in the debate squarely on the side of the *tāza gūyān*. These are presented in the context of various debates (*munāzarāt*), including one between Mulla Firuz and Talib. The issues are the capacity to utter verses and the capacity to understand them (*shi'r gū'ī* versus *shi'r fahmī*). Here Nek Rai seems to anticipate in some respects the position taken by Khan-i Arzu in the eighteenth century.

All this literary training has brought Nek Rai to the age of fourteen, the remaining time having been spent in Jalesar and then perhaps Agra. He now returns for a time to his family's residence in Mathura.[32] But it is finally time to move toward the Deccan, in which direction the emperor himself has already set out. Still, he spends a further six months in Mathura, and his first son is born there; the chronogram for his date is given as *mewa-i bāgh-i dil* ("the fruit of the heart's garden"). Extensive celebrations are held on the occasion, with music and other signs of joy. But this happiness is about to be interrupted, for Nek Rai's father is soon to die. In 1097 H (1685–86 CE) the father's death is recorded with a large number of verses of mourning. This was the end of his carefree youth, writes Nek Rai, and the beginning of serious responsibilities.

Thus a new phase is marked by the father's death, coinciding with his own passage to fatherhood.

> The time of my early youth had passed
> And the time of frolicking too had gone.

The moment has come for him to find employment, but it takes him six months and long reflection before he moves on the matter. The chronogram of his father's death is given as "Lal Bihari left the world like a sigh." At this time the routes from Delhi to Agra were disturbed by bandits and troublemakers. So the powerful noble of Iranian origin (and uncle of Aurangzeb) Shayista Khan was brought there from Bengal as governor in place of Khan Jahan Bahadur. In view of the uncertain conditions, Nek Rai decides to leave Mathura with his mother and other relatives for more safety. A brief notice of Agra follows, its impressive buildings, gardens (a number of which are mentioned by name) and *sarāi*s, the marble tomb of Mumtaz Mahal (the Taj Mahal), and the fact that the town was founded shortly after 900 H (1494–95 CE). The site of Sikandra and Akbar's tomb is also mentioned in passing.

By now we are well into the latter half of the 1680s. It is known, Nek Rai writes, that the Deccani cities of Dar al-Zafar Bijapur and Dar al-Jihad Haidarabad were conquered in the thirty-second regnal year, creating a number of new opportunities. Nek Rai's older brother, Sobha Chand, who is both competent and courageous, has already obtained a job in Bijapur as intendant of the artillery (*topkhāna*) and the *dāgh-o-tashīha* (branding of horses and recruitment of men). Sobha Chand is now the head of the family and about forty years old. It is hence time for Nek Rai to seek his fortune there too, and on 18 Shawwal of the thirty-third regnal year he and his family reach Bijapur. A brief account follows of the journey between Agra and Bijapur, in the course of which Nek Rai himself fell ill. The travel is compared to the Sufi's penance (*chilla*), as the journey was not easy; they had to pass through jungles and mountainous territories, and rains impeded their progress. The itinerary is then detailed.[33]

After Agra, the next large town is Gwaliyar, at a distance of three *manzils*. There are beautiful women to be seen all along the way, as Nek Rai's roving eye notes. On reaching Gwaliyar, he refers to the excellent and high fort there, which is considered to be one of the largest in Hindustan. Those who are imprisoned by the emperor's direct order are kept there. Inside the fort is a large pond. The speech in the place is very sweet, and there are lots of *chamelī* flowers to be seen. The betel leaf (*pān*) in the area is of high quality. From Gwaliyar, the party makes its way to Narwar, where too the *pān* leaves catch his attention, as do the birds. There is a large step well (*bāolī*) outside the town of Narwar, where people gather in the evening. The women here are described as magical, capable of giving even those magical creatures Harut and Marut lessons in sorcery. They come to get water from the *bāolī* with a bucket attached to a rope. The people in general are of an excellent temperament and wear colorful clothes; wearing white is a sign of mourning. If one were to spend time looking at the women, one would lose one's heart several times, writes Nek Rai. But one is not allowed to even touch them with one's hand. Here too is a river that flows below the fort. He believes that this is the fort of Raja Nal, about whom that great poet, the *malik al-shu'arā'* Abu'l Faiz "Faizi," had written in his *masnawī* on *Nal-Daman*.

From Narwar they move on to Sironj, a place with excellent air and a good bazaar with quality grapes. Here too the language is sweet to hear and people wear attractive clothes and are attractive to look at. They cross a small river and move on to Sarangpur. All these places seem to please our traveler, who praises them in poetic terms quite unstintingly, noting that this part of the journey is full of pleasure (*'aish*). Presently they reach Shahjahanpur, a *qasba* with royal buildings, and cross a major hurdle in the form of a river. Having reached the other side, they find the land on both sides prosperous and well populated. The bridge over the river, especially its arches, is excellent. They find that they are now in Dar al-Fath Ujjain, a town remarkable for its prosperous character, a paradise-like city, which is among the most ancient of Hindustan. It was here, according to the Hindus (*ba i'tiqād-i mardūm-i Hunūd*), that the famous and generous Raja Bikramajit (Vikramaditya) had his throne (*ta<u>kh</u>tgāh*), and it is also a sacred city. Stories of this monarch circulate extensively in Hindustan, writes Nek Rai. The artisans of the area are remarkable for their skills, particularly in making jewelry. Among the excellent places in the vicinity of the town is a waterfall (called *ābshār-i Kaliyāda*). This inspires a verse, and Nek Rai writes that in all of Hindustan he has seen no place more beautiful than Kaliyada. From this point on, Sobha Chand is given an imperial escort.

Four *manzils* later they come to the banks of the River Narbada, which is considered to be the frontier of Hindustan and the Deccan (*sarhad-i Hindustān ast wa Dakan*). The waters flow so fast that it can be dangerous for boats. Crossing the river, they reach a place called Baqirpur, saying farewell at last to Hindustan proper. The journey from now on is far more unpleasant, largely on camel back, including the crossing of a pass which leads them to Jahangirpur. The passage is very picturesque, however, and having reached an extensive plain they traverse it to reach Burhanpur. They also pass by the great fort of Asir, a league to the north of Burhanpur; this is located on a high hill and reminds Nek Rai of Daulatabad. Here too imperial prisoners are kept. Burhanpur itself is described as a town with good waters, excellent and handsome people, and a popular bazaar. Through the Fardapur Pass they go on to Aurangabad. This is a rougher and more mountainous route, with dry and rocky ground; many mules are to be seen in these areas, carrying goods from one spot to the other. Eventually, with some difficulty, they reach Aurangabad, some distance from the fort of Daulatabad, a formidable and high spot. Some people in Mughal service come from Bijapur to meet the party there. Two weeks later, with these others, they set out and think of going via Parinda fort, and from there to Sholapur, which is in the region of Bijapur. This is a pleasant spot, reminiscent in many respects of Hindustan, which Nek Rai has already begun to feel homesick for.

Crossing the Bhima River they eventually reach the city of Bijapur. A few verses celebrate their arrival. With the help of his older brother (and the grace of the emperor) Nek Rai is given a job. In this work he spends four years, to the time of the completion of his account, when he is still in his early twenties. By way of conclusion he writes:

> In sum, having completed the journey, the town of Dar al-Zafar was reached. With the aid and intercession of my older brother, who was full of high virtues and ethics, and who held me in affection and benevolence, I was given the honour of the service of the assessment of the expenses of the *parganas* of *sarkār* Haweli Dar al-Zafar and Nusratabad. Until, on account of the convergence of good fortune and the gift of God [*'atā-i wāhib al-'atīyāt*], in the beginning of fortunate Zi-qa'da of the 38th regnal year, my brother was honoured by being appointed the *pīshdast* of the Mir Atish, and this smallest of slaves [of God] in his place was appointed to the intendancy of the *topkhāna* and the *dāgh-o-tashīha* of Bijapur. And a very appropriate *mansab*, in keeping my present stature, was granted.

Nek Rai thanks God for this bounty and ends the text with appropriate verses. The copyist's colophon follows, suggesting that the text and its author's family continued to have a connection with the Deccan.[34]

Conclusion

As we suggested at the outset, the text of the *Tazkirat al-Safar* falls into a larger category of literature, wherein notables and literati from Mughal India wrote first-person accounts in which travel played a more or less important role. We have already mentioned the case of 'Abdul Latif Gujarati from the early seventeenth century, and we could add other near contemporary texts such as those of Mirza Nathan and Shihab al-Din Talish to our list, though these authors also insert wider historical materials into their accounts. Later in the eighteenth century writers such as Anand Ram "Mukhlis" raised this form to an ever higher level of subtlety, since it permitted them to be ironic about their own communities, the political system, and even the monarchy. In each of these works elements of the ethnographic are strongly present, as are ekphrastic aspects, including, in the case of Nek Rai, the description of towns, sites, buildings, and the like. Such descriptions are perhaps characteristic of individuals who inhabited the fringes of the Mughal state since they borrow from the vocabulary of the state (with its own drive to produce gazetteer-like *dastūr al-ʿamal* texts in the seventeenth century). It would probably not be too inappropriate to see in these materials the formation of a north Indian class that was similar to the Chinese literati class, even though the existence of the examination system and its curriculum in imperial China somewhat skewed the nature of that knowledge formation.[35]

We are unable at present to follow the later career of Nek Rai or to determine the extent of success he eventually enjoyed as a *munshī* in the latter decades of Aurangzeb's reign. In any event his trajectory as an individual interests us less than his exemplary character as a member of the Persianized Hindu scribal groups that came increasingly to serve the Mughals in the seventeenth century. We have seen how comfortably he straddles a diversity of cultural and literary heritages; we shall find this same ability to adapt in later characters of the eighteenth century, such as Anand Ram "Mukhlis."[36] Nek Rai is aware that he is not a Muslim, but he is just as comfortable with Chishti saints and their shrines as he is with the story of Rama. The term *composite culture* has been much used and abused in recent years, but arguably one can find it in the life and education of such a *munshī*.

Four key features of his education as suggested in Nek Rai's text immediately spring to mind. The first is an absence, for it is noticeable that he does not speak of the technical aspects that other texts (such as the *Nigārnāma-i Munshī*) insist upon. In view of the post that he eventually came to hold in the Deccan, Nek Rai must have learned *siyāq* and had a course in fiscal literacy (as it were); the affairs of the *dīwān* must have been no mystery to him. Yet nowhere in his account of his education does he even speak of it, as if such banal details were beneath mention. A second aspect is the close relationship between the curriculum of texts that he sets out and that defined for his own son by Chandrabhan "Brahman," which makes it clear that the latter's view was no idealized normative template but a rather practical piece of advice. No doubt different teachers took different routes to these texts, and each student too must have developed his own tastes and preferences. Nek Rai's own fondness for writers such as Wahid Tahir, Ghani Kashmiri, Sa'ib Tabrizi, and Mirza Jalal Asir has already been noted.

A third aspect, in our judgment a crucial one, is the fulcral role of Shaikh Abu'l Fazl in the world of the seventeenth-century *munshī*. Nek Rai admires and imitates the style and also the attitudes of the great *mīr munshī*, and he was surely not alone in this matter, for Abu'l Fazl had come by this time to stand for a point of view in which ecumenical learning and religious pluralism were given a high standing, besides the fact that he (together with his brother, the poet Faizi Fayyazi) also embodied a self-confident Indian claim to the use of the Persian language. A specifically Mughal political and literary tradition thus had come to exist by the mid-seventeenth century, one that differed from its Central Asian and Iranian counterparts, and we must trace this back in part to the late sixteenth century and its usages, when Abu'l Fazl was the great ideologue of the remembered Akbari dispensation.

A fourth aspect is the broader cultural framework within which Nek Rai places the issue of his *Bildung*. If he eschews narrowly technical questions regarding his training and education, it is also clear time and again that the Persian language itself plays a key role in his view of the world. It is through this language, its metaphors and possibilities, that he accedes to and imagines the world around him. The philosophical universe within which he conceives of all matters, including issues of social and religious conflict, is impregnated with Persian and with all the richness of the secular tradition that Indo-Persian represented by the seventeenth century. It is in this sense that we must understand what it meant to become, and to be, a *munshī* in the later Mughal world.

Notes

1. Bayly, *Empire and Information*, 73–78; Fisher, "The Office of Akhbar Nawis."

2. On one such *munshī*, Kishan Sahay from Bihar, who served Antoine Polier, see Alam and Alavi, *A European Experience of the Mughal Orient*, 13–14.

3. A similar figure is that of the *munīm*, on whom see Bayly, *Rulers, Townsmen and Bazaars*, 377–78.

4. Raj, "When Humans Become Instruments."

5. Mirza Khan ibn Fakhr al-Din Muhammad, *Tuhfat al-Hind*. On the *Nigārnāma-i Munshī*, see Nurul Hasan, "Nigar Nama-i-Munshi," 258–63.

6. Balfour, *Inshā'-i Harkaran*; also see Gladwin, *The Persian Munshi*, which includes the *Qawā'id-i Saltanat-i Shāh Jahān* by Chandrabhan "Brahman."

7. See Narayana Rao, Shulman, and Subrahmanyam, *Textures of Time*.

8. These sections draw heavily on Alam, "The Culture and Politics of Persian in Precolonial Hindustan."

9. Taba'taba'i, *Siyar al-Muta'akhkhirīn*, 1:200.

10. Badayuni, *Muntakhab al-Tawārīkh*, vol. 3.

11. Balkrishan, *'Arzdāsht*, British Library, London, Addn. Ms. 16859, cited in Mohiuddin, *The Chancellery and Persian Epistolography under the Mughals*, 41; see also Syed Muhammad 'Abdullah, *Adabiyāt-i Fārsī mein Hindūwon kā Hissa*, 240–43.

12. For an analysis, see Mohiuddin, *Chancellery*, 215–20.

13. Faruqui, *Chandra Bhan Brahman*; for his prose, see Mohiuddin, *Chancellery*, 228–34.

14. Compare Alam, *The Crisis of Empire in Mughal North India*, 169–75, 237–40.

15. Syed Muhammad 'Abdullah, *Adabiyāt*, 121–68.

16. Syed Muhammad 'Abdullah, *Adabiyāt*, 241–43, cites this passage from Chandrabhan's letter. The translation is ours.

17. Even in Bengal the administrative papers prepared and issued in the name of the local Hindu intermediaries were in Persian. Persian *inshā'* even succeeded in influencing Bengali prose. See Acharya, "Pedagogy and Social Learning," 255–72.

18. Badayuni, *Muntakhab al-Tawārīkh*, 2:285. Also see Grandin and Gaborieau, *Madrasa*.

19. Compare Gopal bin Govind's preface to his Persian translation of the *Rāmāyaṇa*, Bibliothèque Nationale de France, Paris, Ms. Blochet, 1, 22.

20. See Athar Ali, "Translation of Sanskrit Works at Akbar's Court." It would seem that in Mughal India, besides the *'aqlīya* and *naqlīya* traditions of the Islamic sciences (*'ulūm*), we need to bear in mind the rise of a third category to include the texts of so-called *hikmat-i 'amalī* ("practical wisdom") through which such materials were included in *madrasa* education.

21. Abdur Rashid, "Some Documents on Revenue Administration during Aurangzeb's Reign," 263–68.

22. Siddiqui, "Khulasat-us-Siyaq," 282–87. This is only one of several similar texts; for another example, see Nandram Kayasth Srivastav, *Siyāqnāma*, and for a survey of such "administrative and accountancy manuals," see Habib, *The Agrarian System of Mughal India*, 470–71.

23. Bhimsen, *Nuskha-i Dilkusha*. Also compare Chandra, *Letters of a Kingmaker of the Eighteenth Century*.

24. Salar Jang Museum and Library, Hyderabad, Accession no. 4519, Mss. No. 7.

25. Kafadar, "Self and Others."

26. See the useful recent reconsideration of this text in Snell, "Confessions of a 17th-Century Jain Merchant."

27. Nek Rai, *Tazkirat al-Safar*, f. 3b.

28. On this Mughal *amīr*, see Athar Ali, *The Apparatus of Empire*, S. 7374, *passim*.

29. Nek Rai, *Tazkirat al-Safar*, fl. 13 b.

30. Nek Rai, *Tazkirat al-Safar*, fl. 18 b.
31. Nek Rai, *Tazkirat al-Safar*, fl. 27a.
32. Nek Rai, *Tazkirat al-Safar*, fl. 34b.
33. Nek Rai, *Tazkirat al-Safar*, fl. 44b.
34. On Kayasthas and Khatris in the eighteenth-century Deccan, see Leonard, *Social History of an Indian Caste*, 23–35.
35. There is a vast literature on this subject, but see the useful overview in Elman and Woodside, *Education and Society in Late Imperial China*.
36. Alam and Subrahmanyam, "Discovering the Familiar."

References

Abdur Rashid, Shaikh, "Some Documents on Revenue Administration during Aurangzeb's Reign." *Proceedings of the Indian History Congress*, fifteenth session (Gwalior: Indian History Congress, 1952).

Acharya, Promesh, "Pedagogy and Social Learning: *Tol* and *Pathsala* in Bengal." *Studies in History*, new series 10, no. 2 (1994), 255–72.

Alam, Muzaffar. *The Crisis of Empire in Mughal North India: Awadh and the Punjab, 1707–1748*. Delhi: Oxford University Press, 1986.

———. "The Culture and Politics of Persian in Precolonial Hindustan." *Literary Cultures in History: Reconstructions from South Asia*, ed. Sheldon Pollock. Berkeley: University of California Press, 2003.

Alam, Muzaffar, and Seema Alavi. *A European Experience of the Mughal Orient: The I'jāz-i Arsalānī (Persian Letters, 1773–1779) of Antoine-Louis Henri Polier*. Delhi: Oxford University Press, 2001.

Alam, Muzaffar, and Sanjay Subrahmanyam, "Discovering the Familiar: Notes on the Travel-Account of Anand Ram Mukhlis." *South Asia Research* 16, no. 2 (1996), 131–54.

Athar Ali, M. *The Apparatus of Empire: Awards of Ranks, Offices, and Titles to the Mughal Nobility (1574–1658)*. Delhi: Oxford University Press, 1985.

———. "Translation of Sanskrit Works at Akbar's Court." *Akbar and His Age*, ed. Iqtidar Alam Khan. New Delhi: Northern Book Centre, 1999.

Badayuni, 'Abdul Qadir. *Muntakhab al-Tawārīkh*. Trans. George S. A. Ranking, W. H. Lowe, and T. W. Haig. 3 vols. Calcutta: Asiatic Society of Bengal, 1884–1925.

Balfour, Francis, ed. and trans. *Inshā'-i Harkaran*. Calcutta: Charles Wilson, 1781.

Bayly, C. A. *Empire and Information: Intelligence Gathering and Social Communication in India, 1780–1870*. Cambridge: Cambridge University Press, 1996.

———. *Rulers, Townsmen and Bazaars: North Indian Society in the Age of British Expansion, 1770–1870*. Cambridge: Cambridge University Press, 1983.

Bhimsen. *Nuskha-i Dilkushā*. Ed. and trans. Jadunath Sarkar and V. G. Khobrekar. Bombay: Maharashtra Department of Archives, 1971.

Chandra, Satish. *Letters of a Kingmaker of the Eighteenth Century*. Delhi: Asia Publishing House, 1972.

Elman, Benjamin A., and Alexander Woodside, eds. *Education and Society in Late Imperial China, 1600–1900*. Berkeley: University of California Press, 1994.

Faruqui, Muhammad 'Abdul Hamid. *Chandra Bhan Brahman: Life and Works, with a Critical Edition of His Diwan*. Ahmadabad: Khalid Shahin Faruqi, 1966.

Fisher, Michael. "The Office of Akhbar Nawis: The Transition from Mughal to British Forms." *Modern Asian Studies* 27, no. 1 (1993), 45–82.

Gladwin, Francis. *The Persian Munshi*. Calcutta: Chronicle Press, 1795.

Grandin, Nicole, and Marc Gaborieau, eds. *Madrasa: La transmission du savoir dans le monde musulman*. Paris: Argument, 1997.

Habib, Irfan. *The Agrarian System of Mughal India, 1556–1707*. 2nd ed. Delhi: Oxford University Press, 1999.

Kafadar, Cemal. "Self and Others: The Diary of a Dervish in Seventeenth-century Istanbul and First-person Narratives in Ottoman Literature." *Studia Islamica* 69 (1989), 121–50.

Leonard, Karen Isaksen. *Social History of an Indian Caste: The Kayasths of Hyderabad*. Berkeley: University of California Press, 1978.

Mirza Khan ibn Fakhr al-Din Muhammad. *Tuhfat al-Hind*. Ed. Nur al-Hasan Ansari. Tehran: Bunyad-i Farhang-i Iran, 1354 Sh. (1975–76).

Mohiuddin, Momin. *The Chancellery and Persian Epistolography under the Mughals: From Babur to Shahjahan, 1526–1658*. Calcutta: Iran Society, 1971.

Nandram Kayasth Srivastav, Munshi. *Siyāqnāma*. Lithograph. Lucknow: Nawalkishor Press, 1879.

Narayana Rao, Velcheru, David Shulman, and Sanjay Subrahmanyam. *Textures of Time: Writing History in South India, 1600–1800*. New York: Other Books, 2003.

Nurul Hasan, S. "Nigar Nama-i-Munshi: A Valuable Collection of Documents of Aurangzeb's Reign." *Proceedings of the Indian History Congress*, fifteenth session. Gwalior: Indian History Congress, 1952.

Raj, Kapil. "When Humans Become Instruments: The Indo-British Exploration of Tibet and Central Asia in the Mid-Nineteenth Century." *Instruments, Travel and Science: Itineraries of Precision from the Seventeenth to the Twentieth Century*, ed. Marie-Noëlle Bourguet, Christian Licoppe, and Hans Otto Sibum. London: Routledge, 2002.

Siddiqui, Noman Ahmad. "Khulasat-us-Siyaq." *Proceedings of the Indian History Congress*, 22nd session. Gauhati: Indian History Congress, 1959.

Snell, Rupert. "Confessions of a 17th-Century Jain Merchant: The Ardhakathānak of Banārasīdās." *South Asia Research* 25, no. 1 (2005), 79–104.

Syed Muhammad 'Abdullah. *Adabiyāt-i Fārsī mein Hindūwon kā Hissa*. Lahore: Majlis-i Taraqqi-yi Adab, 1967.

Taba'taba'i, Ghulam Husain. *Siyar al-Muta'akhkhirīn*. Lucknow: Nawalkishor Press, 1888.

CHAPTER EIGHT

Pages from the Book of Religions

Encountering Difference in Mughal India

ADITYA BEHL

Confronted with an enormous and bewildering diversity of religious movements, artistic and literary expressions, and political structures and fiscal arrangements, historians of Mughal India have tended to produce separate accounts of religious beliefs, of the triumphalist art and architecture sponsored by the Mughal court, of individual rulers, literary, and political figures, and above all of the "steel frame" of the Mughal state as set in place by the emperor Akbar (r. 1556–1605) and its decline and disintegration following the intolerant religious policies of the emperor Aurangzeb (r. 1658–1707).[1] These landmarks do no more than set the external boundaries of a period commonly held to exemplify religious tolerance.[2] However, the historical agents who wrote about their encounters with others' beliefs, ideas, rituals, and practices in this period evince a far more sophisticated understanding of religious difference than the scholarly and commonplace consensus. Faced with the same range of materials and historical agents, the Zoroastrian author of a mid-seventeenth-century Persian text on the religions of Mughal India cites the following quatrain to justify his comparative enterprise:

> The world is a book full of knowledge and justice;
> Its binder is destiny, its binding the beginning and end.
> Its stitching is the true law, and the religions are its pages.
> The community is the pupil, and the messenger the teacher.[3]

The poetic figure of the book of the world, in which the religions form separate pages bound together by destiny, organizes all diversity into the encyclopaedic work that he has compiled. Its beginning and end are the limits of linear time, and the only human agents that appear are the community as the pupil and its leader, the divine messenger. The work in question is the

Dabistān-i Mazāhib (The School of Religions), in which each religion constitutes a teaching (*ta'līm*) of the supreme knowledge that is the subject of the author's investigations. Central to the text of the *Dabistān* is the activity of translation, grounded not in notions of fidelity to an original source, but in the esoteric mysticism of a sect of Parsi holy men. Faced with the multiplicity and incommensurability of religious beliefs, the author strives to find keys to interpretation, the inner meanings of the outer forms that he observes. At play is a sophisticated set of comparative procedures for classifying and understanding religions, outwardly tolerant yet mediated by the esoteric and privileged knowledge of the author's sect of holy men.

The text was discovered in manuscript by Sir William Jones in the 1780s; in his excitement he declared it to contain "more recondite learning, more entertaining history, more beautiful specimens of poetry, more ingenuity and wit, more indecency and blasphemy, than I ever saw collected in a single volume. . . . On the whole, it is the most amusing and instructive book I ever read in Persian."[4] The *Dabistān* was critically edited at the College of Fort William in Calcutta in 1809 and translated into English in 1843 by David Shea and Anthony Troyer, two Orientalists associated with the East India Company and the Royal Asiatic Society.[5] Both these texts, the original Persian and the English translation, represent encounters that attempt to understand cultural and religious difference. While there are Islamic texts devoted to Indian religions before the seventeenth century, most notably Al-Birūnī's *Kitāb al-Hind* and Shahrastānī's *Kitāb al-Milal wa'l-Niḥal*,[6] the Persian *Dabistān-i Mazāhib* differs from the Muslim encyclopaedic tradition in that the author of the *Dabistān* bases his account on fresh interviews and a new reading of scriptural texts. Instead of only repeating the received wisdom of the Arab authors on Indian religions, he frames his newfound knowledge within a rereading of pre-Islamic history grounded in his sect's cosmology and religious practice.

Shea and Troyer's translation of this work, completed in the context of the East India Company's rule, is informed by a vision of the East as superstitious, removed from reason, marked by Oriental despotism, and above all religious as opposed to philosophical and analytical. The translation is informed by a larger Orientalist judgment dividing East and West between the realms of superstitious religion and rational philosophy:

> The Asiatic, from the dawn of his reason, is nourished with the marvellous, trained to credulity, and prepared for mysticism, the bane of practical life; in short, he imbibes from his infancy a superstition from which he never

> frees himself, always prone to interpret every unusual phenomenon as a miracle. No sort of study enables him to correct his first impressions, or to enlighten his ignorance; natural history and experimental philosophy are not cultivated in Asia. . . . He knows no social life embellished by the refinement of mutual sympathy, nor the noble vocations of a citizen who lives—with more than one life in himself, in others, and in the whole community.[7]

Besides the usual complaints of Asiatic superstition, credulity, and prolixity, this Orientalizing judgment contains a key assumption: the absence of rational collective social life in Asia, which militates against the author's ability to represent any religious beliefs critically.

Founding Orientalism as a subject of study involved "a 'genesis amnesia' that systematically obliterated the dialogic conditions of its emergence and the production of its linguistic and textual tools. By turning 'the Orient' into an object of analysis and gaze, Orientalism as an European institution of learning anathematized the Asian pedagogues of its practitioners."[8] Unlocking this "genesis amnesia" involves bringing back to life those figures who helped early Orientalists such as Sir William Jones to make their discoveries, recovering those acts of agency which launched a set of academic disciplines ranging from linguistics to comparative religions. Yet the enterprise immediately involves us in the perplexities of defining the human subject who is at the center of inquiry. The historical agents who provided the cultural and intellectual dialogues necessary for the progress of scholarship, the victims of Orientalism's genesis amnesia, were also members of advanced civilizations who took up the problem of social definition and the constitution of subjectivity.

Who was the author of the *Dabistān,* the interpreter of seventeenth-century Indian religions? This question has preoccupied scholars since Jones's enthusiastic but confused proclamations about the text and its author. Jones wrongly attributed the authorship to the Sufi poet Muhsin Fānī. It has been conclusively proved that this attribution cannot stand, and it is now widely accepted that the author had two names: Amir Zu'lfiqār al-Ḥusainī and the poetic name Mūbad Shāh.[9] This double identity is also maintained in a new dictionary of Indo-Persian literature by Nabi Hadi, who unfortunately does not mention his sources for the following entry:

> *Mūbad Shāh, Zū'lfaqār Ardistāni* (d. 1081/1670) was a Husaini Saiyed from Ardistān, Iran, and emigrated in the reign of Shāh Jahān. . . . [H]e had spent many years of his life as a free-wandering *derwish,* staying in places of worship where shelter was offered. . . . The study of various religions seemed to be his chief passion of life. . . . In order to be discreet and careful, the

author abstained from disclosing his name in the preface of the *Dabistān-i-Mazāhib*.[10]

If this double nomenclature is accepted, then the question arises: Why did the author have two names, one Zoroastrian, one Muslim? Why the need for secrecy? A clue is offered in a casual comment the author makes as he is introducing the different religious sects of India. The quotation emphasizes his community's way of representing itself as well as the importance of secrecy in its religious beliefs:

> It should be known that, as has been mentioned, there are Samrādis, Khodānīs, Rādīs, Shiderangīs, Paikarīs, Milānīs, Alārīs, Shidābīs, Akhshīs, and Mazdakīs, who are in Iran and Turan, and all appear in the dress of Musalmans [*dar libās-i Musalmānī*], although in secret they follow the path of their own religion; in the same way various sects are established in India, but they do not wear the dress of Musalmans.[11]

For fear of persecution the various Zoroastrian sects had double identities in sixteenth- and seventeenth-century Safavid Iran. As a practical strategy they presented themselves as Shi'a Muslims in public and practiced their versions of Zoroastrian religion secretly. In so doing they were using against the Sh'ia Muslims the Sh'ia practice of adopting a public identity different from secretly held religious beliefs (*taqīya*, dissimulation).

The need for secrecy appears to have been a response to the historical situation of the Zoroastrians in Iran at the time. Mary Boyce sums up Safavid policy toward the minority during the reign of Shāh 'Abbās (r. 1587–1628) and the monarchs who followed him:

> Shah 'Abbas . . . had heard tell of the marvellous contents of the books of the "Gaurs" (Zoroastrians), and in particular of one rumoured to have been written by Abraham, containing prophecies of all events to come until the end of time. This he sought persistently, forcing the Zoroastrians to bring him their manuscripts. . . . Naturally the book of Abraham could not be produced, and in the end the king in his frustration had the Dasturan dastur (priest) put to death with several others. . . . After Shah 'Abbas matters grew only worse for them. 'Abbas II (1642–67) moved the Isfahani "Gaurs" to a new suburb (wanting the old one for a pleasure resort); and here under the last Safavid king, Sultan Husayn (1694–1722), they suffered terribly, for soon after his accession he signed a decree for their forcible conversion. A Christian archbishop witnessed the violent measures taken to enforce this; the Zoroastrian temple was demolished, large numbers of the "Gaurs" were

> compelled at sword-point to accept Islam, and the bodies of those who refused stained the river with their blood. A few escaped.[12]

Given these circumstances it is not surprising that Zoroastrians should have begun to migrate from Shiraz and Isfahan in the late sixteenth century and to settle in Mughal India. Moreover it was important for those remaining in Persia to practice dissimulation (*taqīya*) and to present a public identity as Shi'a Muslims to escape persecution and physical violence.

Mūbad Shāh was a follower of Āẕar Kaivān, head of an esoteric sect of Zoroastrians who emigrated from Shiraz around 1580 to the other end of the Persian-speaking world, Patna. As exiled intellectuals and religious leaders the Āẕar Kaivānī Parsis crafted a religious movement that claimed to precede Zoroaster while remaining heavily indebted to Zoroastrian texts and practices. In a pioneering essay on the cosmological frameworks employed by Āẕar Kaivān and his disciples to reread the history of pre-Islamic Iran, Mohamad Tavakoli-Targhi comments on the goals of the movement: "Responding to the threat of physical elimination, the architects of this neo-Mazdean intellectual movement wrote themselves back into history by projecting an Iran-centered universal historical narrative that subordinated the Biblical/Qur'anic mythistory to its own all-encompassing framework. In [their] generative texts . . . human history begins not with Adam, but with the pre-Adamite Mahabad."[13] The *Dabistān* and other key documents, such as the *Desātir*, the falsely archaizing law code of the sect, were produced by intellectuals in exile who used their cosmology and the religious beliefs of Mughal India to craft an intellectual renaissance. The group rewrote Zoroastrian belief and practice creatively and extensively, combining it with ethnographic information derived from their Mughal Indian context, to create this encyclopaedic work on the nature of religious belief.

The *Dabistān* throws into high relief the fundamental issues involved in the enterprise of comparison. As Jonathan Z. Smith has pointed out, "It is axiomatic that comparison is never a matter of identity. Comparison requires the acceptance of difference as the grounds of its being interesting, and a methodological manipulation of that difference to achieve some stated cognitive end."[14] It is this stated cognitive end, the tertium quid in terms of which the comparison is made, that is critical. All too often, as Talal Asad indicates, the category of religion is introduced here as an ahistorical essence, and the subsequent "theoretical search for an essence of religion invites us to separate it conceptually from the domain of power."[15] Although the "comparative-historical method" in its modern incarnation eschews the question of the "truth or falsity of any one religious position," Smith's analyses have been exemplary in demonstrating the ways illicit assumptions such as a pristine originary religion later corrupted or the "true" inner meaning of a doctrinal position have been imported into the arguments of scholars claiming historical objectivity.[16]

With these critiques in mind, we can begin to focus on the clandestine and open agendas of the *Dabistān*, a comparative text that makes its religious assumptions for the small group of Āẕar Kaivān's disciples. A division between exoteric practice and esoteric belief marks the agenda of the author of the *Dabistān*, which is presented throughout in terms of a surface ecumenism and tolerance. Mūbad Shāh gives us a detailed account of Āẕar Kaivānī genealogy and cosmology and the spiritual and ascetic regimen of his group. But he also collects information for comparison, explication, and the presentation of a comprehensive picture of the beliefs and practices of his day. He justifies his enterprise by presenting himself at the end of his work as a mere translator, advancing the claim of transparency and disinterestedness:

> It is well-known that there are five major religions: the Hindus, the Jews, the Magians (Zoroastrians), the Nazarenes, and the Muslims. Each of these five claim that their religion is the true one, and bring forward proofs for the confirmation of their own truth. . . . [T]he author begs to say that he undertook to write this book . . . after looking at all the different religions that came before his eyes. . . . [A]s to the account of each sect, the author wrote down all that had been told to him by the followers and sincere believers of that group, so that not even a trace of partiality or hostility may appear, and in this he performed nothing more than the office of a translator.[17]

This relativist position is, however, belied by Mūbad's use of his own group's beliefs as a standard of comparison. The selection of materials and arrangement of each section draws heavily on the structure of his Āẕar Kaivānī beliefs. Furthermore he and other members of his group disguise themselves as members of other religious groups while remaining Parsis in secret. The tension between the ultimate validity of his own group's esoteric beliefs and the pluralist account of religions that he constructs runs through the text of the *Dabistān*.

Understanding his enterprise requires a much more complex notion of selfhood than the Orientalist notion of finding the "true" author of the text. The clandestine author's journeys through the beliefs of other people reveal also his own beliefs, his own choices and preferences in situations of cultural and religious encounter. Since he is writing for his own small group of believers he begins with an exposition of the genealogy and cosmology of the beliefs of Āẕar Kaivān as the key to understanding religion. His freedom to enter and to explain different systems forces us to reckon with religious belief as a historically complex and interactive phenomenon, confounding Orientalist notions of the irrationality of Asiatic mores and the false credulity of the author of the *Dabistān*. Approaching the *Dabistān* requires filling in the conceptual map that provides the framework for the author's comparative enterprise, explicating his practices of information gathering, evaluating the interpretive strategies that allow him to process this information, and clarifying the relationship of this religious agenda to the political goals of the group.

Crafting the Self

Mūbad Shāh's account of Āẕar Kaivānī practices and of the "ancient" Iranian religion on which they are based involves an elaborate cosmology focused on the planets, the elements of creation, and the angels of the Zoroastrian cosmogony. The opening section of the *Dabistān* traces the history of Zoroastrianism back to the time of first creation. Mūbad Shāh claims that he will present the history of various sects, such as Samrādis, Khodānīs, Rādīs, Shiderangīs, Paikarīs, Milānīs, Alārīs, Shidābīs, Akhshīs, and Mazdakīs. In actuality he presents almost exclusively the beliefs of his own sect, which he calls Sipāsī in its "ancient" incarnation. In his selective account of cosmogony the principle of creation is the Lord (*k͟hudā-yi ta'ālá*), who is first manifested in the form of Āzād Bahman (the Vohu Manah, or "Good Intention," of the Pahlavi *Zend-Avesta*). In the ancient Pahlavi texts Ahura Mazda, or Lord Wisdom, of the Iranians is the supreme divinity who revealed himself and his six Amesha Spentas (bounteous immortals) to Zoroaster. These are held in opposition to a fiendish spirit called Angra Mainyu, who signifies falsehood (*drug*) in the ethical dualism embodied in Zoroaster's teachings.[18]

Mūbad Shāh uses this scheme to construct a godhead common to all the Zoroastrian sects, an almighty Lord who is omnipotent and perfect. The Lord's essence (*ẕāt*) is beyond human comprehension, and his divine will determines all events and actions in the world. According to Mūbad Shāh, the Lord sends the first principle of intelligence, Āzād Bahman, like a ray of the sun from the realm of light (Nūristān). From his light three more rays are emitted, and each in turn becomes an angel linked to a particular heavenly sphere. From Bahman are created all the angels that guard the planets, the elements, and all the spheres of the heavens and earth. Each of the four elements has its own guardian, and the human spirit or soul (*nafs*) has the inborn potential to become a "terrestrial angel" (*zamīnī serūsh*) or an evil spirit through good or bad actions. The text contains a theory of gradual transformation of the human being toward the realm of light through good actions, and of successively worse reincarnations as animals and lower forms of life as the result of bad actions.

The secret practices of the sect are aimed at improving the self's actions so that the person can progress toward the heavenly realm of light and away from the material world. There are several precepts laid down for the practitioner. First, there is to be no slaughter or eating of animals, in particular those animals that do not do violence to others, such as the cow and the horse. However, two loopholes are provided should the slaughter be carried out: the allegorical or esoteric explanation that the slaughter really means the killing of the animal passions of the self, and the karmic explanation that the cow or the horse had harmed a human long ago.[19] Implicit in this theory is the idea, common in seventeenth-century India but not in ancient Zoroastrian belief, of multiple births and existences. The human soul is continually reincarnated through animal, mineral, and botanical lives until it is perfected by carrying out the right actions.

What are these actions? A second set of beliefs has to do with the intake of food and drink, which can be manipulated to achieve the desired results. No alcohol or intoxicating substances should be drunk, for liquor inflames the passions and debases the human being until he reaches an animal state. This is at odds with the celebration of the intoxicating pounded plant *haoma* or *homa* (parallel to the Vedic *soma*) that is found in the ancient Pahlavi texts.[20] Further, according to the Āẕar Kaivānīs, the intake of food can be gradually reduced to receive the benefits of fasting. One of the disciples, Farzānah Bahrām, describes a simple procedure for fasting: "From his usual diet, the practitioner must daily reduce three direms, until he reduces it to the weight of ten direms; he should sit completely alone, and be engaged in meditating."[21] Mūbad Shāh goes on to place this reduction of diet within a larger five-point regimen of ascesis: fasting, silence, staying awake, solitude, and meditation on the divine being (*yazdān*). A number of practices are given of direct invocation, breath control, and physical postures that aid the disciple to transform himself. The practices of invocation are drawn from Sufi practices of *ẕikr* (repetitive meditation). One such practice is a direct calque of the common Sufi *ẕikr* of the negation and affirmation of Allah (*nafi-iṡbāt*). The disciple is to sit with legs crossed, in the position called the lotus seat (*padmāsana*) by yogis, and then do the following:

> With closed eyes and hands placed on his thighs, armpits open and back erect, he should throw his head forward. Then, he should bring the word *nīst* ["there is no"] up from his navel with all his force and raise his head. Then he should turn his head towards his right breast and recite the word *hastī* ["existence"]. Then, reciting the word *magar* ["but/except"], he should bring his head up straight; after this he recites *yazdān* ["God"] and bows his head towards his left breast, the seat of the heart.[22]

The complete formula thus works out to *nīst hastī magar yazdān*, "There is no existence except Yazdān," an adaptation of the Muslim declaration of faith, "There is no God but Allah." The Āẕar Kaivānī use is designed to establish their conception of godhead within the being of the disciple. Although this formula can be used publicly, Mūbad Shāh specifies that its inward meditation is the most effective. In its esoteric repetition three elements must necessarily be present: God, the heart, and the spirit of one's teacher.

After this the disciple is to go on to practice breath control and visualization. After specifying the number of times the disciple has to practice inhalation and exhalation, Mūbad Shāh lays out a scheme for visualizing the body as a stage for the disciple's progress. The scheme is laid out in seven stages or stations (*k̲h̲vān*) and is obviously modeled on the six *cakra*s or nerve centers of yogic practice: "first, the place where one sits; second, above the penis; third, the navel; fourth, the pine-heart; fifth, the windpipe; sixth, between the eyebrows; and seventh, the crown of the head."[23] These stages differ subtly from the yogic model by the inclusion of the windpipe, but the general idea, as in yoga, is that the disciple has to draw his breath up the spinal column through these seven stages. The person who can draw his breath up to the crown of the head through this seven-stage meditative procedure becomes, according to Mūbad, the successor (*k̲h̲alīfah*) to the Lord. At the center of the Āẕar Kaivānī practice of crafting the self is a conception of the subtle body that is immediately comprehensible in the context of seventeenth-century India.

The ultimate goal of all these practices of breath control, visualization, and bodily mortification is to enable the disciple to progress toward the realm of light. Along the way he hears what is called the *āzād āvāy*, or "free sound," within his body. Listening for this sound, which Mūbad specifies is called in Hindi *anāhad* (the "unstruck" sound) and in Arabic the *ṣaut-i muṭlaq* (the absolute sound),[24] was also a common practice among Sufis and the Indian devotional sects. Mūbad compares it to the sound of the bell (*jaras*) that signals the departure of a caravan, and states that one needs to be in complete solitude, at home or in the desert or in the dead of night, to hear it effectively.

Eventually a form appears between the eyebrows. The disciple must concentrate on this form and give himself up to it. After this he must meditate on the heart and leave aside all external things for the inner truth. Finally the searcher reaches the divine being (*īzad*) and contemplates him without the mediation of Arabic, Persian, Hindi, or any other language. The final goal of Āẕar Kaivānī practice, then, is a face-to-face and completely internal encounter with the supreme divinity. The truth of this encounter stands above language.

These secret practices, which form the core of the group's practice, are of a piece with the Āẕar Kaivānī interpretation of scripture and religious precepts. Mūbad Shāh's treatment of the canon of sacred texts detailing the life and principles of Zoroaster reflects similarly a consistent drive toward the esoteric. Zoroaster's life and teachings assume in his account a double meaning: a secret meaning known only to a select intellectual and spiritual elite, and a common exoteric meaning that is publicly available. This exposition of the faith rests on the assumption that the language of the *Zend-Avesta* is enigmatic and allusive and should not be taken literally.[25] The justification for this assumption is Mūbad's reframing of the *Zend-Avesta* as originally a double text: the greater *Zend*, "Mah Zend," and the lesser *Zend*, "Kah Zend." Modern scholars divide the *Zend-Avesta* itself into the archaic *Gāthās*, or hymns of Zoroaster, and the "Younger Avesta" composed by later followers and appropriators of the religion, so that the entire surviving corpus represents stages in the historical development of Zoroastrianism.[26] In Mūbad Shāh's account the greater *Zend* was the now lost original form of the scripture and stated clearly and in plain language (*ṣarīh va bī-ramz*) the laws of the sages. The lesser *Zend*, commonly known as the *Zend-Avesta*, was full of mysterious language and allusions to secret meanings (*ramz va ishārat*).[27]

Zoroaster's life becomes a series of tales to be mined for allusions to secret meanings, and they nearly always accord well with contemporary Āẕar Kaivānī practices. The well-known outlines of the prophet's life—his miraculous birth, his early life as a priest, his enlightenment in the river at the age of thirty, his conversion of King Gushtasp, and his promulgation of the faith he had been granted in the encounter with the supreme Lord[28]—are told through episodes of wonder working, interspersed with comments from members of Mūbad Shāh's group on the hidden significance of particular events. M. N. Dhalla provides a good example of this strategy in a story about Ahriman, or Angra Mainyu, the demonic force of evil:

> The legend that Ahriman appeared at a season festival in the guise of a glutton and devoured everything to the utter confusion of the assembly, until

> he was routed by preparing a dish from the flesh of a certain red cow, mixed with vinegar, garlic, and rue, at the instance of some miraculous advice, may be taken by the masses as literally true. But anyone versed in esoteric wisdom . . . knows that the killing of the red cow stands for the suppression of the sensual appetite, vinegar for the virtue of abstinence, garlic for reflection, and rue for silent meditation. All these would kill Ahrimanian propensities in man. The ignorant invest Ahriman with a personality; but . . . he has no independent existence . . . and is simply the negation of existence.[29]

Here Ahriman signifies only the physical appetites and passions that afflict mortals. According to Āẕar Kaivānī belief, these baser parts of the human being can be eradicated through their regimen for crafting the self. Anyone who can suppress his sensual appetite and engage in abstinence, reflection, and meditation can defeat Ahriman.

Mūbad Shāh rereads the body of Zoroastrian sacred literature from this enigmatic perspective. He describes the parts of the *Zend-Avesta*, the *nasks*, and intersperses them with a fictive account of the creation of the *Desātir* by the prophet Zoroaster. The text to which he devotes the most attention is the *Ṣad Dar* (Hundred Doors). The *Ṣad Dar* was a popular seventeenth-century Zoroastrian work in which the maxims of Zoroaster were arranged as the hundred doors leading to heaven or chapters on practice.[30] It was not originally a Pahlavi text but was composed in Persian in the two or three centuries preceding the *Dabistān* as a guide to correct practice for Indian Parsis.[31] The hundred chapters contain diverse sets of instructions on how to perform all the rituals, duties, and customs appropriate to the Zoroastrian community, backed up by some scriptural sanctions from the *Zend-Avesta*.[32] Mūbad Shah's text does not reproduce the *Ṣad Dar* fully but summarizes the contents of all the hundred chapters and provides a commentary on its meaning to his own religious community.

In his commentary Mūbad Shāh reflects on the advantages of the enigmatic interpretation of these precepts: "The substance of Zoroaster's maxims are hidden in enigmas and signs [*ramz va ishārat*] because with the mass of society [*'avām*] fabulous stories which are far from reasonable excite strong impressions. Moreover, if we tried to give an ignorant person knowledge of the existence and the independence from existence of God, he could not understand."[33] Thus the scriptures offer ordinary people, who are excited by fabulous stories, the chance to apprehend the substance of the laws. The elite few who know how to read esoterically can grasp the real import of the laws and their secret meanings. At odds with this urge toward enigmatic meaning is the broadly tolerant and ecumenical view of other religious groups, sources for many of the Āẕar Kaivānī practices of asceticism and self-mortification. The urge toward secrecy and the privilege accorded to an elitist view of religions are persistent impulses throughout the text that inflect and color Mūbad's way of approaching and interpreting other religions.

Encountering the Other

Based on the evidence in the *Dabistān* Mūbad Shāh (Saiyid Zu'lfiqār al-Ḥusainī) was born in 1025/1616 and traveled throughout northern India, Kashmir, Orissa, and parts of present-day Afghanistan until the year 1063/1652. As Simon Digby has pointed out, the Āẕar Kaivānīs' material circumstances remain obscure, though it is not impossible that they engaged in overland trade. For a small group, they were extremely highly placed. Mūbad Shāh met "influential Mughal officials and literary men, while important religious figures—among them Guru Har Gobind, Chidrūp Gosā'īṉ, and Sarmad—as well as an assortment of Catholic priests, Tibetan lamas, *sanyasis, bairagis* and Kashmiri and south Indian Brahmans were prepared to give time to his society and questioning."[34] But how did he and the other disciples of Āẕar Kaivān present themselves to these people? We know already that Mūbad Shāh and other members of the sect traveled under two different identities, Parsi and Shi'a Muslim. The variation in identities is most apparent in the travels of Parrah Kaivān, one of the inner circle of Āẕar Kaivān's disciples, who used to travel about in the guises of many different religious sects. He had encounters with members of each religious system:

> Parrah Kaivān, a Yazdānī, is one of the accomplished saints, and manifests himself in the guise of every sect. When in that of a Vairāgī, he was in Gujarat for the sake of a pleasure walk, and he saw some Vairāgīs [who had been branded on the arm from a Krishna pilgrimage]. He asked them, "Why this wound?" They responded, "It is the mark of Vishnu; whoever has it is recognized as his own by Vishnu." Kaivān Parrah said, "When the soul is separated from the body, they burn the corpse; no mark of it remains, but the soul is not perishable, and has no mark. How then will Vishnu recognize it?" When he came to Ahmedabad . . . he saw a muezzin who shouted out the call to prayer from the top of a mosque. When he came down, Kaivān Parrah asked, "Have you received an answer?" The muezzin said, "From whom?" Kaivān replied, "From him to whom you have been calling."[35]

The assumption of such religious disguises, while holding on in secret to the superiority of Āẕar Kaivānī belief, seems to have been characteristic of the sect. What is significant here is that the disguised Parsi presents himself, chameleon-like, in the dress of a group that takes him to be one of their own. But he is able to use the disguise to point out to them the error of their ways and to push them toward a perspective closer to his own.

On the surface the Āẕar Kaivānīs appeared to pass as Shi'a Muslims, generic holy men, or members of the religious groups that they encountered. They also endorsed the validity of every religious sect by focusing on the common truth and exhorting disciples to remove all religious distinction from their minds before engaging in meditation and visualization exercises. This tolerant and cosmopolitan surface covers a variety of interpretive strategies for judging all religious beliefs against the Āẕar Kaivānī system. Mūbad Shāh does not anywhere state his definition of religion in the *Dabistān* but describes sects as cohesive groups somewhat like his own: the followers of a particular religious leader, historical or contemporary, or a community that has a revealed scripture. After his extensive account of the beliefs of his own group he gives us eleven more lessons, one devoted to each major religious group: the Hindus (a section devoted to all the Indian sects), the Tibetans, the Jews, the Christians, the Muslims, the Ṣādiqiyya (an early breakaway Shi'a group, followers of a prophet named Musaylima), the Vāḥidiyya (followers of another Muslim holy man, Vāhid Maḥmūd), the Roshaniyyah (a militant Afghan sect organized by a Miyān Bāyezid in the reign of Akbar), the Ilāhiyya (the religion devised by the Mughal emperor Akbar), the wise men or philosophers of every religious tradition, and the Sufis. Each lesson is divided into *naẓar*s (insights) into the smaller sects which Mūbad Shāh groups within each major religion.

Mūbad Shāh's comparative project uses all the strategies that are available to the interpreter who must work at the borders of disparate religious worldviews. When he takes up the larger social collectivities that represent themselves forcefully around him, he is challenged by the radical differences encoded in their beliefs. His responses describe an arc from similarity to incommensurable difference and include sociological exegesis, causal explanation, translation and interpretation of scripture, and finally undecidability between competing truth claims. How is one to make sense of Mūbad Shāh's attempt to write about every religion that he encountered or could know from books? His primary strategy seems to be classification, from the larger twelve-fold division into major religions to the smaller "insights" into particular sects. Classification for Mūbad Shāh *is* interpretation. He breaks up the information that he collects into a very specific set of topoi, mostly recognizable as Āẕar Kaivānī in provenance. Within each subsection he tries to focus in turn on a group's description of the godhead, their account of the creation of the world, the physiological division of the world into elements, the origins of the religion and its leaders, its stages of evolution, its cosmology and units of time, any scriptural or textual evidence, anecdotes of members of the sect whom the author met, and exegetical solutions to understanding the information presented. When the information available is too scanty to allow him to follow this procedure, he contents himself with whatever he can glean and his judgment of the value of the belief or practice.

Mūbad Shāh's framework for understanding other religions can thus be boiled down to a set of comparative standards that reflect his own background and the Āẕar Kaivānī predilection for esoteric meaning. First he needs to determine whether or not a group believes in divinity, which he understands as absolute truth (*ḥaqq*). His definition rests on the Āẕar Kaivānī distinction between essence and appearance, and if there is any way in which he can read a group's description of their notion of divinity as conforming to a common essence, he will do so. Second, he wants to know if they have any prophets or religious leaders so that he can determine the nature of religious authority within the group. Third, he focuses on the group's sources of belief, with a special emphasis on whether or not their religion is revealed. In his definition a revealed religion possesses a heavenly (*āsmānī*) book with an ancient pedigree and the special linguistic features of a scriptural text. The *Desātir*, the Āẕar Kaivānīs' own heavenly book, is written in an archaizing language that is based on Persian. If the group does have such a scripture, Mūbad Shāh translates the parts to which he has access and which appear to express important tenets of belief.

Next he provides us with a causal explanation for the origin of a particular faith. The cause can be historical or genealogical and reflects the Āẕar Kaivānīs' interest in rewriting the past to explain their own position and to express their political stance in relation to the Safavids. Another sort of explanation of religious belief is his sociological exegesis of the current activities of a group. Here he focuses on whatever information he can get about their strength, their activities, their status, and even the hierarchical structure of the group. However, there are cases where he comes up short against the problem of religious incommensurability and the impossibility of deciding between conflicting truth claims. This is where his system for comparison founders, but not before teaching us a great deal about strategies for making religious distinctions.

Let us see how this system for comparing religions works in practice. The second section of the *Dabistān* treats the religious beliefs of the Hindus. The author treats all the religious sects that originated in India; in Mūbad Shāh's usage, the term *Hindu* does not mean a member of an organized unified religion called Hinduism, in which sense it is sometimes misunderstood at present. Instead *Hindu* seems to have a geographical valence, that is, the religions and sects of the people of India (*ahl-i Hind*), which are innumerable. This section is divided into twelve chapters: the Pūrva Mīṃāṃsā school of philosophy, Puranic accounts of creation, the Smārtas, Vedānta, Sāṃkhya, the yogis, the Śāktas, the Vaiṣṇavas, the Cārvākas, the school of logic (Tarka Śāstra), the Buddhists, and the other systems of belief in India. The selection is aimed at comprehensive coverage and is the result of fairly extensive reading, interviewing, and observation. Mūbad includes the six standard *darśana*s, or schools of philosophy, the cosmogony of the Puranas, the worshippers of Viṣṇu and Śiva-Śakti, the legal or orthodox school, an ancient materialist sect, the Buddhists, and miscellaneous religious groups contemporary to him (under which he includes the *bhakti* or devotional sects and the Sikhs).

The organization of these sections follows roughly the comparative method and the Āẕar Kaivānī topoi sketched out above. Thus the first chapter makes the followers of the Mīmāṃsā school of philosophy into the analogues of the ancient Zoroastrian sages that the Āẕar Kaivānīs venerated.[36] The school was based on the careful reading of the Vedic corpus and isolated the principles of logic, inference, and causality out of the materials on ritual sacrifice. Mūbad Shāh's synthetic presentation of their doctrines does not correspond to any standard picture of the school. First he notes two points of difference from his own religion: "The entire world is not ruled by the orders of a real Lord, and he has no permanence or true existence. Whatever good or evil, reward or punishment, happens to created beings, is entirely the result of their actions. . . . Without action, there are no effects of action."[37] He does, however, understand that the school distinguished between actions enjoined by the Veda (which were to be performed) and those forbidden in the sacred texts (which were to be avoided). He goes on to explain the origins of the gods Brahmā, Viṣṇu, and Śiva through his own theory of action: they were only human beings who reached their divine status through right action. If a human being acted like Brahmā he could eventually reach Brahmā's station and displace him for the next age. This is a characteristic kind of causal explanation for Mūbad Shāh and recurs frequently. Regarding action, he seems to be laboring under a misconception: while the Mīmāṃsakas did focus on action, it was ritual action and its consequences that interested them, not a general theory of *karma*. According to Mūbad Shāh, for the Mīmāṃsakas action led to advancement to higher states or degradation, and except for their lack of belief in the one true Lord they were to be respected as spiritual seekers. He concludes that they were very much like his own Yazdānī, or "ancient," school.

Once he has discovered ancient philosophers in India, Mūbad Shāh turns his attention to their account of creation, so important to his own Zoroastrian sect. He draws on the second chapter of the tenth-century *Bhāgavata Purāṇa,* a text widely used by Vaiṣṇava sects to present the ten incarnations of Viṣṇu. The strategy he uses to understand cosmogony is familiar from his account of Zoroastrian belief. He gives us a detailed account of creation and the incarnations, then a set of esoteric meanings that are attached to the events of creation or the various forms of Viṣṇu. He gives to the progenitor of the world the name *ḥaqq,* the absolute truth, rather than Brahmā or another divinity from the Puranas. In his view the various worlds (*lokas*), incarnations, and created beings are manifested in contingent form (*mumkināt*) and cannot share in the essence (*gauhar*) of the creator.[38] The creator deems incarnation necessary so that created beings can know him and satisfy their desires and natures. As a reading of the notion of *avatāra,* this seems far-fetched because of the distinction Mūbad makes between human and divine essences.

He proceeds to give us the secret meaning of incarnation according to Shidosh, one of the major disciples of Āẕar Kaivān. Shidosh uses a Sufi gloss to explain that "the first intellect is the knowledge of Allah, and the universal soul is the life of Allah, and here they have used his attributes and distinguished between them. Brahmā is his creative power, and when they mention his old age, this just implies his perfection. . . . Viṣṇu means the attribute of divine love . . . and the *avatāras* are rays or reflections of divine essence."[39] Thus the living power of God and his various attributes are represented in concrete form in individual gods. But Mūbad Shāh is not entirely satisfied with this explanation of the meaning of *avatāra* or incarnation and asks Shidosh what the esoteric meanings of Viṣṇu's incarnation as a fish or a tortoise could be. They decide that by the fish, the Hindus must mean the guardian angel of water, and by the tortoise, the angel of the earth. The boar must mean animal passions, and when the waters of lust submerged the earth Viṣṇu as the boar conquered with the tusks of goodness a demon signifying iniquity.[40] They conclude that the Indians believe that every class of being has its own angel, with a female energy or wife. They worship these angels, but the true object of their worship is the Lord and not these created beings. To worship the images of all these gods and goddesses is to participate in the rituals of an angelology akin to Mūbad Shāh's own Zoroastrian angelology.

Mūbad Shāh has thus far painted a picture of Indian religions using many of the principles of the Āẕar Kaivānīs to discover the esoteric meanings of the faiths he encounters. The source for his philosophical speculations may be a work covering the different *darśanas* or systems of Indian philosophy (a *sarva-darśana-saṃgraha*), though he does not cite an explicit text, translation, or interpreter. He is motivated by a double agenda: to include as much diversity of belief as he can encompass within his classificatory scheme and to discover the truth of all these systems. Mūbad Shāh himself follows his spiritual guide's lead by describing Āẕar Kaivān's non-Zoroastrian disciples as *yagānah-bīn*, "seers of the one (truth)."[41] In this sense his work is not just a valuable account of the religions of seventeenth-century India, but also a journey in search of the single standard of comparison that is embodied in the Āẕar Kaivānī notion of the one truth.

It is not surprising that he finds monistic philosophical systems most sympathetic to his own view of religion. His presentation of Vedānta comprehends very well the bases of Śaṅkara's *māyā-vāda* (doctrine of the illusory nature of the world).[42] He begins by calling the Vedāntins Sufis and searchers into truth (*muḥaqqiq*), then defines their religion as focused on a truly existing Being who is knowledge in essence. The world is a likeness or exemplification (*tamṣīl*) of that divine knowledge, which he describes as "the ruler of lives and the greatest of the souls, the most sacred essence; these are his qualities. They call that sacred essence and that greatest existing being the *paramātmā* [the supreme spirit]."[43] He is sympathetic to the Vedāntic interpretation of the *parama-vākya*, or supreme sentence of the Upaniṣads, *tat tvam asi* ("You are that"), as the assertion of the identity of the subject and object of mystical practice, quoting Sufi verses from Shabistarī's *Gulshan-i Rāz* in support.

Mūbad Shāh moves beyond scriptural and philosophical interpretations to present actual ethnographic encounters and eyewitness accounts of religious practices and customs. Thus he describes meetings with eminent personages as well as descriptions of sects less familiar to him, such as the Śākta worship of Śiva and the goddess that he witnessed in eastern parts of India. Here he gives a connected account of the six *cakras* or nerve centers of the body in Śākta or Tantric practice, a structure that is roughly homologous with the Āẕar Kaivānī conception of the zones of the body. He describes the goddess as enthroned at the place in the head above all of these, and then details the Tantric procedures for propitiating her. These are divided into external and internal worship and include details of sexual and other practices such as the consumption of fish, flesh (both animal and human), alcohol, and cannabis. Despite their dissimilarity to his own sect's ascetic regimen, he presents for the most part as fair an account as he can manage. Indeed Mūbad Shāh makes a concerted effort in all the sections of the *Dabistān* to present the social and ritual life of contemporary religious groups. The categories he uses are drawn from his own Āẕar Kaivāni group; this helps him if there is a historical founder or leader of the group, a textual tradition, and ascetic practices such as fasting or breathing exercises.

For instance, in his sympathetic account of the Sikhs he presents the life of Guru Nānak, the Sikh theological and religious vision, travels with Guru Har Gobind, and the Sikh doctrine of the reincarnation of Nānak in the person of the guru. The life of Guru Nānak draws on information provided to him by members of Guru Har Gobind's entourage and thus repeats episodes from the birth stories (*janamsākhīs*) of Nānak. He begins with Nānak's basic views on religious diversity, coinciding with the Āẕar Kaivānī position: "Although Nānak praised the Muslims, and also the *avatāras* and gods and goddesses of the Hindus, he knew them all to be created beings and not creators; he denied their descent and their mingling with humans. They say he had the string of beads [*tasbīh*] of the Muslims in his hand and the sacred thread of the Hindus around his neck."[44] Here Mūbad Shāh presents Nānak's acceptance of the symbols and practices of both Hindus and Muslims, but also his denial of the external truth of these faiths in favor of the composite religion based on inspiration and devotion that he devised. As an Āẕar Kaivānī it makes perfect sense to him to adopt the exoteric symbols and stories of a religious group while crafting his own set of beliefs.

As an ethnographer Mūbad Shāh presents sympathetically the theological doctrine of the successive reincarnations of Guru Nānak into the persons of the succeeding Gurus and something of the Sikh use of mythological materials from sources such as the *Yoga-Vasiṣṭha*. As a member of an organized and devoted religious group himself he is also able to compose a shrewd sociological account of the religious organization of the Sikhs into *masnads*, or regional centers, and of the tithes collected by them and in general of the material base of the group. He is alert to the caste base of the group among the Jats, as well as to the social division of labor among the Sikhs between commerce, farming, and various occupations. Additionally he is able to present the succession struggles in the community and describe the military strength and readiness of Guru Har Gobind's camp. He knew Har Gobind personally and describes several meetings with him. As an ethnographer it is easier for him to assimilate the social reality of a small group of persecuted religious leaders and their attempts at presenting a coherent theology.

Mūbad Shāh's difficulties arise not when there is a similarity between the social organization of the group on which he is reporting and his own group of Parsi holy men, but when he has to represent radical difference. When he comes to the Cārvāka (materialist tradition of Indian philosophy), he misunderstands them to believe in the constructed nature of all reality, the Buddhist notion of *skandhas*. He presents this idea as best as he can, but then founders on the absence of any living principle to worship or any maker of the universe. He is unable to say much except to present their scathing critique of Vedic sacrifice and to report the Cārvāka mockery of Brahmanic religion. For example, on seeing a Brahman with a sacred thread around his neck, the Cārvākas say, "The cow cannot be without a rope."[45] The difficulty Mūbad has with the different religious emphasis of the Cārvākas points to two further topoi in his enterprise that are problematic: strategies for translation and strategies for thinking about a standard for measuring incommensurable religious systems.

The first crux, how he translates different religious systems into Persian, is best illustrated by a concrete example. The section on Judaism contains a fairly accurate translation of Genesis 1:1–4:8, framed by the knowledge he obtained about the Jews from his friend Sarmad, a Jewish *faqīr* from Iran. Sarmad had fallen in love with a Hindu boy named Abhay Chand and camped at his door until the boy's parents hid him away from Sarmad. As a consequence, "a sense of madness over-powered him so [that] he took to nudity. After some time the boy joined Sarmad. . . . Long hair had grown over his body and long nails on his fingers."[46] They traveled widely in Iran and India in search of illumination and were close friends of Mūbad Shāh. Sarmad's famous Persian quatrains express some of his mystical views about the negation and affirmation of the godhead in passionate verse.[47] He was ultimately executed in Delhi in 1660 or 1661 for his extreme religious views by order of Emperor Aurangzeb.

While Mūbad's translation follows the book of Genesis closely, the interpretation of Judaism that he develops from the text is idiosyncratic. He quotes Sarmad as authoritative on the following points about Judaism:

> According to the Yahūds, God the Almighty is corporeal and has a body. His body is like that of mankind, and in its image; and after a while he is scattered just as a ray of light disappears. In the Mosaic Torah and Holy Book it is written that the spirit is the subtle form of the body, and that in human form it is the locus of divine manifestation, and that punishment and reward are experienced in this form. Human life lasts 120 years; man's whole life may be considered one day, which, when he dies, is followed by night; his body assumes partly the form of a mineral, partly the form of a vegetable, and partly that of an animal; when 120 years have passed, night comes

> to an end and morning shines forth; if an atom of his bodily dust be in the east and another atom in the west, they unite in one place, and life is renewed for another hundred years, as we have said, when night falls again.[48]

Mystical Jewish sects with esoteric interpretations of the holy books are not unknown, but this passage clearly has more going on under the surface. Sarmad's idea that the Tetragrammaton is anthropomorphic, and the corollary that God is to be approached through the body, has obvious links with Sufi ideas about God's beauty and possibly Sarmad's own mystical and romantic involvement. God's creation of Adam and Eve in Genesis can be interpreted through this anthropomorphic lens and justifies Sarmad's practice of mystical absorption.

Mūbad Shāh had Sarmad and Abhay Chand check the Persian translation against the original text to ensure its accuracy. Its exegesis, however, is another matter. Since there was no exegetical community present to check the collective interpretation of Sarmad, Abhay Chand, and Mūbad Shāh, it stood unchallenged as a Zoroastrian view of Judaism. Their enterprise implies a larger problem in the interlinguistic transfer of meaning: How is one to find adequate synonyms for translating metaphysical concepts and complex belief systems? Since the translator has to come to terms with the authority, or lack thereof, of the exegetical community that frames the text for him, the larger implication is that exegetical communities determine the meanings of a religious scripture. Clearly the fixing of the terms of synonymy in translation is linked to this determination of meaning. At stake is the place of a religious community in the world, as well as the authority of the exegetes, issues that surface in their most radical form when there are many competing claims to truth within the same historical context.

The translation and its exegetical frame reflect also Mūbad Shāh's attempt to find his own religion in Sarmad's views. The idea of animal, vegetable, and mineral reincarnation bears a suspicious resemblance to the Āẕar Kaivānīs' own theory of action and reincarnation. Certainly Sarmad's renunciation of the world and practice of absorption are very similar to the Āẕar Kaivānī regimen. Mūbad's encounters with difference are also confirmations of his own beliefs, using the key of esoteric meaning. His practice of carrying out ethnography is validated by his attempt to translate all religions into a truth beyond linguistic divisions. The ultimate sanction for this stance is the Āẕar Kaivānī mystical goal of absorption in divinity, represented as existing in a space beyond Arabic, Persian, Hindi, or any other language. Yet there is a curious tension throughout his enterprise, for according to the Āẕar Kaivānīs, they already have the secret key to interpreting religion. What then is the need for constructing the unique and encyclopaedic account of all religions that constitutes the *Dabistān*? And what happens when members of competing religious systems actually encounter each other? Whose secret key is the most valid?

One of the last sections of the *Dabistān* is devoted to the Dīn-i Ilāhī, the syncretic state religion started by the Mughal emperor Akbar. Mūbad Shāh gives what is supposed to be a verbatim report of the debates at Akbar's imperial court, although these debates took place several years before his own birth. In these staged theological debates (*munāẓirah*) members of various religious sects appear before Akbar, who is himself presented as the *Khalīfatullah,* or successor to God, the Messiah of the new Islamic millennium. A typical dispute is as follows:

> A learned Brahman from the Deccan, Shaikh Bhāvan by name, conceived an enmity towards his own people, became a Muslim and obtained this name. He had the fourth Veda with him. In some of the precepts in that book, it appeared that the letter *lām* appeared many times and therefore it appeared like the declaration *lā-ilah ill-Allah* ("There is no God but God"). There was also the statement that whoever does not declare this, does not obtain salvation. In another place it was written that it was permissible to eat cow's flesh under certain circumstances, and also that a corpse should be buried rather than burnt after death. Thus, the Shaikh was triumphant over the Brahmans. But Nāin Javet (an Āẕar Kaivānī) related that he asked him to translate this passage, and that its meaning proved to be contrary, and the opposite of the sentence, "There is no God but God." The restriction against cow's flesh was also contrary to the custom of the Muslims, and about the burial of corpses he gave a different account from that which is lawful among the Muslims. The emperor and all the others laughed at the Brahman, and his majesty said, "Look at these Muslims and Hindus, who among all these arguments did not think to ask what was the meaning of this passage," and praised Nāin Javet exceedingly.[49]

While this is one of the briefer debates, it shows the existence of some critical standards to judge the religious claims of the men who came to Akbar's court. It is one of a long series of disputes between Shi'as and Sunnis, Christians and Muslims, Jews and Christians, and all the Indian sects. The topics include claims to divinity, the validity of miracles, incarnation, and prophecy, and ultimately the truth value of any religious system. The Āẕar Kaivānī interventions underscore their tendency to point out the logical inconsistencies in others' beliefs and to claim the superiority of their own secret knowledge. The final arbiter in these debates is Akbar himself, who in most cases dismisses the disputants. This is because they reach an aporia that cannot be resolved without addressing the lack of a common yardstick to measure all religions.

Here the author comes face to face with the incommensurability of religious beliefs but does not solve the problem decisively. Emperor Akbar's solution is to put forward his own eclectic state religion, an imperial cult using vegetarianism and sun worship and focused on accepting the emperor as a spiritual guide and the perfected man (*insān-i kāmil*). Mūbad Shāh does not adopt this as a final solution, even though some aspects of the Dīn-i Ilāhī are close to Zoroastrian practice. Instead he ends his consideration of all the major religious systems with a double move. On the one hand he confirms the truth of each of the five major religions—Hinduism, Judaism, Zoroastrianism, Christianity, and Islam—and on the other he represents himself as an unbiased translator,[50] a claim that his text undoes on virtually every page. Although he does not resolve the problem of incommensurability among truth claims, his account opens up interesting critical possibilities. Since he accepts that there is no adequate synonymy between religious concepts, there are no right or wrong translations, only more or less adequate ones. His avoidance of fixity and his presentation of plural worldviews means that many correct translations are possible, but not a uniquely true one. Yet he has insisted throughout on the single truth of all religions as the yardstick of comparison. What knowledge formations enable this seemingly self-contradictory position, and how are they related to our understanding of comparison? To begin to answer these questions, we must turn to the larger historical scene within which the *Dabistān* is articulated and to the political agenda of the Āẕar Kaivānīs.

The Future According to Āẕar Kaivānī

We have come back squarely to the issue of comparison, which in its modern guise claims objectivity but smuggles in a clandestine and ahistorical essence of religion as the comparandum. Mūbad Shāh performs the opposite move, which is to disguise his name and identity but to present as self-evident the oneness and equality of all religions. If all religions are equal and one, then there is no contradiction in digging further for the truth and presenting the hidden meanings that underlie religious discourse. It is clear that these declarations of ecumenism and religious unity are part of the Mughal discursive scene, the larger assumptions taken for granted by all the historical agents concerned with problems of the definition of truth, religious community, and appropriate spiritual action. Unfortunately they have been taken at face value by modern historians of religion. As the *Dabistān* amply demonstrates, one has merely to scratch the surface to find much more complex phenomena at play. Instead of a single equivalence that sanctions the translation of religious belief, we must be prepared to posit multiple equivalences that undercut the universalizing language of comparative religions. The disguised ethnographer must erase himself from his own encyclopaedia to translate the truths of other religious systems, but the assumptions and frameworks of his core "technology of the self" are not so easily jettisoned. Mūbad Shāh does not depart from the orthodoxy of his sect, and his comparative method is not divorced from the marginal political stance of the group.

Āẕar Kaivānī metaphysical and social attitudes were directed toward upholding a system of ascesis, based on their hidden key to religious belief. Far from embodying the Orientalist stereotypes of credulity, irrationality, and superstition, the group created a system of mortification and esoterica that was a complex response to their own historical situation. They constructed a pre-Islamic genealogy for their group that rewrote history to glorify their religious movement in exile, claiming an anteriority and superiority to all the religions of Persia and India. This "pristine religious origin," jealously guarded by the members of the sect, justifies the comparative enterprise of the *Dabistān*. It is clear from the text that the Āẕar Kaivānīs saw themselves as inheritors of a civilization that was wiped out by a religion of the black-clad and foretold a political second coming that would free them from their shackles. In their version of the life of Zoroaster, Mūbad Shāh describes an elaborate dream vision of the prophet's in which he saw (among other things in a journey through Hell) a tree with seven branches.[51] Each branch, signifying an age of the world, was made of a different metal, ranging from gold and silver to tin, steel, and an alloy of iron. The seventh branch, made of iron alloy, signifies the current age of affliction.

During this seventh age, at the end of a period of just rule, royal power passes into the hands of the heretic Mazdak, betrayer of the true faith. Then a "group of the black-clad people, oppressors of the poor, without name, honour, or excellence, friends of tumult and evil, deceitful and hypocritical and cheating, bitter-hearted as aloes but with honey on their tongues, dark-faced betrayers of bread and salt . . . will destroy the fire-temples. . . . The sons and daughters of the free nobles will fall into the hands of this group, and the virtuous and powerful will become their servants."[52] The remedy for this calamity and for the extinguishing of the fires of the true faith is a future king of the ancient lineage of Kay. The dialogue between Zoroaster and the Lord goes on to describe a number of natural disasters, such as fire, flood, and storms, that would engulf the world. These disasters would herald the coming of a mighty king named Bahrām, who would gather together the armies of India and China and liberate the Persian homeland from the tyranny of the black-clad Muslims. His ascension to the throne would also mean the rekindling of the sacred fires and the revitalization of the ancient institutions described elsewhere in the *Dabistān*.[53]

This utopian vision cannot be separated from the ascetic regimen of right action articulated by the holy men of the group. Driven to India by the political situation under which the sect had to labor in Iran, the followers of Āẕar Kaivān modified their beliefs in accordance with their environment. As Mūbad Shāh puts it, for the ancient Sipāsīs going insane, suffering distress for one's young children, and suffering calamities are the result of one's own past character and actions. However, he spells out what the sect thought about undeserved punishment: "But when these things happen to a wise man, if it seems undeserved, it comes from the tyranny of the earthly king; the king of the time to come will question him about his actions."[54] Not all bad things that happen are deserved, and tyranny need not be accepted fatalistically. The phenomenon of unjust oppression, such as the Safavid targeting of the religious minorities, allows also the articulation of a military and political agenda through the account of Zoroaster's visionary journey.

Unrealistic as it may seem within the imperial cosmopolitanism of the Mughal scene, Mūbad Shāh's dream of what could be serves also to sharpen our own understanding of tolerance as characteristic of the period between Akbar and Aurangzeb. Mūbad Shāh's strategies of distinction—naming, describing, classifying, judging, translating, and polemicizing—are found in a range of contemporary materials that also try to construct what's around them into an orderly universe. Encyclopaedism is a response to a multireligious scene, not a dualistic one, in which hierarchies of judgment, taxonomy, and exclusion coexist uneasily with a surface ecumenism that has been too easily accepted by modern historians of religion. The ecumenical claim of the unity of all religions thus constantly undoes itself in the face of irreducible differences between the truth claims of large religious systems. However, the claim also creates a space in which the self can demonstrate itself, can discover what otherness means in the trajectory of self-discovery. Until we can understand how these pluralist perceptions of selves and others work, we cannot make larger generalizations about the shape of belief and practice in a historical period, nor indeed understand what tolerance really meant in the intellectual and religious culture of an age.

Notes

A longer version of this essay appears in David Haberman and Laurie Patton, eds., *Notes from a* Maṇḍala*: Essays in the History of Indian Religions in Honor of Wendy Doniger* (Newark: University of Delaware Press, 2010). I am grateful to Peter Brown, Simon Digby, Stephen Dunning, Anthony Grafton, and Christian Novetzke for their insightful and probing responses to different versions of this essay; they have helped me to frame the argument more effectively and sharply. I am also grateful to the Shelby Cullom Davis Center for Historical Studies at Princeton University for supporting the research and writing of this paper, as well as for the year-long intellectual fellowship which made it possible.

1. On religious beliefs, see, for instance, Rizvi, *Religious and Intellectual History of the Muslims in Akbar's Reign,* or his *Muslim Revivalist Movements in Northern India in the Sixteenth and Seventeenth Centuries.* Good discussions of art and architecture include Beach, *The Imperial Image*; Welch et al., *The Emperor's Album*; Brand and Lowry, *Akbar's India.* Some classic instances on individual figures are V. A. Smith, *Akbar the Great Mogul*; Naik, *'Abdu'r-Rahīm Khān-i-Khānān and His Literary Circle*; Alam Khan, *Mirza Kamran.* For a nuanced discussion and detailed references on Akbar, see the introduction to M. Alam and Subrahmanyam, *The Mughal State.* This anthology of representative essays also contains a more extensive bibliography than it is possible to give here. On Aurangzeb, see Sharma, *The Religious Policy of the Mughal Emperors.* More recent studies that attempt to displace the historiography of decline include M. Alam, *The Crisis of Empire in Mughal North India*; Bayly, *Rulers, Townsmen, and Bazaars*; Gommans, *The Rise of the Indo-Afghan Empire.*

2. Most recently by Amartya Sen, in his *The Argumentative Indian,* but the notion of Mughal India as a tolerant place has a long genealogy, in part rooted in historical materials from this period.

3. *Dabistān-i Maẓāhib,* ed. Malik, 1:4.

4. Sir William Jones, "Letter to John Shore, 24 June 1787," in Cannon, *The Letters of Sir William Jones,* 739.

5. *Dabistān-i Māẓāhib, editio princeps*; *The Dabistán, or School of Manners.*

6. Al-Birūnī's *Kitāb al-Hind* is translated by Edward C. Sachau as *Alberuni's India.* Compare Habibullah, "An Early Arab Report on Indian Religious Sects"; Jeffery, "Al-Bīrūnī's Contribution to Comparative Religion"; Jahn, "On the Mythology and Religion of the Indians in Mediaeval Moslem Tradition"; Friedmann, "Medieval Muslim Views of Indian Religions." On Shahrastānī's *Kitāb al-Milal wa'l-Niḥal,* see Lawrence, *Shahrastānī on the Indian Religions,* and more generally Waardenburg, *Muslim Perceptions of Other Religions.*

7. *Dabistán, or School of Manners,* clxxxv–clxxxvi.

8. Tavakoli-Targhi, "Orientalism's Genesis Amnesia," 1.

9. For the full working-out of this scholarly controversy, see the longer version of this essay in *Notes from a* Maṇḍala. [Ed. A hitherto unknown manuscript of the *Dabistān* has recently been found by Dr. Karim Najafi of the Iran Cultural House (New

Delhi). It was completed in 1060 Hijri (1650 CE), and carries marginal comments in the hand of the author himself. Although this manuscript (like the others) does not contain the author's name in the preface or introductory sections, there are several clear indications of it spread through the text. It establishes that the author was named Mirza Zu'lfiqar Azar Sasani, known by the penname of "Mubad" (we find no Husaini, no Ardistani, and not even a Shah, and Amir becomes Mirza). While the disciple for whom the manuscript was prepared, Majduddin Muhammad, was a Muslim, it remains possible that Mirza Zu'lfiqar his master was either a Zoroastrian or felt close to that faith. The elevated phrases and terms used to describe him suggest that Mirza Zu'lfiqar had the reputation already in his lifetime of a great scholar and a near saintly figure. (I thank Muzaffar Alam and Sanjay Subrahmanyam for this information.)]

10. Hadi, *Dictionary of Indo-Persian Literature*, 360–61.

11. *Dabistān*, 1:188. All translations from the text are my own, using the R. R. Malik edition.

12. Boyce, *Zoroastrians*, 181–82.

13. Tavakoli-Targhi, "Contested Memories," 150.

14. J. Z. Smith, *To Take Place*, 14.

15. Asad, *Genealogies of Religion*, 29.

16. Smart, "Comparative-Historical Method." See in particular J. Z. Smith's 1988 Louis Henry Jordan Lectures in Comparative Religion at the School of Oriental and African Studies (University of London), published under the title *Drudgery Divine: On the Comparison of Early Christianity and the Religions of Late Antiquity*.

17. *Dabistān*, 1:367.

18. Gershevitch, introduction, 9.

19. *Dabistān*, 1:23.

20. See Boyce, *Textual Sources for the Study of Zoroastrianism*, 55–56. For a comprehensive discussion of the controversial identity of the *haoma* or *soma* plant, see Doniger O'Flaherty, "The Post-Vedic History of the Soma Plant."

21. *Dabistān*, 1:27.

22. *Dabistān*, 1:27.

23. *Dabistān*, 1:28.

24. *Dabistān*, 1:28.

25. *Dabistān*, 1:111.

26. For a summary account, see Nigosian, *The Zoroastrian Faith*, 46–70.

27. *Dabistān*, 1:112.

28. For a scholarly account, see Boyce, *A History of Zoroastrianism*, 181–91, as well as Herzfeld, *Zoroaster and His World*, vols. 1–2. For a fascinating ancient textual fragment, the Cow's Lament on the birth of Zoroaster, see Malandra, *An Introduction to Ancient Iranian Religion*, 35–39.

29. Dhalla, *Zoroastrian Theology*, 315.

30. See "Sad Dar," in West, *Pahlavi Texts*, 253–361.

31. For the different versions of the *Ṣad Dar*, as well as the dating of the manuscripts of the work, see West, introduction to *Pahlavi Texts*, xxxvi–xxxvii.

32. For Zoroastrian rituals and practices, see Modi, *The Religious Ceremonies and Customs of the Parsees*; Choksy, *Purity and Pollution in Zoroastrianism*.

33. *Dabistān*, 1:111.

34. Digby, "Some Asian Wanderers in Seventeenth Century India," 254. For a detailed itinerary, see 254–55.

35. *Dabistān*, 1:180.

36. There is a slight terminological confusion here. Mūbad Shāh gives the name of this school as Būdh Mīmāṃsā, which could be a reference either to the Pūrva or Uttara Mīmāṃsā schools. The translators Shea and Troyer construe Būdh as *būdah* (was) and hence translate the name as Pūrva Mīmāṃsā, the "earlier" school, but I am not convinced. Buddha Mīmāṃsā makes no sense as the name of a school of philosophy. I have chosen therefore to render it simply as Mīmāṃsā. For the history of the Mīmāṃsā school, see Dasgupta, *A History of Indian Philosophy*, vol. 1.

37. *Dabistān*, 1:122.

38. *Dabistān*, 1:127.

39. *Dabistān*, 1:128.

40. *Dabistān*, 1:129.

41. *Dabistān*, 1:45.

42. For an account of Śaṅkara and the Advaita Vedānta school, see Potter, ed., *Encyclopedia of Indian Philosophies*.

43. *Dabistān*, 1:147.

44. *Dabistān*, 1:198.

45. *Dabistān*, 1:183.

46. Rizvi, *Muslim Revivalist Movements in Northern India in the Sixteenth and Seventeenth Centuries*, 343, n. 4. On Sarmad's nudity, see also Bernier, *Travels in the Mogul Empire*, 317.

47. See Asiri, *Rubā'iyāt-i Sarmad*.

48. *Dabistān*, 1:217.

49. *Dabistān*, 1:298.

50. *Dabistān*, 1:367.

51. *Dabistān*, 1:87–89.

52. *Dabistān*, 1:88.

53. *Dabistān*, 1:89–90.

54. *Dabistān*, 1:24.

References

Alam, Muzaffar. *The Crisis of Empire in Mughal North India: Awadh and the Punjab, 1707–1748*. Delhi: Oxford University Press, 1986.

Alam, Muzaffar, and Sanjay Subrahmanyam, eds. *The Mughal State, 1526–1750*. Delhi: Oxford University Press, 1998.

Alam Khan, Iqtidar. *Mirza Kamran: A Bibliographical Study*. Bombay: Asia Pub. House, 1964.

Al-Birūnī. *Alberuni's India*. Trans. Edward C. Sachau. 1910. Delhi: Munshiram Manoharlal, 1989.

Asad, Talal. *Genealogies of Religion: Discipline and Reasons of Power in Christianity and Islam*. Baltimore, Md.: Johns Hopkins University Press, 1993.

Asiri, F. M., ed. *Rubā'iyāt-i Sarmad*. Santiniketan: Visva-Bharati, 1950.

Bayly, C. A. *Rulers, Townsmen, and Bazaars: North Indian Society in the Age of British Expansion*. Cambridge: Cambridge University Press, 1983.

Beach, Milo Cleveland. *The Imperial Image: Paintings for the Mughal Court*. Washington, D.C.: Freer Gallery of Art, Smithsonian Institution, 1981.

Bernier, François. *Travels in the Mogul Empire, A.D. 1656–1668*. 1934. Trans. Vincent A. Smith. Delhi: Low Price Publications, 1989.

Boyce, Mary. *A History of Zoroastrianism*. Vol. 1. Leiden: Brill, 1975.

———, ed. and trans. *Textual Sources for the Study of Zoroastrianism*. Manchester: Manchester University Press, 1984.

———. *Zoroastrians: Their Religious Beliefs and Practices*. London: Routledge and Kegan Paul, 1979.

Brand, Michael, and Glenn D. Lowry. *Akbar's India: Art from the Mughal City of Victory*. New York: Asia Society Galleries, 1985.

Cannon, Garland, ed. *The Letters of Sir William Jones*. Oxford: Clarendon Press, 1970.

Choksy, Jamsheed K. *Purity and Pollution in Zoroastrianism: Triumph over Evil*. Austin: University of Texas Press, 1989.

Dabistān-i Mazāhib. Ed. R. R. Malik. Tehran: Tahuri, 1984.

Dabistān-i Mazāhib, editio princeps. Calcutta, 1809.

The Dabistán, or School of Manners. Trans. David Shea and Antony Troyer. Paris: Printed for the Oriental Translation Fund of Great Britain and Ireland, 1843.

Dasgupta, Surendranath. *A History of Indian Philosophy*. 5 vols. Cambridge: Cambridge University Press, 1952–57.

Dhalla, M. N. *Zoroastrian Theology*. New York: AMS Press, 1972.

Digby, Simon. "Some Asian Wanderers in Seventeenth Century India: An Examination of Sources in Persian." *Studies in History*, new series, 9, no. 2 (1993), 247–64.

Doniger O'Flaherty, Wendy. "The Post-Vedic History of the Soma Plant." *Soma: Divine Mushroom of Immortality*, ed. R. Gordon Wasson. New York: Harcourt Brace Jovanovich, 1971.

Friedmann, Yohannan. "Medieval Muslim Views of Indian Religions." *Journal of the American Oriental Society* 95 (1975), 214–21.

Gershevitch, Ilya. Introduction to *The Avestan Hymn to Mithra*. Cambridge: Cambridge University Press, 1959.

Gommans, Jos J. L. *The Rise of the Indo-Afghan Empire, c. 1710–1780*. Leiden: Brill, 1995.

Habibullah, A. B. M. "An Early Arab Report on Indian Religious Sects." *History and*

Society: Essays in Honour of Professor Niharranjan Ray, ed. D. Chattopadhyaya. Calcutta: K.P. Bagchi, 1978.
Hadi, Nabi. *Dictionary of Indo-Persian Literature*. New Delhi: Indira Gandhi National Centre for the Arts, 1995.
Herzfeld, Ernst. *Zoroaster and His World*. Princeton, N.J.: Princeton University Press, 1947.
Jahn, K. "On the Mythology and Religion of the Indians in Mediaeval Moslem Tradition." *Mélanges d'orientalisme offert à Henri Massé*. Teheran: Impr. de l'université, 1963.
Jeffery, A. "Al-Bīrūnī's Contribution to Comparative Religion." *Al-Bīrūnī Commemorative Volume*, ed. Hakim Mohammed Said. Karachi: Oxford University Press, 1979.
Lawrence, Bruce B. *Shahrastānī on the Indian Religions*. The Hague: Mouton, 1976.
Malandra, William W., trans. and ed. *An Introduction to Ancient Iranian Religion: Readings from the Avesta and the Achaemenid Inscriptions*. Minneapolis: University of Minnesota Press, 1983.
Modi, J. J. *The Religious Ceremonies and Customs of the Parsees*. Bombay: J.B. Karani's Sons, 1937.
Naik, C. R. *'Abdu'r-Rahīm Khān-i-Khānān and His Literary Circle*. Ahmedabad: Gujarat University, 1966.
Nigosian, S. A. *The Zoroastrian Faith: Tradition and Modern Research*. Montreal: McGill-Queen's University Press, 1993.
Potter, Karl H., ed. *Encyclopedia of Indian Philosophies: Advaita Vedānta up to Śaṃkara and His Pupils*. Princeton, N.J.: Princeton University Press, 1981.
Rizvi, S. A. A. *Muslim Revivalist Movements in Northern India in the Sixteenth and Seventeenth Centuries*. Agra: Agra University, 1965.
———. *Religious and Intellectual History of the Muslims in Akbar's Reign, with Special Reference to Abu'l Fazl, 1556–1605*. Delhi: Munshiram Manoharlal, 1975.
Sen, Amartya. *The Argumentative Indian: Writings on Indian History, Culture and Identity*. New York: Allen Lane, 2005.
Sharma, Sri Ram. *The Religious Policy of the Mughal Emperors*. New York: Asia Publishing House, 1972.
Smart, Ninian. "Comparative-Historical Method." *The Encyclopedia of Religion*, ed. Mircea Eliade. New York: Macmillan, 1987.
Smith, Jonathan Z. *Drudgery Divine: On the Comparison of Early Christianity and the Religions of Late Antiquity*. Chicago: University of Chicago Press, 1990.
———. *To Take Place: Toward Theory in Ritual*. Chicago: University of Chicago Press, 1987.
Smith, Vincent A. *Akbar the Great Mogul*. Oxford: Clarendon Press, 1917.
Tavakoli-Targhi, Mohamad. "Contested Memories: Narrative Structures and Allegorical Meanings of Iran's Pre-Islamic History." *Iranian Studies* 29, nos. 1–2 (1996), 149–75.
———. "Orientalism's Genesis Amnesia." *Comparative Studies of South Asia, Africa and the Middle East* 16, no. 1 (1996), 1–14.

Waardenburg, Jacques, ed. *Muslim Perceptions of Other Religions: A Historical Survey.* New York: Oxford University Press, 1999.

Welch, Stuart Cary, Annemarie Schimmel, Marie L. Swietochowski, and Wheeler M. Thackston. *The Emperor's Album: Images of Mughal India.* New York: Metropolitan Museum of Art, 1987.

West, E. W., trans. *Pahlavi Texts. Sacred Books of the East,* vol. 24, ed. F. Max Mueller. Oxford: Clarendon Press, 1885.

CHAPTER NINE

"If There Is a Paradise on Earth, It Is Here"

Urban Ethnography in Indo-Persian Poetic and Historical Texts

SUNIL SHARMA

Sir Sayyid Ahmad Khan's encyclopaedic work on the archaeological monuments of Delhi, *Āsār as-Sanādīd* (Monuments of Rulers), was first published in 1847. This important work, which documented both Muslim and non-Muslim buildings as well as surviving inscriptions in all languages found on them, "was very much in the traditional mould of Indo-Muslim histories in which poetry, panegyric and moral and political instruction were fused, and the cultural diversity of Hindustan lavishly praised."[1] In an effort to present his history as a more objective work, with scholarly citations and devoid of traditional rhetorical flourishes and quotations from Persian poets, Sir Sayyid prepared a second edition in 1854, and subsequently a third edition. The *Āsār as-Sanādīd* displays a profound attachment to the city and organic connectedness to urban life, which is a prominent feature of other historical and poetic works of the tradition; in the words of C. A. Bayly, "This sense of luminosity of place, and of the pleasures and ease of the erstwhile great and culturally diverse empire of Hindustan, was a compelling motif for much of the poetic-cum-historical work of the eighteenth and early nineteenth centuries."[2] The roots of this mode of writing stretch back to the nascent period of Indo-Persian literature, but such texts especially flourished in the early modern period in South Asia. An exploration of the archaeology of the image of the flourishing and multicultural city in selected Indo-Persian poetic and historical texts demonstrates that the so-called hyperbolic and flowery descriptions of urban places that began appearing in the sixteenth century are replete with valuable proto-ethnographic information and represent a system of knowledge that was transmitted by the use of a special language and certain literary topoi and genres.

Persian literature contains a large corpus of topographical works in verse and prose whose subject is the praise of places that were thriving centers of power, commerce, and culture at a particular moment in history. In Indo-Persian literature authors made use of three interconnected literary topoi to encompass the totality of Indian society: description of the richness of India in all things, from its flora and fauna to the inhabitants; celebration of the architectural achievements of a particular city; and an expression of wonder and praise for a city's bazaars and its comely and industrious tradesmen. Over time these topoi were innovatively employed by authors of a variety of verse and prose texts, ranging from *masnavīs*, biographical dictionaries, and historical chronicles to travelogues. The earliest author to combine all three topoi is Amīr Khusraw (1253–1325), whose intimate connection with the Khalji and Tughlaq sultans of Delhi coincided with a flurry of building activity in that city and provided him ample topographical material for his poetry. In the first long narrative poem that Amīr Khusraw composed, in 1289, and dedicated to Sultan Kay Qubād (r. 1287–90), *Qirān as-Sa'dayn* (The Conjunction of Two Auspicious Planets), he lavishly praised the newly constructed congregation mosque and public water tank (*hawz*), concluding his depiction of the flourishing state of the capital, Delhi (*hazrat-i Dihlī*), with a lyrical paean to the Hindu lads of the city. His own pride in being a Muslim in India inspired him to glorify the land as an integral but unique part of the world of Islam; it also inspired him to write in 1318 a long verse panegyric of epic proportions, known as the *Sultān-nāmah* (Book of the Sultan), or more popularly as *Nuh Sipihr* (Nine Spheres).[3] Section 2 of this work includes descriptions of a palace and a congregational mosque completed in Delhi during the reign of Sultan Mubārak Shāh (r. 1316–20). Having found a novel way to praise the city, he includes a series of apostrophes, whereby different cities of the Islamic world state their superiority over Delhi, only to find that their claims are shot down one by one by the voice of fate in favor of the Indian capital city. The best known and most often quoted part of this work is a poetic ethnography of India in the third section that includes laudatory discourses in verse on India's climate, flora and fauna, learning, and religious systems, as well as a list of languages spoken in the vast expanse of the country.[4] Later Indo-Persian authors discussed below were familiar with these aspects of Amīr Khusraw's works, and his presence inhabited their writings right until the nineteenth century.

The second episode on a grand scale of Indo-Persian writing about buildings and cities took place in the Mughal period, especially during the reigns of Akbar (r. 1556–1605) and Shāh Jahān (r. 1628–58), whose court poets were actively involved in the project of translating the emperor's vision of a newly commissioned building or city into the discursive realm of the arts.[5] A Persian poet's all-encompassing gaze would often settle on the beautiful inhabitants of a place, who embodied a city's vigor and vitality, and the picture would be transformed into lyrical verse with ethnographic overtones. Working within a strictly defined system of poetics, but one which allowed for literary innovation and artistic freedom, poets described relationships between cities and their inhabitants in the metaphoric language of love. The *shahrāshūb* or *shahrangīz* (city disturber, i.e., someone who enlivens a city by his beauty) was a favorite literary topos that was often treated as a full-fledged poetic genre and employed by Persian poets in an ingenious (homo)erotic weave of commerce and love in the form of addresses to different craftsmen and youths marked as belonging to special social or religious groups.[6] Beginning in the sixteenth century such poems were written about every major urban center in the Persianate world, that is, the Iranian, Central and South Asian, and Ottoman regions, either as longer narratives in the *masnavī* form or as short unconnected poems organized around a city and patron. These poetical vignettes of bazaar life provide valuable historical information about crafts, guilds, and the jargon of various trades. In the Persianate courtly literature produced in India for the Mughals and Deccani rulers there seem to be no works of this genre in an independent form; it is more usual to find hybrid texts in which authors combine various modes of historical and lyrical writing in both prose and poetry. Unlike their Iranian counterparts, Indo-Persian writers following Amīr Khusraw were apparently more intent on emphasizing the multicultural fabric of their societies in sweeping tones rather than providing detailed information about specific crafts.

The metaphoric and aesthetic link between physical beauty, commerce, and crafts against the backdrop of a specific location can be seen at the very threshold of Mughal presence in India, as in Bābur's dismissive summation of the land of Hindustan: "There is no beauty in its people, no graceful social intercourse, no poetic talent or understanding, no etiquette, nobility, or manliness. The art and crafts have no harmony or symmetry." But he is sufficiently impressed by the wealth and variety in the commercial sphere there to add a few lines later, "The one nice aspect of Hindustan is that it is a large country with lots of money. . . . Another nice thing is the unlimited numbers of craftsmen and practitioners of every trade."[7] When the Mughals began to construct new cities or redesign old ones according to their preferences and needs, there was no dearth of positive elements that were observed and recorded by writers. A historian from the period of Akbar, Muhammad ʿĀrif, in his *Tārīkh-i Akbarī* or *Tārīkh-i Qandahārī*, written in 1580–84, describes the construction in the capital city (*dār al-khilāfat*) of Agra in 1563–64 in these words:

> A flourishing city [*shahr*], the abundant delicacies of which at the time of abundance and hope could provide security for all the inhabitants under the expanse of heaven, and the spring-habitations of the benefits of its monuments [*athar*] could encompass the inhabitants of the regional districts, through the abundance of pools [*iyad*] and the multitude of pastures. . . . Such a number of varied vessels and fords have been floated on the river for the need of so many people as cannot be accommodated in geometrical figures of the imagination. The multitude of foreigners from all sorts of nations, from the corners of the four sides of the world, have gathered, for trade and fulfillment of necessities, in such a country that the dar al-khilafat of Agra has become all of India.[8]

The abundant public spaces of this city reflect the diversity, vastness, and richness of the multitudinous inhabitants of not only the local setting, but the world at large. An emphasis on and propagation of this view of India was integral to the success of the imperial vision of the ruling polity.

Abū al-Fazl, the court historian of Akbar, echoes the same view regarding the capital, as he states in his section on the province (*sūbah*) of Agra, "It is filled with people from all countries and is the emporium of the traffic of the world."[9] In his detailed ethnography of Hindustan, which is part of the *Ā'īn-i Akbarī* (Institutes of Akbar), Abū al-Fazl analyzes the causes of misunderstanding and quarrels between different religions in India; he states that the first cause is "the diversity of tongues and the misapprehension of mutual purposes, and thus the alloy of ill-will is introduced and the dust of discord arises."[10] Āsaf Khān Ja'far Beg, the author of the second half of the *Tārīkh-i Alfī*, the "History of the Millennium" written in 1592 to mark a momentous point in Islamic history, explains in his introduction Akbar's motives in having this history written: "His heart is inclined to justice, desiring that the followers of different faiths, each having become acquainted with the truth and reality of the others' religion, should act with restraint and abandon bigotry."[11] All the arts would be at the service of the ruler's desire to celebrate the rich diversity of the people, where the city was a synecdoche for the entire country or world.

At a time when Mughal historians were writing universal histories another monumental work, *Haft Iqlīm* (Seven Climes), was completed in 1594 by the Iranian émigré Amīn Ahmad Rāzī at the court of Akbar. This prose work uses the framework of the cosmos to write a dictionary (*tazkirah*) of the world combining history, geography, and literary biography with copious quotations of poetry. Under each clime the author introduces the major cities there, beginning with a description of its architectural monuments and moving on to biographical sketches of its notables, especially those who are involved in the practice of Persian poetry. In the case of Delhi, which lies in the third clime, he mentions its impressive architecture and the intimate association of Sufis to the city.[12] Interestingly he reserves his enthusiasm in the lyrical vein for the city of Ahmadabad in the second clime, which was an important center of poetic production. This suggests that the metaphorical response of a writer to a particular city is not a random choice, but rather a deliberate one meant to convey an impression about that city. In the following glowing report on Ahmadabad, the writer attempts to convey something specific about the connection between beauty and place:

> It is the capital of Gujarat, and takes precedence in fame over all the provinces of India due to the pleasantness [*latāfat*] and quality [*kayfiyat*] of its inhabitation [*ābādānī*]. It is exceptional in the craftsmanship and elegance of its architecture and buildings. If it is said that no other city exists in the entire world with such greatness and beauty, it would not be an exaggeration. Its bazaar, in contrast to those of other Indian cities, is extremely vast and neat, where perfectly decorated shops of two or three stories have been constructed, and all its inhabitants, both female and male, are so cute and lovely that they take the life of one who looks at them—and bestow life when they speak![13]

An important commercial center at this time, Ahmadabad had become the focus of a lively group of literati under the patronage of the Mughal general 'Abd al-Rahīm Khān-i Khānān. The first major stop for many poets coming from Iran, it was certainly worthy of being praised. Amīn Ahmad Rāzī's use of the lyrical mode of description that would seem to belong in a *ghazal* rather than a prose text suggests that this type of coded language is the normative discourse in poetry and prose to convey positive characteristics of a place and paint a verbal picture that could be appreciated by his audience. Although biographical dictionaries document the names and achievements of actual individuals from the past and present, such representations of the general population were also included in them.

A similar treatment of the subject of urban life can be found in Persian texts written in or about the Deccan, although these have received much less scholarly attention. The Mughal ambassador to the Deccan, an Iranian by the name of Asad Beg, while visiting the court of Ibrāhīm 'Ādilshāh (r. 1579–1626) at Bijapur in 1603–6 describes the bazaars of the newly constructed city of Nauraspur in *Vaqā'i-i Asad Beg* as "filled with wine and beauty, dancers, perfumes, jewels of all sorts . . . and viands. In one street were a thousand bands of people drinking, and dancers, lovers and pleasure-seekers assembled; none quarreled or disputed with another, and this state of things was perpetual. Perhaps no place in the wide world could present a more wonderful spectacle to the eye of the traveler."[14] In a later period such poetic responses of wonder and novelty are seen again in Indo-Persian and Urdu travelogues that describe encounters with foreign places and peoples.

In the realm of Persian poetry in the Deccan the *Sāqīnāmah* (Book of the Saqi), the work of Nūruddīn Muhammad Zuhūrī (d. 1616), active at the Nizāmshāhī and 'Ādilshāhī courts at Ahmadnagar and Bijapur, is partly a verbal panorama of a new city (*shahr-i naw*) built on the outskirts of Ahmadnagar.[15] Zuhūrī's tour of the city begins with the private spaces of the assembly (*majlis*) and tavern (*maykhānah*), both felicitous places where wine drinking takes place, and moves out to public sites in the city such as the fort, almshouse, bathhouse, garden, and bazaar. Describing the bathhouse Zuhūrī observes, "Iraqis and Indians are as friendly to each other as basil and wild rose in a garden" (*'irāqī u hindī bi-ham dūstān / chu rayhān u nasrīn bi-yak būstān*).[16] He lapses into a rapturous lyrical mode to describe the flourishing marketplace as he passes through. He begins with a rhetorical question and response: "What can I say of the ways of the bazaars? / They are not bazaars, but fresh rose gardens" (*chih gūyam zi ā'īn-i bāzārhā / na bāzārhā tāzah gulzārhā*). He goes on to boast that "the city is bejeweled by groups of skilled ones" (*zi īl-i hunar shahr dar zīvar ast*) who ensnare people by their professional ability and physical beauty. Included in his catalogue of people of the bazaar are tradesmen such as a *kamāngar* (archer), *bazzāz* (grocer), a *sabbāgh* (dyer), a *'attār* (druggist), a *talāgar* (goldsmith), *javāhirfurūsh* (jeweler), and a *sarrāf* (money-changer). He then moves on to a higher echelon of inhabitants of that city, the *ahl-i 'ilm* (men of learning), *hakīmān* (physicians), and *ahl-i nujūm* (astronomers), concluding this section with an appreciation of the high status of poets in the city. Zuhūrī dwells much less on the amorous qualities of the professionals and plays on the commercial aspects of their activities.

A quarter of a century after Zuhūrī composed his influential work in the Deccan, Mughal court poets were busy composing poems that would adequately convey the breadth of Shāh Jahān's architectural vision, especially with respect to the construction of the city of Shāhjahānābād. The poetry of this period includes a vast amount of topographical material, including palaces, gardens, and baths, and also records photographic instances of urban life,[17] although the published material does not represent the totality of what was produced at the time because many literary manuscripts still remain unpublished.

The poet laureate Abū Tālib Kalīm Kāshānī (d. 1650) takes pride of place among the literati who were involved in the emperor's project. Kalīm's descriptive *masnavī* on the city of Akbarabad (Agra) and the garden of Princess Jahānārā also uses the *shahrāshūb* topos in the manner of Zuhūrī's work. The panoramic tour of the city of Akbarabad in verse that includes descriptions of the magnificent building complexes sponsored by the emperor Shāh Jahān is the verbal equivalent of what painters of this time were doing in miniatures. Kalīm begins his poem by praising first the land of Hindustan and then the city of Akbarabad, which is so vast that a thousand Egypts exist in its every street. He emphasizes the diverse composition of its inhabitants:

chunīn shahrī bih 'ālam kas nadīdah ast
kih dar vay haft iqlīm āramīdah ast
zi har kishvar dar ū khalq aramīdah
ta'adī rā na dīdah na shanīdah[18]

Nobody has seen such a city in the world
In which the seven climes repose.
People from every country reside here,
Who have neither heard of nor seen any harm.

He then leads the reader through the bazaar of Akbarabad, where he encounters an *'attār* (druggist), a *bazzāz* (cloth seller), a *sarrāf* (money-changer), a *jawharfurūsh* (jeweler), a *khayyāt* (tailor), a *zargar* (goldsmith), a *shaykhzādah* (shaikh's son), and a *sipahzādah* (soldier boy); in addition, there are specifically Indian tradesmen such as a *mahājan* (merchant), a *tanbolī* (*pān* seller), and a *dobī* (washerman), as well as handsome Rajput and Pathan lads. Writing from the point of view of an elite Iranian poet in India, Kalīm regards the composition of the exotic Indian metropolis and reports on this new and exciting world to an audience composed of both Indians and non-Indians in the same manner that romances and travel narratives about India were consumed by literary audiences in the Islamic world.

Kalīm's poetic descriptions of city life can be read alongside the prose work of Shāh Jahān's *munshī* Chandarbhān "Barahman" (d. 1662). In the second section of his *Chahār Chaman* (Four Meadows), a compendium of miscellaneous prose writings on courtly life, he sketches vignettes of life in the exalted camp (Urdū-i mu'allā) of Delhi. He comments on the bustling lives of the merry inhabitants of the walled city, especially the merchants in the bazaar who have "cash, gems, goods and objects from every land filling the shops." When he writes about the capital city Shahjahanabad (Dār al-hukūmat), Barahman quotes the famous line, frequently used by writers of this period but wrongly attributed to Amīr Khusraw, "If there is a paradise on earth / It is here, it is here, it is here" (*agar firdaws bar rū-yi zamīn ast / hamīn ast u hamīn ast u hamīn ast*). After describing the architectural wonders, he turns to the bazaar:

'Irāqī u Khurāsānī zi hadd-i bīsh / nihādah pīsh-i khud sarmāyah-yi khvīsh
Firangī az Firangistān rasīdah / navādir az banādir bīsh chīdah . . .
chih shahrī ānkih Misr az vay nishānī / Harāt az kūchah-yi ū dāstānī
bih ma'mūrī u ābādī chunān ast / kih dar har kūchah-ash sad Isfahān ast
nishastah har taraf gawharfurūshī / bar āvardah zi daryāhā khurūshī
fitādah har taraf sad la'l-i rakhshān / buvad dar har dūkān kān-i Badakhshān
bar āyad az barā-yi imtihānī / matā'-i haft kishvar az dūkānī[19]

Innumerable Iraqis and Khurasanis have put forward their capital.
Europeans from Europe are here collecting many rarities from the ports . . .
What a city!—Egypt is but a sign of it; Harat a mere tale of its streets.
It is so fully inhabited that there are a hundred Isfahans in its every street.
Pearl-sellers are everywhere, causing the seas to roar.
Glimmering rubies everywhere, a Badakhshan mine in every shop!
The merchandise of seven countries [climes] can be examined in one shop.

Similar praise is reserved for the imperial cities of Akbarabad (Dār al-khilāfat) and Lahore (Dār al-saltanat) as he passes over the other provinces in a more cursory manner. At all times he emphasizes the variety of people, buildings, animals, and merchandise to be found in India.

This portion of Chandarbhān's work was cited by another Hindu *munshī*, Sujān Rā'i, in the historical chronicle *Khulāsāt at-Tavārīkh* (Compendium of Histories), composed around 1696 in the time of the Mughal emperor Aurangzeb (r. 1658–1707). Writing in the new style of Indo-Persian historiography that appeared in the seventeenth century Sujān Rā'i traces the history of India all the way from the rule of King Yudhishthira in mythical times up to the reign of Aurangzeb, just as earlier Persian historians of Iranian lands had begun writing Iran-centered histories that took up the story of the world from a pre-Islamic Iranian moment and continued chronologically down to their times. This kind of historical writing privileges a narrative centered on one place, in this case Delhi. Right at the outset, while praising the creator of the universe, he emphasizes the multifariousness of the world; phrases describing the world as a variety of pictures (*'anvā-i nuqūsh*), numerous creations of nature (*ikhtirāt-i gūnāgūn-i khalqat*), and the diverseness of creation (*rangārang-i āfrīnish*) announce the tenor of his writing. This is followed by a portrayal of the vastness of India, with its variety of places and flora and fauna, and discourses on forms of Indian learning, where he mentions the canonical Sanskrit texts. The different classes of holy men of India (*darvīshān-i Hind, sanyāsīs, jogīs, bairāgīs*, etc.) are also portrayed in a more detached and less hyperbolic style than that of previous authors of proto-ethnographic works, as are the divisions of four *āshramas* and four sects (*firqah*) of Hindus. In praising Indians he asserts that anybody can become an Indian: the Zangi, Firangi, Arabistani, Irani, and Turani are all Indians by choosing to reside here. A significant portion of the work is a gazetteer of the empire, beginning with the capital, Delhi, and Shahjahanabad, where he cites Amīr Khusraw's verses on the monuments of the city. In his topographical survey of the major buildings, Sujān Rā'i's choice of expressions, such as the variety of palaces (*'anvā-i qusūr*), types of habitations (*aqsām-i amākin*), numerous settlements (*gūnāgūn nashīman*), and multitude of porticoes (*chandīn ayvān*), paints a dizzying picture of the cosmopolis and is meant to convey a sense of diversity. He then surveys the inhabitants of the city:

> People from Rum, Zang, Sham, Firang, Angrez, Holland[?], Yemen, Arabistan, Iraq, Khurasan, Khvarazm, Turkestan, Kabul, Zabulistan, Khita, Khotan, Chin Machin, Kashghar, Qalmastan, Tibet, Kashmir and all the provinces of India have made their homes in this city. They have learnt the ways and speech of [this land], for the origins of the language of Hindustan are right here. They are occupied with their work and professions. Their inhabitation is as harmonious as phrases are in prose and the way they live as fitting as verses in a poem.[20]

After describing the briskness of the bazaar and the variety of goods to be found there, he quotes the verses of Chandarbhān about the people of the city. He concludes that Egypt, Istanbul, Qazvin, and Isfahan, despite their greatness, cannot compete with the great and vast Mughal capital. It is significant that the information embedded in this work passed into the vernacular sphere with a "translation" into Urdu of the first part of the chronicle, excluding the section on Muslim rulers of India, under the title *Ārā'ish-i Mahfil* (Embellishment of the Assembly). This translation was made by the *munshī* Mīr Sher 'Alī "Afsos," who had left Mughal patronage for service at Fort William College under Lord Wellesley and John Gilchrist.[21] Three editions of this book were published in Calcutta, in 1808, 1848, and 1863, and it became an important text in the new curriculum for the training of administrators who would be equipped with knowledge about India.[22]

The emphasis on the diversity of the population of a city continued to be a feature of works written in the seventeenth century, both in Persian but now also in vernacular compositions. A remarkable poem about the important port city Surat was composed by the Urdu poet Muhammad Valī (d. 1720). It is a celebration of the city's beauties and its bustling commercial life and is written in a tone that is squarely in the tradition of Indo-Persian praise of urban life. His verbal sweep of the demography of the city results in a description of its charming and industrious inhabitants, Hindus, Muslims, Parsis, and also Europeans, each person contributing to the city's overall prosperity. Valī is effusive in his praise for the city; if in its nobility it can be compared to Mecca, it is still superior to all other places. Ultimately it is the diversity of the population that makes it great:

vahāñ sākin itte haiñ ahl-i mazhab / ke gintī meñ na āveñ un kī mashrab
agarche sab haiñ vo abnā-i Adam / vale bīnish meñ rangārang-i 'ālam[23]

People of so many religions live here
That one cannot count their sects.
Although all are the children of Adam,
In appearance they represent the diversity of the world.

Although the eighteenth century saw the production of a vast body of Urdu literature about the decline of cities that celebrates nostalgia and gloom, in Persian the positive mode lasted well into the nineteenth century.[24] A work emphasizing the biographical element that would also contribute to Sir Sayyid's ethnography is the *Muraqqa'-i Dihlī* (Picture Album of Delhi) by the Navāb Dargāh Qulī Khān (d. 1766), who was based in Hyderabad and accompanied Nizām al-Mulk Āsaf Jāh to Delhi during the years 1738–41. His short prose work paints a lively and happy picture of Delhi, with descriptions of groups of individuals from Sufis to courtesans.

Valī's sighting of the Europeans in Surat resonates in the wonder-filled response of the poet Asadullāh Khān Ghālib (1797–1869) to the colonial city of Calcutta.[25] He remarks in a Persian poem that the inhabitants there are "from every land and of every vocation" (*az har diyār u az har fan*).[26] Ghālib's fixation with beauty and occupations is a familiar element by now, but in its form and treatment of the subject his poem is entirely new and displays a precociously modern sensibility, also mirrored in early Persian and Urdu European travelogues that "often used the conventional symbols and metaphors of women from classical Persian poetry in describing Europeans,"[27] but more specifically drew directly on the body of city literature and especially the *shahrāshūb* topos. One early Indo-Persian visitor to Europe in 1800–1802, Mīrzā Abū Tālib Khān (d. 1806), admires the architecture, gardens, beautiful inhabitants, and markets of London and other British cities, and at every possible opportunity he bursts into familiar verses extolling the beauty of places and particular women whom he meets. Interestingly Abū Tālib does not have the same reaction when he visits France, Italy, and parts of the Ottoman Empire.

Persian texts from the nineteenth century, a time when Urdu was rapidly becoming the primary language of South Asian Islamicate culture, inevitably draw attention to themselves for being innovative and using traditional forms in distinctive ways. One such text of a class that maintained a structural connection to the ethnographic literature discussed in this paper, *Tashrīh al-Aqvām* (Concise Account of Peoples), has the underpinnings of a colonial-period ethnography. This illustrated work was completed by the Eurasian colonel James Skinner (1778–1841) in Hansi in 1825, probably with the assistance of a *munshī*, and bears a dedication to Sir John Malcolm. It is a survey of the occupational groups and religious mendicants under the headings of four castes of the Hindus (the four original castes, mixed castes, castes derived from Vishvakarma, and miscellaneous castes), various groups of Hindu, Jain, and Sikh mendicant and religious orders, as well as Muslim *qavvāl*s and *faqīrs*.[28] The headings are reminiscent of *shahrāshūb* poems in the wide range of occupations covered. The writer employs both Persian words, such as *rīsmānsāz* (rope maker), *khishtpaz* (brick maker), and *zargar* (goldsmith), along with Indic words such as *bhangī* (sweeper), *kumhār* (potter), and *baid* (physician). An account of the history of the house of Timur is provided, from the scion of the dynasty down to the time of Akbar II (r. 1806–37), while the final section treats Muslim families and tribes in Avadh and the Panjab. Thus, like some other works in this genre, it also provides a panorama of the social fabric of India under the power of a ruling polity, here the late Mughals. A more scientific ethnographic approach in this work distinguishes it from the other panegyric texts discussed above; the emphasis is on the origins of the groups described rather than on the poetics of the tradition, and a single metropolis is not the focus of the poet's vision but rather a larger sociogeographic region. The choice of Persian and the historical material found here distinguish the *Tashrīh al-Aqvām*, but actually it is also in the class of early Western-style visual ethnographic representations in eighteenth-century East India Company school paintings, and in the nineteenth century in works such as Frederic Shoberl's *Hindoostan* (1822) in the series *World in Miniature* and John Forbes Watson and John Kaye's *The People of India* (1868–75). Skinner also wrote another work, *Tazkirat al-'Umarā* (History of Notables), both of which "were vital keys to local society, reproducing the material of Persian histories in a form which could be used by revenue officers."[29]

Sir Sayyid Ahmad Khan's *Āsār as-Sanādīd* was thus the culmination of a long and complex tradition of topographical literature in prose and poetry. The first edition of this work was dedicated to Sir Thomas Metcalfe, resident of Delhi. The second edition was thoroughly reworked with the assistance of Edward Thomas, following Sir Sayyid's admission to the Archaeological Society of Delhi. In this edition the fourth chapter on Delhi and its inhabitants, written in the manner of a traditional Persian *tazkirah,* was dropped but then reinserted into the next edition.[30] It contains a brief description of the climate of Delhi and its people,[31] followed by biographies of 119 individuals: religious groups of Muslims under such specific headings as *kabā'ir-i mashā'ikhīn, rasūl-shāhīs, majzūb, hukamā-yi kirām,* and *'ulamā-i dīn*; Koran reciters and memorizers (*qurrā* and *huffāz*); singers (*bulbulnavāyān*); calligraphers (*khushnavīsān*); and maestros of music (*arbāb-i mūsīqī*). The shift from a description of types or classes of people to naming particular individuals belonging to those groups in the *Āsār as-Sanādīd* can be viewed as a hybridization of various genres, such as the *tazkirah,* and topoi *shahrāshūb* from classical literature. The early photographic ethnographic series *People of India* was commissioned by the government of India and includes photographs of specific named individuals as well as anonymous people who represent a particular ethnic group or category.

Although the texts discussed in this paper are not the only ones that represent the tradition, neither are they random instances; these works either had a wide circulation and made an impact on the worldview held by people in their own day, or in retrospect appear to be located at a pivotal juncture in the course of the literary and social history of urban India in the early modern period. Indo-Persian authors maintained a particular attitude to their environment and material world. The full extent of their inherent intertextuality is difficult to gauge without surveying the tradition as a whole, due to the canonical status of certain authors and the use of a metaphoric register that remained unchanged for centuries. For instance, in the nineteenth-century *Āsār as-Sanādīd,* while describing the people of Delhi, Sir Sayyid Ahmad Khan quotes directly from Amīr Khusraw's *Qirān as-Sa'dayn*, one of the first panegyrical works on Delhi. When it comes to literary genres, the hybrid nature of Indo-Persian texts, and their degree of intertextuality, has not been fully appreciated, allowing works to be categorized in predetermined ways that may have limited use for modern readers. Each author securely adheres to the textual tradition, even if his response to his surroundings is novel in its own way and to a great extent informed by his choice of language and genre. One should also be mindful of the origins of the author and his own position in the society he is describing; the numerous Iranian poets settled in India in the sixteenth and seventeenth centuries were genuinely responding to the novelty of their situation and locale, while Indian writers have an effusive patriotic zeal that informs their writing. There is also the qualitative difference between works in this genre that were produced in other parts of the Persianate world; for instance, the *shahrāshūb* topos remains a universally fundamental device and even enters the realm of travel and wonder literature, but in Indo-Persian works we notice a greater interest in the diverse religious and mendicant groups that are found in India. The description of a city functioned as a focus around which a variety of information could be transmitted, whether to celebrate the here and now or, increasingly over time, to remind oneself of what was glorious about the past and its relevance to the present.

Notes

1. Juneja, Introduction, 10. Also see the articles by C. M. Naim and Christian Troll on this work.

2. Bayly, *Empire and Information*, 197.

3. For further details about this work, see the introduction by Mohammad Waheed Mirza to the Persian edition of this text, *The Nuh Sipihr of Amir Khusraw*. English translations of these sections are found in Nath and Gwaliari, *India as Seen by Amir Khusrau*.

4. This unabashed manner of praising India was a subgenre in Indo-Muslim literature. One example of this is an unusual work that appeared in the eighteenth century. In 1763–64 Āzād Bilgrāmī, writing in Arabic, produced a work called *Subhat al-marjān fī āthār Hindustān* (Coral Rosary of the Monuments of Hindustan), whose first chapter is on the excellence and eminence of India. The editor of this work, Fazlur Rahman Nadwi, writes, "When Āzād wrote this chapter, he incorporated everything, whether real or fantastic, history or mythology which tended to prove the greatness of India" (v. 1, 17). For a translation of this text, see Ernst, "India as a Sacred Islamic Land," 556–64.

5. Poetic works produced in an Iranian milieu are discussed in Losensky, "The Palace of Praise and the Melons of Time." A related subgenre of this poetry is the substantial body of Mughal works in verse and prose on the gardens and natural beauty of Kashmir by Munīr Lāhorī, Fānī Kashmīrī, and others, where descriptions of nature, buildings, and people engaged in various occupations occur together.

6. For an overview of this topic, see de Bruijn, "Shahrangīz, 1." Also see Sharma, "The City of Beauties in the Indo-Persian Poetic Landscape," where I have dealt more specifically with the literary aspects of this topic.

7. Bābur, *The Baburnama*, 350–51.

8. Brand and Lowry, *Fatehpur-Sikri*, 292.

9. Abū al-Fazl, *The Ain-i-Akbari*, v. 3, 190–91.

10. Abū al-Fazl, *The Ain-i-Akbari*, v. 2, 3.

11. Quoted in Rizvi, *Religious and Intellectual History of the Muslims in Akbar's Reign*, 254.

12. The connection between sacred space and biographical literature is discussed in Hermansen and Lawrence, "Indo-Persian Tazkiras as Memorative Communications." Bayly's view on the genre of *tazkirahs* is that "these collective biographies demonstrate the diverse origins of the people who were known to the Indian critical public. . . . Just as Islam was supposed to pay no attention to class in matters of faith, so in matters of style and literary excellence, men from humble backgrounds could achieve great fame" (*Empire and Information*, 195). However, biographical dictionaries vary greatly in their orientation and usually have a particular agenda that determined the inclusion and exclusion of individuals.

13. Ahmad Amīn Rāzī, *Haft Iqlīm*, v. 1, 80–81.

14. Quoted in Alam and Subrahmanyam, "Discovering the Familiar," 136.

15. This work was dedicated to Burhan Shah II (r. 1591–95) in Ahmadnagar. When Zuhūrī moved to the Bijapur court he incorporated parts of it in his *Seh nasr* in praise of Nauraspur, the new city of Sultan Ibrahim II (r. 1580–1627).

16. Zuhūrī, *Sāqīnāmah-yi Zuhūrī*, 115–27.

17. For an art historical point of view on this subject, see Koch, *Mughal Art and Imperial Ideology*, xxvi.

18. Kalīm, *Dīvān-i Abū Tālib Kalīm Hamadānī*, 142.

19. Barahman, *Chahār Chaman*, 87.

20. Sujān Rā'i, *The Khulasatu-t-Tawarikh*, 30.

21. Bayly writes about this work, "Though [Afsos] recognised distress and decline, the overall tone seems to have been panegyric" (198). For the place of this work in the rise of Urdu historiography, see Habibullah, "Historical Writing in Urdu," 482; Bredi, "Remarks on *Ārāish-e mahfil*."

22. The role of *munshī*s in this context is discussed by Alam and Subrahmanyam in their paper in this volume.

23. Valī, *Kullīyāt-i Valī*, 378.

24. Poets such as Mīrzā Muhammad Rafī' Saudā (d. 1781), Qā'im Chāndpurī

(d. 1793), Mīr Taqī Mīr (d. 1810), and Nazīr Akbarābādī (d. 1830) wrote satirical poems painting a bleak picture of urban society with the collapse of commerce and social hierarchies, as a result of the weakening Mughal power and Nādir Shāh's invasion of Delhi. Since networks of patronage had broken down, these poems mock the professions that are in decline. On this type of poetry, called "elegiac historiography" by Bayly, see Rahman, "Shahrangīz, 3"; Petievich, "Poetry of the Declining Mughals," 104. For a happy portrayal of city life in Urdu poetry, see Behl, "Poet of the Bazaars."

25. See Hyder, "Ghalib and His Interlocutors," for a discussion of Ghalib's *Chirāgh-i dair* that forms the backdrop of the poet's arrival in Calcutta: "Inscribed in lyrical language here is a symbolic microcosm of cosmopolitanism, the accommodation of uniqueness, beauty, nature, religion, sorrow, accountability, and binaries and syntheses of difference" (468–69).

26. Ghālib, *Dīvan-i Ghālib Dihlavī*, 389.

27. Tavakoli-Targhi, *Refashioning Iran*, 56.

28. For more information on this work, see Dover, "The Cultural Significance of Col. James Skinner"; Losty, *The Art of the Book in India*, 148, 152.

29. Bayly, *Empire and Information*, 167. Poetic works continued to be written alongside the new ethnographic works about Indian society in the nineteenth century. For example, the *Kāshī-istut* (Praise of Kashi) by Matan Lāl 'Āfrīn' uses both the traditional topos of a description of a flourishing city and a scientific ethnographic description. The author, in his praise of Banaras, includes the usual representative inhabitants along with a detailed classification of the city's Brahmans. I am grateful to Stefano Pellò for drawing my attention to this text.

30. Further discussion of this can be found in Juneja, Introduction, 10–12, and footnotes on 88–89; also in the introduction to the recent edition by Khaliq Anjum, v. 1, 142–53.

31. Here the author declares, "What I have described is free from exaggeration and hyperbole. In truth, the people here are unlike those of any other part of the world [*iqlīm*]. Every individual is a collection of a thousand virtues and a bouquet of a hundred thousand skills" (Ahmad Khan, *Āsār as-Sanādīd*, v. 2, 13).

References

Abū al-Fazl. *The Ain-i-Akbari*. Trans. H. S. Jarrett. Delhi: Taj, 1989.

Ahmad Amīn Rāzī. *Haft iqlīm*. Ed. Javād Fāzil. Tehran: 'Alī Akbar 'Ilmī, 1961.

Alam, Muzaffar, and Sanjay Subrahmanyam. "Discovering the Familiar: Notes on the Travel-Account of Anand Ram Mukhlis, 1745." *South Asia Research* 16, no. 2 (1996), 131–54.

Āzād Bilgrāmī. *Subhat al-marjān fi āthār Hindustān*. Ed. Fazlur Rahman Nadwi. Aligarh: Ma'had al-Dirāsāt al-Islāmīyah, Jāmi'at 'Alīkarh al-Islāmīyah, 1976–80.

Bābur. *The Baburnama: Memories of Babur, Prince and Emperor*. Trans. Wheeler M. Thackston. New York: Oxford University Press, 1996.

Barahman, Chandarbhān. *Chahār Chaman*. Ed. Yunus Jafferey. New Delhi: Iran House, 2004. (Special edition no. 22 of the journal *Qand-i Pārsī*.)

Bayly, Chris. *Empire and Information: Intelligence Gathering and Social Communication in India, 1780–1870*. Cambridge: Cambridge University Press, 1999.

Behl, Aditya. "Poet of the Bazaars: Nazīr Akbarābādī, 1735–1830." *A Wilderness of Possibilities: Urdu Studies in Transnational Perspective*, ed. Kathryn Hansen and David Lelyveld. New Delhi: Oxford University Press, 2005.

Brand, Michael, and Glenn Lowry, eds. *Fatehpur-Sikri: A Sourcebook*. Cambridge, Mass.: The Aga Khan Program for Islamic Architecture at Harvard University and MIT, 1985.

Bredi, Daniela. "Remarks on *Ārāish-e mahfil* by Mīr Sher 'Alī Afsos." *Annual of Urdu Studies* 14 (1999), 33–54.

de Bruijn, J. T. P. "Shahrangīz, 1. In Persian." *Encyclopaedia of Islam*, 2nd ed. Leiden: Brill, 2010. Brill Online (accessed 18 April 2010), http://www.brillonline.nl/subscriber/entry?entry=islam_COM-1026.

Dover, Cedric. "The Cultural Significance of Col. James Skinner." *Calcutta Review* 134, no. 1 (1955), 17–23.

Ernst, Carl W. "India as a Sacred Islamic Land." *Religions of India in Practice*, ed. Donald S. Lopez Jr. Princeton, N.J.: Princeton University Press, 1995.

Ghālib. *Dīvan-i Ghālib Dihlavī*. Ed. Muhsin Kiyānī. Tehran: Rawzanah, 1997.

Habibullah, A. B. M. "Historical Writing in Urdu: A Survey of Tendencies." *Historians of India, Pakistan and Ceylon*, ed. C. H. Philips. London: Oxford University Press, 1961.

Hermansen Marcia K., and Bruce B. Lawrence. "Indo-Persian Tazkiras as Memorative Communications." *Beyond Turk and Hindu: Rethinking Religious Identities in Islamicate South Asia*, ed. David Gilmartin and Bruce B. Lawrence. New Delhi: India Research Press, 2002.

Hyder, Syed Akbar. "Ghalib and His Interlocutors." *Comparative Studies of South Asia, Africa and the Middle East* 26, no. 3 (2006), 462–75.

Juneja, Monica. Introduction to *Architecture of Medieval India: Forms, Contexts, Histories*. Delhi: Permanent Black, 2001.

Kalīm. *Dīvān-i Abū Tālib Kalīm Hamadānī*. Ed. Muhammad Qahramān. Mashhad: Āstān-i Quds-i Razavī, 1990.

Koch, Ebba. *Mughal Art and Imperial Ideology*. New Delhi: Oxford University Press, 2001.

Losensky, Paul. "The Palace of Praise and the Melons of Time: Descriptive Patterns in 'Abdī Bayk Šīrāzī's *Garden of Eden*." *Eurasian Studies* 2 (2003), 1–29.

Losty, Jeremiah P. *The Art of the Book in India*. London: British Library, 1982.

Matan Lāl 'Āfrīn'. *Kāshī-istut*. Lucknow: Naval Kishor, 1873.

Mirza, Mohammad Waheed. Introduction to *The Nuh Sipihr of Amir Khusraw*. London: Oxford University Press, 1950.

Naim, C. M. "Syed Ahmad and His Two Books Called 'Asar-al-Sanadid." *Modern Asian Studies*, forthcoming (2010).

Nath, R., and Faiyaz Gwaliari. *India as Seen by Amir Khusrau*. Jaipur: Historical Research Documentation Programme, 1981.

Petievich, Carla. "Poetry of the Declining Mughals: The *Shahr Ashob*." *Journal of South Asian Literature* 25, no. 1 (1990), 99–110.

Rahman, Munibur. "Shahrangīz, 3. In Urdu." *Encyclopaedia of Islam*, 2nd ed. Leiden: Brill, 2010. Brill Online (accessed 18 April 2010), http://www.brillonline.nl/subscriber/entry?entry=islam_COM-1026.

Rizvi, Saiyid Athar Abbas. *Religious and Intellectual History of the Muslims in Akbar's Reign, with Special Reference to Abu'l Fazl (1556–1605)*. Delhi: Munshiram Manoharlal, 1975.

Sayyid Ahmad Khan. *Āsār as-Sanādīd*. Ed. Khaliq Anjum. Delhi: National Council for Promotion of Urdu Language, 2003.

Sharma, Sunil. "The City of Beauties in the Indo-Persian Poetic Landscape." *Comparative Studies of South Asia, Africa and the Middle East* 24, no. 2 (2004), 73–81.

Sujān Rā'i. *The Khulasatu-t-Tawarikh*. Ed. M. Zafar Hasan. Delhi, 1918.

Tavakoli-Targhi, Mohamad. *Refashioning Iran: Orientalism, Occidentalism and Historiography*. New York: Palgrave, 2001.

Troll, Christian. "A Note on an Early Topographical Work of Sayyid Ahmad Khan: *Asar al-Sanadid*." *Journal of the Royal Asiatic Society* 2 (1972): 137–39.

Valī. *Kullīyāt-i Valī*. Ed. Nūr al-Hasan Hāshmī. Lucknow: Uttar Pradesh Urdu Academy, 1981.

Zuhūrī. *Sāqīnāmah-yi Zuhūrī*. Lucknow: Matba' Mustafā'ī Muhammad Mustafā Khān, 1844.

CHAPTER TEN

Early Persianate Modernity

MOHAMAD TAVAKOLI-TARGHI

The post-Orientalist reexamination of modernity has been at the center of recent historical *reactivations* of modern times.[1] The conventional Enlightenment story treats modernity as a peculiarly European development and as a byproduct of "Occidental rationalism."[2] Viewed from within this hegemonic paradigm, non-European societies were "modernized" as a result of Western impact and influence.[3] Thus Westernization, modernization, and acculturation were conceived as interchangeable concepts accounting for the transition of "traditional" and non-Western societies.[4] These assertions have been reevaluated by scholars examining the cultural genealogies and etiologies of modernity.[5] Locating the West in a larger global context beginning with the Age of Exploration, Stuart Hall suggests, "The so-called uniqueness of the West was, in part, produced by Europe's contact and self-comparison with other, non-western, societies (the Rest), very different in their histories, ecologies, patterns of development, and cultures from the European model."[6] Demonstrating the critical importance of "the Rest" in the formation of Western modernity, Hall submits that "without the Rest (or its own internal 'others'), the West would not have been able to recognize and represent itself as the summit of human history."[7] Hall's revised conception of modernity allows for an expanded framework of analysis encompassing what I call the formative role of *heterotopic* experiences for the Age of Exploration in the formation of the *ethos* of modernity.

In contrast to *utopias*, the imaginary places in which human societies are depicted in perfect forms, Michel Foucault explored *heterotopias* as alternative real spaces. As existing loci beyond the everyday space of experience, heterotopias "are something like counter-sites, a kind of effectively enacted utopia in which the real sites, all other real sites that can be found within the culture, are simultaneously represented, contested, and inverted." These loci of alterity served the function of creating "a space of illusion that exposes every real space . . . a space that is other, another real space, as perfect, as meticulous, as well arranged as ours is messy, ill-constructed, and jumbled." Calling the latter type a "compensatory" heterotopia, Foucault speculated that "on the level of the general organization of terrestrial space" colonies might have "functioned somewhat in this manner."[8] He offered as historical examples the regulated colonies established by Jesuits and Puritans. Similarly sixteenth-century reports of European exploration of exotic heterotopias deepened the Renaissance "humanists' understanding of human motives and action" and enlarged their framework of understanding. "As late as the 18th century," according to Stephen Toulmin, "Montesquieu and Samuel Johnson still found it helpful to present unusual ideas by attributing them to people in a far-off land like Abyssinia or Persia."[9]

The attribution of "unusual ideas to people in a far-off land" was not merely a "literary device."[10] For instance, the physical presence of the Persian ambassador Muhammad Riza Bayk (d. 1717) in France in 1715–16 provided the pertinent context for the imaginary scenarios informing the "unusual ideas" and the central question of *Persian Letters*: "How can one be Persian?"[11] As spectacles and as native informants of exotic heterotopias, travelers like Muhammad Riza Bayk inspired native European spectators, who in turn provided them with a space of self-recognition and self-refashioning. Considering the material significance of the Rest in the formation of Western modernity, such attributions can be considered residues of a genesis amnesia in European historiography. Such a historiographical amnesia has made possible the construction of a coherent and continuous medieval and modern Western civilization. As Maria R. Menocal has demonstrated, the "European Awakening" was "an Oriental period of Western history, a period in which Western culture grew in the shadow of Arabic and Arabic-manipulated learning."[12]

What Toulmin calls the "counter-Renaissance" search for certainty constituted European modes of self-refashioning as archetypically universal, rational, and modern.[13] This dehistoricizing universalist claim enabled European rationalists to obliterate the heterotopic context of their self-making and thus constitute themselves as the originators of modernity and rationality. This *amnesiac* or *forgetful* assertion gained hegemonic currency and thus constituted non-Western modernity as Westernization. The universalist claims of European Enlightenment blackmailed non-European modernity and debilitated its historiography by engendering a tradition of historical writing that used a dehistoricized and decontextualized European rationality as its scale and referent. Iranian historians and ideologues, like their Indian and Ottoman counterparts, thus developed a fractured conception of historical time that viewed contemporary European societies ahead of their own time. This conception of historical time was similar to the time-distancing devices of European and American anthropologists who denied *coevalness* to their contemporary non-Western societies.[14] Such a *schizochronic* conception of history informed the nationalist historiography of Iranian modernity, a historiography that assumed the noncontemporaneity of the contemporaneous Iranian and European societies. In a recent expression of this dehistoricizing and detemporalizing presumption, Daryush Shayegan, a leading Iranian critic, invited his readers "to be rational for once." Claiming to "stay with the facts," he argued, "For more than three centuries we, the heirs of the civilizations of Asia and Africa, have been 'on holiday' from history. We succeeded so well in crystallizing time in space that we were able to live outside time, arms folded, safe from interrogation."[15] Informed by the same temporal assumption of nonsimultaneity with Europe, Riza Davari, an Iranian philosopher who has set himself the task of transcending Western humanism, asserted, "The past of the West is our future."[16]

The temporal comprehension of these engagé critics is genealogically related to the ironic and self-Orientalizing rhetorical argument of an early twentieth-century Iranian constitutionalist who contended that if Adam, the forefather of humanity, returned today, he would be pleased with his Iranian descendants who have preserved his mode of life for many millennia, whereas his unfaithful European children have totally altered his traditions and mode of life. With the exception of a short-lived ancient cultural efflorescence, this rhetorical argument is similar to the Hegelian postulate of the fundamental similarity of the ancient and contemporary Persian mode of life, a postulate which Hegel shared with his contemporary Orientalists. Such Hegelian and Orientalist temporal assumptions have been reinforced by Iranian historiographical traditions that equate modernity with Westernization.

Nationalizing Iran

Recognized as the heterotopia of modernity and scientific rationality, Europe was constituted as the horizon of expectation for the Iranian passage to modernity. Thus European history, as the *future past* of the desired present, functioned as a normative scenario for the prognosis or forecasting of the future Iran. This anticipatory modernity introduced a form of historical thinking that narrated Iranian history in terms of the European past. By universalizing that past, scholars have misrecognized historical deviations from the European norm as abnormalities and national illnesses.[17] Thus the development of feudalism, capitalism, the bourgeoisie, the proletariat, democracy, freedom, scientific rationality, and industry in "well-ordered" Europe informed the diagnoses of their lack, absence, retardation, and underdevelopment in Iran.[18] In other words, alternative non-European historical processes have been characterized as the absence of change and as unhistorical history. For instance, John Malcolm (1769–1833), the author of an influential Orientalist *History of Persia* (1815), which was translated into Persian in 1876, observed:

> Though no country has undergone, during the last twenty centuries, more revolutions than the kingdom of Persia, there is, perhaps, none that is less altered in its condition. The power of the sovereigns, and of the satraps of ancient times; the gorgeous magnificence of the court; the habits of the people; their division into citizens, martial tribes, and savage mountaineers; the internal administration; and the mode of warfare; have continued essentially the same: and the Persians, as far as we have the means of judging, are at the present period, not a very different people from what they were in the time of Darius, and the Nousheerwan.[19]

In a more concise statement, Hegel (1770–1831) similarly asserted, "The Persians . . . retained on the whole the fundamental characteristics of their ancient mode of life."[20] This dehistoricizing assumption—that is, the contemporaneity of an early nineteenth-century mode of life with that of ancient times—informed both Orientalist and nationalist historiographies that constituted the heightened period of European colonialism and imperialism as the true beginning of rationality and historical progress in Iran. Whereas a progressive conception of time informed the modern European historiography from the late eighteenth century to the present, the accounts of modern Iran, like that of other non-Western societies, were unanimously based in a regressive conception of history. Thus the passage to modernity was constituted as a radical break with the "stagnant" and eternally recurring Iranian mode of life.

Malcolm viewed Islam and "the example of the prophet of Arabia and the character of some of the fundamental tenets of his faith" as the most prominent factors "in retarding the progress of civilization among those who have adopted his faith." These "retarding" factors explained why "every country inhabited by Mahomedans [never] attained a state of improvement which can be compared with that enjoyed by almost all those nations who form the present commonwealth of Europe." He concluded his recounting of the Iranian past with a reflection on its future. "The History of Persia, from the Arabian conquest to the present day," he claimed, "may be adduced as a proof of the truth of these observations: and while the causes, by which the effects have been produced, continue to operate, no material change in the condition of that empire can be expected." Malcolm wondered whether "the future destiny of this kingdom" could be altered with "the recent approximation of a great European power." The experience of the Ottomans, who were "wrapt up in the habits of their ancestors and . . . for ages resisted the progress of that civilization with which they were surrounded," did not seem promising to him. Thus the proximity with European powers and the "consequent collision of opposite habits and faith, was more likely to increase than to diminish those obstacles which hitherto prevented any very intimate or social intercourse between Mahomedan and Christian nations."[21] This prognosis, a forerunner of the "clash of civilizations,"[22] was grounded in the epistemological differentiation of the progressive Christian "commonwealth of Europe" and the stagnant "Mahomedan nations" of Asia.

Along with the global hegemony of the West, this binary opposition became an ever more significant component of an Iranian national historiography venerating progress, development, and growth. A celebratory history of Europe provided the normative manual for deciphering the abnormalities of Iran's past and for promoting its modernization, that is, Westernization. Ervand Abrahamian, the author of one of the most sophisticated accounts of modern Iran, offers a paradigmatic view of the nineteenth century, a view that is embedded in Persian historical writings.[23] "Traditional Iran," in his estimation, "in sharp contrast to feudal Europe, . . . had no baronial rebellions, no magna carta, no legal estates, and consequently no representative institutions." These and other *lacks* constitute the foundation for explaining a series of reformist failures of the nineteenth-century Qajars: "The attempt to construct a statewide bureaucracy failed. . . . The Qajars were equally unsuccessful . . . in building a viable standing army . . . [and] even failed to recapture the full grandeur of the ancient shah-in-shahs." By narrating a failed version of European history, this progressive historian of Iran assumes a typically Orientalist vantage: "The Qajar dynasty was an epitome of ancient oriental despotism; in fact, it was a failed imitation of such absolutism."[24] Such a characterization is a common feature of Orientalist, nationalist, and also Marxist historiography of nineteenth-century Iran.[25] The opening paragraph of Guity Nashat's *The Origin of Modern Reforms in Iran* is likewise a testimony to the centrality of Europe in the horizon of expectation for "traditional" Iran: "In 1870 a young Iranian of modest background, Mirza Huseyn Khan, was presented with an opportunity to regenerate Iran. During the next ten years he introduced regulations that were designed to transform the country's traditional political, military, and judicial institutions to resemble Western models. He also attempted to introduce Western cultural innovations and Westernized modes of thought."[26] Viewed as a Western model used to transform traditional societies, the modern, as in this case, was commonly understood "as a known history, something which has already happened elsewhere, and which is to be reproduced, mechanically or otherwise, with a local content."[27] As a mimetic plan, Iranian modernity, like its non-Western counterparts, can at best be hailed as a "project of positive unoriginality."[28] An eternally recurring Iranian premodernity was thus superseded by an already enacted Western modernity.

Viewing modernity as a belated reduplication of Western models, historians of Iran often invent periodizations that are analogous to standard European historical accounts. Recognizing Descartes's *Discours sur la Méthode* and Newton's *Principia* as two founding texts of modern thought in Europe, Iranian historians have the same expectations for the Persian rendering of these texts. In a modularized periodization of the Iranian "discovery of the West" and the "dissemination of European 'new learning,'" Mangol Bayat, a historian of modern Iran, writes that a Persian translation of Descartes's *Discours* was commissioned by Arthur Gobineau and published in 1862.[29] Referring to I'tizad al-Saltanah's *Falak al-Sa'adah* (1861), she adds that only one year earlier Isaac Newton and the idea of heliocentricity had been "introduced to the Iranian public."[30] This periodization concerning the introduction of modern European philosophical texts is similarly advanced by Faraydun Adamiyat, Elie Kedouri, Nikki Keddie, Jamshid Bihnam, and Aliriza Manafzadeh.[31] Adamiyat, a pioneering historian of Iranian modernity, contended that *Falak al-Sa'adah* and the Persian translation of *Discours* provided the "context for rational transformation" (*zaminah-'i tahavvul-i 'aqlani*) of nineteenth-century Iran. To dramatize the historical significance of Descartes's translation, he speculated that all copies of an 1853 edition of the text might have been burned.[32]

In these narratives the Comte de Gobineau, a French diplomat in Tehran and an infamous anti-Semite,[33] was credited as the initiator of the rationalizing tasks of translating Descartes's generative text of European modernity into Persian. Although Gobineau commissioned this translation, he doubted whether Iranians and other Asians were capable of absorbing modern civilization.[34] Like Gobineau, Iranian nationalist historians of scientific modernity often assumed that "the defense of geocentricism was of greatest importance for Muslim traditional scholars, just as it was for the medieval church."[35] In such accounts the endeavor for modernity was often depicted as a contention between the rational European astronomy and the irrational Muslim astrology. For example, Bayat wrote that I'tizad al-Saltanah "rose in defence of Newton and other European scientists' theories, and he declared obsolete the 'knowledge of the ancients.'"[36] Likewise Arjomand argued that I'tizad al-Saltanah's work was "the first book of its kind, aimed at combating the belief in traditional astronomy and astrology and bringing what might be termed scientific enlightenment to 19th-century Iran."[37]

Recounting the contentions for scientific rationality, historians of modern Iran often selected scholars who endorsed astrology and opposed heliocentrism as Muslim representatives, ignoring those who did not fit into this schema. By claiming that the Persian publication of Descartes in the 1860s was the beginning of a new age of rationality and modernity, these historians provided a narrative account that accommodated and reinforced the foundational myth of modern Orientalism, a myth that constitutes the West as ontologically and epistemologically different from the Orient.[38] This Orientalist problematic was validated by a nationalist historiography that constitutes the period prior to its own arrival as a time of decay, backwardness, and despotism. By deploying the foundational assumptions of Orientalism for the enhancement of its own political project, Iranian nationalist historiography participated "in its own Orientalizing."[39] As self-designated vanguards of modernity and national homogenization, both official and counter-official Iranian nationalists naturalized and authenticated Orientalists' temporal assumption of the noncontemporaneity of the contemporaneous Iranian and European societies.[40]

Homeless Texts

In the mid-seventeenth century a purely self-congratulatory view of European civilization as the paragon of universal reason and the concurring "blackmail of the Enlightenment" had not yet been formed.[41] Similarly Europe's Oriental-Other had not yet been dehistoricized as only traditional, static, and unchanging, and Muslims were not viewed as antiscientific. More significantly historical thinking had not yet been confined to the boundaries of modern nation-states. It is during this period that an alternative account of a Persianate modernity can be retrieved. Predating the consolidation of modern nation-states and the co-optation of modernity as a state-legitimating ideology, following Foucault, modernity may be envisaged as an ethos rather than a well-demarcated historical period.[42] By envisaging modernity as an ethos historians of Iran and India may imagine a joint fact-finding mission that would allow for reactivating what the poet Mahdi Akhavan Salis (1928–90) aptly recognized as "stories vanished from memory" (*qissah-ha-yi raftah az yad*).[43] These vanished stories may be retrieved from a large corpus of texts made homeless with the emergence of *history with borders*, a convention that confined historical writing to the borders of modern nation-states.

The convention of history with borders has created many *homeless texts* that have fallen victim to the fissure of Indian and Iranian nationalism. Although abolished as the official language of India in the 1830s, the intellectual use of Persian continued and Persian publications in nineteenth-century India outnumbered those produced in other languages. Publishers in Calcutta, Bombay, Lucknow, Kanpur, Delhi, Lahore, Hyderabad, and other cities in the Indian subcontinent also published more Persian books than their counterparts in Iran. Many of the literary and historical texts edited and published in India achieved canonical status in neighboring Iran. Rammohan Roy, the acclaimed "father of modern India," was in fact the editor of one of the first Persian newspapers, *Mir'at al-Akhbar* (founded in 1822). This Indo-Iranian intellectual symmetry continued until the end of the nineteenth century, when a Persian newspaper, *Miftah al-Zaffar* (founded in 1897), campaigned for the formation of Anjuman-i Ma'arif, an academy devoted to the strengthening of Persian as a scientific language.[44] Whereas the notion of Western civilization provided a safety net supplementing European national histories, no common historiographical practice captures the residues of the colonial and national conventions of historical writing that separates the joint Persianate literary culture of Iran and India—a literary culture that is irreducible to Islam and the Islamic civilization. A postcolonial historiography of Indian and Iranian modernity must begin to reactivate the concurring history that has been erased from memory by colonial conventions and territorial divisions.

The conventional account of Persianate acquaintance with the Cartesian notion "I think, therefore I am" differs radically from an account retrievable from the *Travels* of François Bernier (b. 1620), a French scholar who resided in India for a few years. Approximately two hundred years prior to Arthur de Gobineau, Danishmand Khan Shafi'a Yazdi (1578–1657), a Mughal courtier and Iranian émigré who was aware of current intellectual developments in Europe, dared to be wise (in Kant's sense of *sapere aude*) and commissioned Bernier to translate into Persian the works of Descartes (1560–1650), William Harvey (1578–1657), and Jean Pecquet (1622–74).[45] Bernier, a student of the philosopher Gassendi and a recipient of a doctorate in medicine in 1652, who is also considered a founding figure of modern Orientalism,[46] was an employee of Mirza Shafi'a, who was granted the title *danishmand* (scholar, scientist) for his intellectual endeavors. Bernier reported, "[I] explain[ed] to my Agah [master] the recent discoveries of Harveus and Pecquet in anatomy . . . [and] discours[ed] on the philosophy of Gassendi and Descartes, which I translated to him in Persian (for this was my principal employment for five or six years)." Illustrating the intellectual audacity and curiosity of Danishmand Khan, Bernier wrote, "My Navaab, or Agah, Danech-mend-khan . . . can no more dispense with his philosophical studies in the afternoon than avoid devoting the morning to his weighty duties as Secretary of State for Foreign Affairs and Grand Master of the Horse. Astronomy, geography, and anatomy are his favourite pursuits, and he reads with avidity the works of Gassendi and Descartes."[47] Danishmand Khan was known by his contemporaries to have espoused and "disseminated many of the innovating principles of that [European] community" (*aksari az ahkam-i tahrifat-i an jama'at tikrar minimud*) and desired to know "European sciences" (*'ilm-i ahl-i farang*) at a time when Europe was still plagued with religious wars.[48] His sustained interest in European intellectual developments is evident from his securing of a promise from Bernier "to send him the books from *ferngistan* [Europe]."[49] It was within the dynamic intellectual community around Danishmand Khan that Bernier became familiar with Persian translations of classical Sanskrit texts, including the *Upanishads*, which he brought back to Paris.[50] But the writings of Danishmand Khan and his cohorts who trained Bernier, this pedagogue of the "educated society in the seventeenth century" in Europe, have remained virtually unknown. This is in part because of the stereotypical perception of the period of the Indian Mughal emperor Aurangzeb's rule (1658–1707) as the age of Muslim bigotry and medieval decline. Confined within the grand narratives of "historical stages" and counter-colonial Hindu nationalism, historians of "medieval" India have mostly found facts of decline, all too often the only facts that they have searched for. During the same period François Martin, a friend of Bernier who visited Iran in 1669, observed that Persians "love the sciences, particularly mathematics." Contrary to received ideas, Martin reported, "It is believed that they [the Persians] are not very religious."[51] Likewise Pietro della Valle (1586–1652) could still confide that the Persianate scholar Mulla Zayn al-Din Lari, who has remained unknown to historians of Iran, "was comparable to the best in Europe."[52]

The scholarly efforts of Raja Jai Singh (1688–1743) provide another precolonial example of Persianate scholars' engagement with the modern sciences. Jai Singh built the observatories of Delhi, Banaras, and Jaipur, and based on new observations prepared the famous Persian astronomical table *Zij-i Muhammad Shahi* of 1728.[53] After the initial draft of his astronomical calculations he sent a mission to Portugal in 1730 to acquire new observational equipment and to inquire about recent astronomical findings. The mission, which included Father Emmanuel de Figueredo (1690?–1753?) and Muhammad Sharif, returned with an edition of Phillipe de La Hire's *Tabulae Astronomicae* from 1702.[54] Mubashshir Khan provides a brief account of Jai Singh's scientific mission in his *Manahij al-Istikhraj*, an eighteenth-century guide to astronomical observation and calculations. He reported that Mirza Muhammad 'Abid and Mirza Khayr Allah were two "Muslim engineers" who assisted Raja Jai Singh in the building of observatories. He had met Mirza Khayr Allah, who explained to him how Jai Singh, with the assistance of "Padre Manuel," acquired European observational equipment and a copy of La Hire's *Tabulae*. La Hire's calculations were used by Jai Singh in a revised edition of his *Zij-i Muhammad Shahi*.[55] This astronomical table, which was well known to eighteenth-century Iranian scholars, has remained virtually unknown to historians of Iran.[56] It is significant to note that almost a century earlier Shah 'Abbas II (1642–66) sent a mission to Rome to learn European painting techniques. The delegation included Muhammad Zaman "Paulo," whose "macaronic" style left a long-lasting imprint on representational art in both India and Iran.[57]

Works of Tafazzul Husayn Khan (d. 1800), well known to his Iranian friends and associates, are among other homeless texts that are elided from both Indian and Iranian annals of modernity. Hailed as an *'allamah* (arch-scholar), he was an exemplary figure of the late eighteenth century who interacted closely with the first generation of British Orientalists in India and actively promoted vernacular inquiry into modern science. In the 1780s he translated Newton's *Principia*, Emerson's *Mechanics*, and Thomas Simpson's *Algebra*.[58] In his obituary in 1803 the *Asiatic Annual Register* remembered Tafazzul Husayn Khan as "both in qualities and disposition of his mind, a very remarkable exception to the general character of Asiatic genius." Taking an exception to William Jones's assessment that "judgment and taste [were] the prerogative of Europeans," the obituary stated, "But with one, at least, of these proud prerogatives, the character of Tafazzul Husayn unquestionably interferes; for, a judgment at once sound, clear, quick, and correct, was its indistinguishable feature."[59] To document the accomplishments of this "Asiatic" who had "cultivated ancient as well as modern European literatures with ardour and success . . . very uncommon in any foreigner," the *Asiatic Annual Register* published letters received from Ruben Burrows (1747–92), David Anderson, and Lord Teignmouth (1751–1834).[60] Lord Teignmouth remarked of Tafazzul Husayn Khan, "Mathematics was his favourite pursuit, and perceiving that the science had been cultivated to an extent in Europe far beyond what had been done in Asia, he determined to acquire a knowledge of European discoveries and improvements; and with this view, began the study of the English language." He further noted that in two years Tafazzul Husayn Khan "was not only able to understand any English mathematical work, but to peruse with pleasure the volumes of our best historians and moralists. From the same motives he afterwards studied and acquired the Latin language, though in a less perfect degree; and before his death had made some progress in the acquisition of the Greek dialect."[61] Well acquainted with Tafazzul Husayn Khan, the British Orientalists Sir William Jones, Richard Johnson, and Ruben Burrows utilized his knowledge of classical Indo-Islamic sciences in their Orientalogical endeavors.[62]

Mir 'Abd al-Latif Shushtari (1758–1806), a close associate of Tafazzul Husayn Khan, who traveled to India in 1788, provided a synopsis of European modernity, modern astronomy, and new scientific innovations in his *Tuhfat al-'alam* (1216 H/1801).[63] Shushtari constituted the year 900 of Hijrah (1494–95 CE) as the beginning of a new era associated with the decline of the caliphate (*khilafat*) of the pope (*papa*), the weakening of the Christian clergy, the ascent of philosophy, and the strengthening of philosophers and scientists. Referring to the English Civil War, he explained the decline of religion. While both philosophers and rulers affirmed the unity of God, they viewed prophecy, resurrection, and prayers "as entirely myths" (*hamah ra afsanah*).[64] He also explained the views of Copernicus and Newton on heliocentricity and universal gravitation. Shushtari rejected the astrological explanations of "earlier philosophers" (*hukama-yi ma taqaddam*) and found affinities between the contemporary British scientific views and the "unbounded rejection of astrologers in the splendid *Shari'ah*" (*kah hamah ja dar Shari'at-i gharra' takzib-i munajimin varid shudah ast*). Critical of the classical explanation of tides, as recounted by 'Abd Allah Jazayiri (d. 1173 H/1760) in *Tilism-i Sultani*, he offered a Newtonian account, relating the tides to gravitational actions of the sun and moon on oceanic waters.[65] Accordingly he explained why the magnitude of the high tides in Calcutta differed from that of the coastal cities of the Persian Gulf. Writing some fifty years prior to I'tizad al-Saltanah's *Falak al-Sa'adah* (1861), Shushtari viewed Newton as a "great sage and distinguished philosopher" (*hakim-i a'zam va filsuf-i mu'azzam*) and ventured that in view of Newton's accomplishments all the "the golden books of the ancients" (*gawharin namah-ha-yi bastaniyan*) are now "similar to images on water" (*nimunah-'i naqsh bar ab ast*).[66] Shushtari's critical reflections on European history and modern sciences was appreciated by Fath 'Ali Shah (r. 1797–1834), who assigned the historian Marvazi Vaqayi' Nigar (d. 1834) the task of editing an abridged edition of *Tuhfat al-'Alam*, which is known as *Qava'id al-Muluk* (Axioms of Kings).[67] Given Shushtari's competence in both classical and modern astronomy, a periodization of Iranian "scientific modernity" that lionizes I'tizad al-Saltanah's *Falak al-Sa'adah* (1861) as its harbinger needs serious reconsideration. This is particularly important since I'tizad al-Saltanah was familiar with *Qava'id al-Muluk*.[68]

Aqa Ahmad Bihbahani Kirmanshahi (1777–1819), an Iranian Shi'i scholar and a friend of Shushtari who visited India between 1805 and 1810, devoted a chapter of his travelogue, *Mir'at al-Ahval-i Jahan Namah* (1810), to "the classification of the universe according to the school of the philosopher Copernicus."[69] In the introduction he explained, "Eminent philosophers are so numerous in Europe that their common masses [*avvam al-nas*] are inclined philosophically and pursue mathematical and natural sciences."[70] Like many other Muslim scholars, Bihbahani linked the "new views" (*ara'-i jadidah*) of Copernicus to those of ancient Greek philosophers but emphasized that "most of his beliefs are original" (*mu'taqidat-i u aksari tazigi darand*).[71] He explained favorably the heliostatic system, the sidereal periods for the rotation of planets around the sun, the daily axial and annual orbital revolutions of the earth, and the trinary rotations of the moon. This Muslim theologian found no necessary conflict between Islam and modern astronomy.[72]

The corpus of homeless texts of modernity includes Mawlavi Abu al-Khayr's concise account of the Copernican solar system, *Majmu'ah-'i Shamsi* (1807), which appears to have been known in Iran.[73] Like the works of Tafazzul Husayn Khan, *Majmu'ah* is a product of dialogic interaction between Persianate scholars and British colonial officers. Among topics discussed in the *Majmu'ah* are the movements of the earth, the law of inertia, the planetary motions, and universal gravitation. In the introduction Mawlavi Abu al-Khayr noted that his book was based on English-language sources and was translated "with the assistance" (*bi-i'anat*) of Dr. William Hunter (1718–83).[74] It is significant to note that Hunter had introduced Raja Jai Singh's *Zij-i Muhammad Shahi* to the English reading public in an article appearing in *Asiatic Researches* (1799).[75] It is likely that Mawlavi Abu al-Khayr had assisted Hunter in understanding and translating this highly technical Persian text.

During the first three decades of the nineteenth century numerous other texts on modern sciences were written in Persian that do not appear in accounts of Iranian and Indian modernity.[76] Muhammad Rafi' al-Din Khan's treatise on modern geometry and optics, *Rafi' al-Basar* (1250 H/1834), was one such text. The author was informed by English sources brought to his attention by Reverend Henry Martyn (1781–1812), a renowned Christian missionary and a translator of the Bible into Persian.[77] With an increased mastery of modern science, Persianate scholars became active in the production of scientific knowledge. In *A'zam al-Hisab*, a treatise on mathematics completed in 1814, Hafiz Ahmad Khan A'zam al-Mulk Bahadur (d. 1827) took issue with the Scottish astronomer James Ferguson on reckoning the difference between the Christian and the Muslim calendar. Aware of the self-congratulatory views of Europeans, "particularly among the people of England," A'zam al-Mulk wrote a treatise on astronomy, *Mir'at al-'Alam* (1819), in order to "disprove" the assertion that Muslims were "uninformed of mathematics and astronomy."[78] Based on Copernican astronomy and informed by the most recent observations and discoveries at the Madras Observatory, this treatise likewise remains homeless and among those not yet included in the Indian and Iranian nationalist accounts of modernity.

Orientalism's Forgotten Texts

Similar to the capitalist process of commodification and reification,[79] histories of Orientalism have concealed the traces of creativity and agency of the intellectual laborers who produced the works that bear the signature of pioneering Orientalists. The archives of unpublished Persian texts commissioned by eighteenth- and nineteenth-century British Orientalists reveal this underside of Orientalism. Having examined the works of the British who commissioned these unpublished works, it appears to me that they authored books that closely resemble their commissioned Persian works.

Sir William Jones (1746–94), who is viewed as the founder of British Orientalism as well as "one of the leading figures in the history of modern linguistics,"[80] relied heavily on the intellectual labor of numerous Persianate scholars. He labeled these scholars "my private establishment of readers and writers."[81] They included Tafazzul Husayn Khan (d. 1801), Mir Muhammad Husayn Isfahani, Bahman Yazdi, Mir 'Abd al-Latif Shushtari (1757–1806), 'Ali Ibrahim Khan Bahadur, Muhammad Ghaus, Ghulam Husayn Khan Tabataba'i (1727–1814?), Yusuf Amin (1726–1809), Mulla Firuz (d. 1833), Mahtab Rai, Haji Abdullah, Sabur Tiwari, Siraj al-Haqq, and Muhammad Kazim. In addition Jones was assisted by many pundits, including Radhacant Sarman. In one letter he specified, "My pendits must be *nik-khu, zaban-dan, bid-khwan, Farsi-gu* [well tempered, linguist, Vedantist/Sanskrit-reader, and Persophone]."[82] As the manager of an extensive scholarly enterprise Jones appropriated as his own the finished works that were the products of the intellectual capital and labor of these Indian scholars.

Jones's connection to Persianate scholars predated his arrival in India in 1783. Mirza I'tisam al-Din, an Indian who traveled to England between 1766 and 1769, reported that during his journey to Europe he helped to translate the introductory section of the Persian dictionary *Farhang-i Jahangiri*, which was made available to Jones when he composed his academic bestseller, *A Grammar of the Persian Language* (1771). Munshi I'tisam al-Din recounted:

> Formerly, on ship-board, Captain S[winton] read with me the whole of the Kuleelaah and Dumnah [*Kalilah va Dimnah*], and had translated the twelve rules of the Furhung Jehangeree [*Farhang-i Jahangiri*], which comprise the grammar of the Persian language. Mr. Jones having seen that translation, with the approbation of Captain S[winton], compiled his Grammar, and having printed it, sold it and made a good deal of money by it. This Grammar is a very celebrated one.[83]

With the publication of "The Sixth Discourse: On the Persians" (1789), Jones was recognized as "the creator of comparative grammar."[84] While he continues to be lionized for his remarks concerning the affinity of languages,[85] the Persian-Indian scholars and texts that informed Jones's work have remained unknown.

A few decades prior to Jones the Persian lexicographer and linguist Siraj al-Din Khan Arzu (1689–1756) wrote a comprehensive study of the Persian language, *Muthmir* (Fruition), discerning its affinity with Sanskrit.[86] Textual evidence indicates that Jones was familiar with this work and used it in writing the lecture that gained him recognition as "the creator of the comparative grammar of Sanskrit and Zend."[87] In his study of phonetic and semantic similarities and differences in Persian, Arabic, and Sanskrit and the interconnected processes of Arabization (*ta'rib*), Sanskritization or Hindianization (*tahnid*), and Persianization (*tafris*) in Iran and India, Arzu was fully aware of the originality of his own discernment of the affinity of Sanskrit and Persian. He wrote, "Amongst so many Persian and Hindi [Sanskrit] lexicographers and researchers of this science [*fann*], no one except *faqir* Arzu has discerned the affinity [*tavafuq*] of Hindi and Persian languages."[88] Arzu was amazed that lexicographers such as 'Abd al-Rashid Tattavi (d. 1658), the compiler of *Farhang-i Rashidi* (1064 H/1653) who had lived in India, had failed to observe "so much affinity between these two languages."[89] The exact date of the completion of Arzu's *Muthmir* has not been ascertained. But it is clear that he used the technical term *tavafuq al-lisanayn* (the affinity/concordance of languages) in his *Chiragh-i Hidayat* (1160 H/1747), a dictionary of rare Persian and Persianized concepts and phrases.[90] In this dictionary he offered examples of words common to both Persian and Hindi (Sanskrit).[91] Since Arzu died in 1756, *Muthmir* must have been written prior to that date. His works on the affinity of Sanskrit and Persian certainly predated the paper published in 1767 by Father Coeurdoux, who inquired about the affinity of Sanskrit and Latin.[92]

The traces of Persianate texts can be found in many other works of British Orientalists. For instance, Charles Hamilton's *Historical Relation of the Origin, Progress, and Final Dissolution of the Rohilla Afghans* (1787) corresponds closely to Shiv Parshad's *Tarikh-i Fayz Bakhsh* (1776).[93] Similarly W. Francklin's *History of the Reign of Shah-Aulum, the Present Emperor of Hindustan* (1798) is comparable in content and form to Ghulam 'Ali Khan's *Ayi'in 'Alamshahi*.[94] Likewise a large set of Persian-language reports on Tibet provided the textual and factual foundations for Captain Samuel Turner's *An Account of an Embassy to the Court of the Teshoo Lama in Tibet Containing a Narrative of a Journey through Bootan, and Part of Tibet* (1800). The most fascinating of these textual concordances is William Moorcroft's *Travels in the Himalayan Provinces of Hindustan and Panjab*.[95] Moorcroft is recognized as "one of the most important pioneers of modern scientific veterinary medicine" and "a pioneering innovator in almost everything he touched." In 1812 he commissioned Mir 'Izzat Allah to journey from Calcutta to the Central Asian city of Bukhara. Along the way Mir 'Izzat Allah collected invaluable historical and anthropological information which he recorded in his "Ahval-i Safar-i Bukhara."[96] His findings provided the factual foundations for the "pioneering" *Travels in the Himalayan Provinces* of Moorcroft. A preliminary inquiry indicates that Moorcroft might not have personally made the recounted journey that is praised for its "accuracy of historiographical and political observations."

Based on these and other collated texts, it seems that in its formative phase European students of the Orient, rather than initiating original and scientific studies, relied heavily on the research findings of native scholars. By rendering these works into English the colonial officers in India fabricated scholarly credentials for themselves, and by publishing these works under their own names gained prominence as Oriental scholars back home.[97] The process of translation and publication enabled the Europeans to obliterate the traces of the native producers of these works and thus divest them of authorship and originality, attributes which came to be recognized as the distinguishing marks of European scholars of the Orient. In many of these cases European scholars differentiated their works by adding the scholarly apparatuses of footnotes and references, citations that were already available in the body of the commissioned texts.

In some other cases scholarly competition helped to preserve the name of the original authors. For instance, Mirza Salih Shirazi served as a guide for the delegation led by Sir Gore Ouseley (1770–1844), the British ambassador extraordinary and plenipotentiary, who visited Iran between 1811 and 1812.[98] Mirza Salih kept records of the journey of this delegation, which included the leading Orientalists William Ouseley (1767–1842), William Price, and James Morier (1780–1849).[99] Mirza Salih composed a set of dialogues in Persian which were published in William Price's *A Grammar of the Three Principal Oriental Languages*.[100] Price wrote, "While we were at Shiraz, I became acquainted with Mirza Saulih, well known for his literary acquirements: he entered our train and remained with the Embassy a considerable time, during which, I prevailed upon him to compose a set of dialogues in his native tongue, the pure dialect of Shiraz."[101] In his *Travels in Various Countries of the East, More Particularly Persia*, published thirteen years earlier, William Ouseley had cited an "extract from some familiar Dialogues, written at my request by a man of letters at Shiraz."[102] This extract was the opening of the "Persian Dialogues" written by Mirza Salih.[103] Both Ouseley and Price claimed that the "Dialogue" was written at their request.[104] These competing claims may account for the preservation of the name of Mirza Salih as its author. In the introduction to the "Dialogue" Price humbly noted, "Having myself no motive but that of contributing to the funds of Oriental literature, and of rendering the attainment of the Persian language to students; I have given the Dialogues verbatim, with an English [*sic*] translation as literal as possible."[105] Mirza Salih also assisted Price in the research for his *Dissertation*.[106] Ouseley likewise credited Mirza Salih for providing him with a "concise description and highly economiastick" narrative on historical and archaeological sites used in his *Travels in Various Countries*. Having relied on Mirza Salih's contribution, Ouseley referred to part of the work as "the result of our joint research."[107] Oddly enough Mirza Salih is remembered only as a member of the first group of Iranian students sent to England in 1810, who were supposedly in need of "instruction in reading and writing their own language."[108]

The obliteration of the intellectual contributions of Persianate scholars to the formation of Orientalism coincided with the late eighteenth-century emergence of authorship as a principle of textual attribution and accreditation in Europe. The increased significance of authorship is attributed to the Romantic revolution and its articulation of the author "as the productive origin of the text, as the subjective source that, in bringing its unique position to expression, constitutes a 'work' ineluctably its own."[109] With the increased cultural significance of innovation (*inventio*), European interlocutors constituted themselves as the repositories of originality and authorship. It was precisely at this historical conjuncture that contemporary works of non-European scholars began to be devalued and depicted as *traditio*. This rhetorical strategy authorized the marginalization of Persianate scholarship at a time when the existing systems of scholarly patronage in Iran and India were dislocated. Without stable institutional and material resources that authorized the Persianate scholars, Orientalists were able to appropriate their intellectual works. The institutionalization of Orientalism as a field of academic inquiry and its authorization of original sources enabled European scholars to effectively appropriate the works of their non-Western contemporaries, who were denied agency and creativity.

The challenge of postcolonial historiography is to rehistoricize the processes that have been concealed and ossified by the Eurocentric accounts of Orientalism. This challenge also involves uncovering the underside of Occidental rationality. Such a project must go beyond a Saidian critique of Orientalism as "a systematic discourse by which Europe was able to manage—even produce—the Orient politically, sociologically, militarily, ideologically, scientifically, and imaginatively." Said's *Orientalism* provided the foundation for immensely productive scholarly works on European colonial agency, but these works rarely explore the agency and imagination of Europe's Other, who are depicted as passive and traditional. This denial of agency and *coevality* to the Rest provided the ground for the exceptionality of the West. By reconstituting the intertextual relations between Western texts and their repressed Oriental master texts, the postcolonial historiography can reenact the dialogical relations between the West and the Rest, a relationship that was essential to the formation of the ethos of modernity. The reinscription of the homeless texts into historical accounts of Orientalism is essential to this historiographical project.

Decolonizing the Historical Imagination

The preceding analysis calls for the decolonization of historical imagination and the rethinking of what is commonly meant by South Asian and Middle Eastern modernity. By anticipating a period of decline that paved the way for British colonization, historians of Mughal India have searched predominantly for facts that illustrate the backwardness and the disintegration of this empire. Mughal historiography in this respect has a plot structure similar to the late Ottoman history. In both cases the dominant themes of *decline* and *disintegration* are based on a projection about the rise and progress of Europe. In a similar manner historians of modern Iran inherited historiographical traditions that militate against the construction of historical narratives about the pre-Constitutional or pre-Pahlavi times as anything but an age of ignorance (*bikhabari*), stagnation, and despotism. Anticipating the coming of the Constitutional Revolution of 1905–9 historians have crafted narratives of intolerable conditions as instigators of the revolution.[110] The title of Nazim al-Islam Kirmani's paradigmatic account of the revolution, in which he participated between 1910 and 1912, *Tarikh-i Bidari-i Iraniyan* (The History of the Awakening of Iranians), reveals this prevalent assumption of prerevolutionary dormancy. To legitimate the Pahlavi dynasty (1926–79) as the architect of Iranian modernity and progress, Pahlavi historians likewise depicted the Qajar period as the dark age of Iranian history. These two Iranian historiographical traditions have been informed by, and in turn have informed, Orientalist accounts of Qajar shahs as absolute Oriental despots and Islam as a barricade against rationalization and secularization. Inscribing the history of Europe on that of India and Iran, both Indian and Iranian historians have deployed a regressive conception of time that constitutes their respective histories in terms of lacks and failures.

These bordered histories have rendered homeless texts that yield a different account and periodization of Persianate modernity. Historians of modern India often view Persian as a language only of the "medieval" Muslim Mughal court and thus find it unnecessary to explore the Persian texts of modernity.[111] Historians of Iran also consider unworthy Persian texts produced outside of the country, viewing Persian as a solely Iranian language. The conventional Persian literary histories, moreover, regard poetry as a characteristically Iranian mode of self-expression. With the privileged position of poetry in the invented national *mentalité*, the prose texts of the humanities are devalued and scholarly efforts are infrequently spent on editing and publishing nonpoetic texts. Thus a large body of historically significant prose texts of modernity have remained unpublished. This willful marginalization of prose is often masked as a sign of the prominence of poetry as an intrinsically Iranian mode of expression. These factors account for the elision of texts produced in India, which are stereotypically considered either linguistically faulty or as belonging to the corpus of the degenerate "Indian-style" (*sabk-i Hindi*) texts. Consequently Persian-language texts documenting precolonial engagement with the modern sciences and responding to European colonial domination have remained nationally homeless and virtually unknown to historians working within the confines of modern Indian and Iranian nationalist paradigms. This has led to several historiographical problems. Exclusion of these homeless texts from national historical canons, on the one hand, has contributed to the hegemony of Eurocentric and Orientalist conceptions of modernity as something uniquely European. On the other hand, by ignoring the homeless texts both Indian and Iranian historians tend to consider modernity only under the rubric of a belated Westernization. Such a conception of modernity reinforces the exceptionality of Occidental rationality and corroborates the programmatic view of Islamic and Oriental societies and cultures as static, traditional, and unhistorical.

This historical imagination is simultaneously grounded on two problematic conceptions of historical time: the presupposition of the noncontemporaneity of the contemporaneous Western and Oriental societies and the dehistoricizing supposition of the contemporaneity of the noncontemporaneous early nineteenth-century and ancient modes of life. With the onset of Westernization, consequently, the premodern repetition of ancient modes of life is replaced with the repetition of Western modernity. These dehistoricizing assumptions continue to provide the historiographical foundations for both Orientalist and nationalist accounts of Indian and Iranian modernity.

Notes

For a detailed study of issues in this essay, see Mohamad Tavakoli-Targhi, *Refashioning Iran: Orientalism, Occidentalism and Nationalist Historiography* (Basingstoke, England: Palgrave Publishers in association with St. Antony's College, Oxford University, 2001).

1. For instance, see Dussel, "Eurocentrism and Modernity," and *The Underside of Modernity*; Chakrabarty, *Provincializing Europe*.

2. Weber, *The Protestant Ethic*, 25; Habermas, *The Philosophical Discourse of Modernity*, 1.

3. For instance, see Lewis, "The Impact of the West"; Binder, "The Natural History of Development Theory."

4. According to G. E. Von Grunebaum:

> Acculturation, or more precisely Westernization, in the Near and Middle East has gone through distinct typical phases. After the shock caused by the discovery of inadequacy, there followed an almost complete surrender to foreign values and (not infrequently misunderstood) aspirations; then with Westernization partially realized, a recoiling set in from the alien, which however, continues to be absorbed greedily, and a falling back on the native tradition; this tradition is restyled and, in some instances, newly created with borrowed techniques of scholarship to give respectability to

> the results. Finally, with Westernization very largely completed in terms of governmental reforms, acceptance of the values of science, and adoption of Western literary and artistic forms, regained self-confidence expresses itself in hostility to the West and in insistence upon the native and original character of the borrowed product. (*Modern Islam*, 248)

5. Hall, "The West and the Rest"; Roberts, *The Triumph of the West*, particularly 194–202.

6. Hall, "The West and the Rest," 187. The dichotomy the West and the Rest was originally formulated by Marshall Sahlins in his *Culture and Practical Reason.*

7. Hall, "The West and the Rest," 221.

8. Foucault, "Of Other Spaces," 24, 27.

9. Toulmin, *Cosmopolis*, 28.

10. For instance, see Shklar, *Montesquieu*, 30.

11. The first edition of *The Persian Letters* was published in 1721. In Letter 91, documenting this evocative pertinent context writing about Muhammad Riza Bayk, Montesquieu noted, "There has appeared a personage got up as a Persian ambassador, who has insolently played a trick on the two greatest kings in the world" (172–73).

12. Menocal, *The Arabic Role in Medieval Literary History*, 2. Explaining the scholarly resistance to this view of European awakening, Menocal writes, "The tenor of some of the responses to the suggestion that this Arab-centered vision might be a viable historiographical reconstruction for the West has occasionally been reminiscent of the reactions once provoked by Darwin's suggestion (for so was the theory of evolution construed) that we were 'descended from monkey'" (3).

13. According to Toulmin, "In four fundamental ways . . . 17th-century philosophers set aside the long-standing preoccupation of Renaissance humanism. In particular, they disclaimed any serious interest in four different kinds of practical knowledge: the oral, the particular, the local, and the timely" (*Cosmopolis*, 30).

14. Johannes Fabian defines the denial of coevalness as "a persistent and systematic tendency to place the referent(s) of anthropology in a Time other than the present of the producer of anthropological discourse" (*Time and the Other*, 31).

15. Shayegan, *Cultural Schizophrenia*, 12.

16. Davari Ardakani, *Shimmah'i az Tarikh-i Gharbzadigi-i Ma*, 88.

17. For a study of the past as illness, see Tavakoli-Targhi, "Going Public."

18. For instance, see Ashraf, "Historical Obstacles to the Development of a Bourgeoisie in Iran"; Alamdari, *Chira Iran aqab mand va gharb pish raft?*

19. Malcolm, *The History of Persia*, 2:621. For the Persian translation, see Malcolm, *Tarikh-i Iran.*

20. Hegel, *The Philosophy of History*, 188.

21. Malcolm, *The History of Persia*, 2:622, 2:623, 2:624.

22. Huntington, *The Clash of Civilizations?*

23. For instance, see Ahmad Ashraf's renowned work, *Mavani'-i Tarikhi-i Rushd-i Sarmayahdari dar Iran.*

24. Abrahamian, *Iran between Two Revolutions,* 35, 38, 39, 40, 47.

25. Katouzian, *The Political Economy of Modern Iran,* 7–26, 298–300; Katouzian, "Arbitrary Rule"; Abrahamian, "Oriental Despotism"; Abrahamian, "European Feudalism and Middle Eastern Despotisms," 135.

26. Nashat, *The Origins of Modern Reform in Iran.*

27. Meagan Morris, "Metamorphoses at Sydney Tower," *New Formations* 11 (summer 1990), 10, cited in Chakrabarty, "Postcoloniality and the Artifice of History," 17.

28. Morris, "Metamorphoses at Sydney Tower," 10, cited in Chakrabarty, "Postcoloniality and the Artifice of History," 17.

29. Bayat, *Iran's First Revolution,* 36.

30. Bayat, *Iran's First Revolution,* 37.

31. Kedourie, *Afghani and Abduh,* 44–45; Keddie, *Sayyid Jamal al-Din "al-Afghani,"* 197–99; Jamshid Bihnam, *Iraniyan va Andishah-'i Tajaddud,* 32–34; Manafzadeh, "Nukhustin Matn-i Falsafah-'i Jadid-i Gharbi bah Zaban-i Farsi."

32. Adamiyat, *Andishah-'i Tarraqi va Hukumat-i Qanun,* 17, 18.

33. On Arthur de Gobineau's anti-Semitism, see Pulzer, *The Rise of Political Anti-Semitism in Germany and Austria.*

34. Gobineau, *Les religions et philosophies dans l'Asie centrale,* 98, 110–14; Gobineau, *Toris ans en Asie,* 322–23, 330–36.

35. Arjomand, "The Emergence of Scientific Modernity in Iran," 15.

36. Bayat, *Iran's First Revolution,* 37.

37. Arjomand, "Emergence of Scientific Modernity," 17.

38. According to Edward Said, "Orientalism is a style of thought based upon an ontological and epistemological distinction made between 'the Orient' and (most of the time) 'the Occident.' Thus a very large mass of writers . . . have accepted the basic distinction between East and West as the starting point for elaborate theories, epics, novels, social descriptions, and political accounts concerning the Orient, its people, 'mind,' destiny, and so on" (*Orientalism,* 2–3).

39. Writing about the developments in the Middle East after the Second World War, Said observed, "Despite its failures, its lamentable jargon, its scarcely concealed racism, its paper-thin intellectual apparatus, Orientalism flourishes today in the forms I have tried to describe. Indeed, there is some reason for alarm in the fact that its influence has spread to 'the Orient' itself: the pages of books and journals in Arabic (and doubtless in Japanese, various Indian dialects, and other Oriental languages) are filled with second-order analyses by Arabs of 'the Arab mind,' 'Islam,' and other myths" (*Orientalism,* 322).

40. Elaborating on the function of intellectuals in self-Orientalizing, Said wrote:

> Its role has been prescribed and set for it as a "modernizing" one, which means that it gives legitimacy and authority to ideas about modernization, progress, and culture that it receives from the United States for the most part. Impressive evidence for this is found in the social sciences and surprisingly enough, among radical intellectuals whose Marxism is taken

> wholesale from Marx's own homogenizing view of the Third world. . . . So if all told there is an intellectual acquiescence in the image and doctrines of Orientalism, there is also a very powerful reinforcement of this in economic, political, and social exchanges: the modern Orient, in short, participates in its own Orientalizing. (*Orientalism*, 325)

41. Foucault, "What Is Enlightenment?," 312.

42. Foucault, "What Is Enlightenment?," 309–10.

43. Salis, "Akhir-i Shahnamah," 85.

44. "Anjuman-i Ma'arif," *Miftah al-Zaffar*, 22 March 1899, 182–83.

45. Bernier, *Travels in the Mogul Empire*, 324–25. Danishmand Khan, also known as Muhammad Shafi', was born in Iran and went to Surat, India, in 1646. Shah Jahan appointed him a *bakhshi* (military paymaster) and granted him the title Danishmand Khan. Alamgir appointed him governor of Shah Jahan Abad, or New Delhi, where he died in 1670. William Harvey was a lecturer at the Royal College of Physicians and discovered the circulation of blood. Jean Pecquet discovered the conversion of chyle into blood.

46. Schwab, *The Oriental Renaissance*, 142–46.

47. Bernier, *Travels in the Mogul Empire*, 324–25, 352–53.

48. Shahnavaz Khan, *Ma'asir al-Umara*, 2:32.

49. François Bernier to M. Caron, 10 March 1663, in Martin, *François Martin Mémoires*, 548.

50. This Persian translation of the *Upanishads* was rendered into French and Latin by Anquetil-Duperron in 1801–2.

51. Martin, *François Martin Mémoires*, 441–42.

52. Pietro della Valle, *Viaggi di Pietro della Valle* (Brighton, 1843), 326–28, cited in Arjomand, "Emergence of Scientific Modernity," 7; Gurney, "Pietro Della Valle."

53. On the *Zij-i Muhammad Shahi*, see Hunter, "Some Account of the Astronomical Labours of Jaha Sinha." The article includes the preface of the Zij and its English translation. Also see Kaye, *The Astronomical Observatories of Jai Singh*; Forbes, "The European Astronomical Tradition"; Sharma, "Jai Singh"; Mercier, "The Astronomical Tables of Rajah Jai Singh Sawa'i"; Khan Ghori, "Development of Zîj Literature in India."

54. La Hire (1640–1718), *Tabulae astronomicae . . .* was originally published in 1687.

55. Muhammad 'Ali Mubashshir Khan, *Manahij al-Istikhraj*, unpublished manuscript, Kitabkhanah-'i Astan-i Quds-i Razavi, no. 12302. On the influence of La Hire, see Sharma, "Zïj Muhammad Shahi and the Tables of de La Hire."

56. Many copies of *Zij-i Muhammad Shahi* are available in Iranian libraries. One of the earliest editions is reported to be extant in the library of Madrasah-'i 'Ali-i Sipahsalar, which was renamed after the 1979 Revolution Madrasah-'i 'Ali-i Shahid Mutahhari. See Razaullah Ansari, "Introduction of Modern Western Astronomy in India during 18–19 Centuries," 364. Astan-i Quds-Razavi's copies are dated 1240 H/1824 in Yazd and 1242 H/1826.

57. 'Abbas Mazda, "Nufuz-i Sabk-i Urupa'i dar Naqashi-i Iran," particularly 61; Ardakani, *Tarikh-i Mu'assasat-i tamadduni-i Jadid dar Iran,* 1:234. The claim that Muhammad Zaman traveled to Europe is refuted by the Russian Orientalist Igor Akimushkin. See Soudavar, "European and Indian Influences," 379 n.16. For a critical evaluation of the controversy over Muhammad Zaman's career, see Ivanov, "Nadirah-'i Dawran Muhammad Zaman."

58. Shushtari, *Tuhfat al-'Alam va Zayl al-Tuhfah,* 363–67. See also "An Account of the Life and Character of Tofuzel Hussein Khan," Characters, 1–8, quote on 1.

59. "An Account of the Life and Character of Tofuzel Hussein Khan," section on Characters, 1. Spelled Tofuzzel Hussein in the original.

60. Ruben Burrows was supposed to write "notes and explanations" to Tafazzul Husayn Khan's translation of Newton's *Principia.* According to the *Asiatic Annual Register,* "The translation was finished, but it has not been printed; and we believe Mr. Burrows never added the annotations he mentions." See "An Account of the Life and Character of Tofuzel Hussein Khan," Characters, 7. Mir 'Abd al-Latif Shushtari noted that Tafazzul Husayn Khan acquired his knowledge of European philosophy (*hikamiyat-i farang*) from Burrows (*Tuhfat al-'Alam,* 371). On Ruben Burrows, see Burrows, "A Proof that the Hindoos Had the Binomial Theorem." Tafazzul Husayn Khan, who "wrote the Persian language with uncommon elegance," was appointed by Hastings to accompany David Anderson to Mahajee Scindiah. According to David Anderson, Husayn Khan learned English from his "brother, Mr. Blaine," and European mathematics and astronomy "from his communication with the learned Mr. Broome." In 1792, upon a friend's request, Anderson asked Tafazzul Husayn Khan to inquire about "the ancient astronomy of the Hindus." All quotes are from a letter by David Anderson published in "An Account of the Life and Character of Tofuzel Hussein Khan," Characters, 2–3.

61. "An Account of the Life and Character of Tofuzel Hussein Khan," Characters, 8.

62. For Husayn Khan's acquaintance with Sir William Jones and Richard Johnson, see "An Account of the Life and Character of Tofuzel Hussein Khan," Characters, 4.

63. Cole, "Invisible Occidentalism."

64. Shushtari, *Tuhfat al-'Alam,* 252, 255.

65. Shushtari, *Tuhfat al-'Alam,* 36–40, 299–315, particularly 36, 38.

66. Shushtari, *Tuhfat al-'Alam,* 303, 307. For an alternative interpretation of this passage, see Cole, "Invisible Occidentalism," 11–12. As it relates to the state of astronomical knowledge, Shushtari mentioned meeting the ninety-year-old Mir Masih Allah Shahjahanabadi, who resided in Murshidabad and had spent most of his life mastering astronomy. He reports studying *Zij-i Muhammad Shahi,* the observations of Chayt Singh, and other astronomical texts which were in the possession of Mir Masih. It would be important to locate the works these two scholars. See Shushtari, *Tuhfat al-'Alam,* 374.

67. Mirza Muhammad Sadiq Vaqyi' Nigar, *Qava'id al-Muluk,* Iranian National Library, Tehran, MS no. F/1757.

68. See I'tizad al-Saltanah's biographical note on Vagayi'nigar in his *Iksir al-Tavarikh*, 274–77.

69. Bihbahani Kirmanshahi, *Mir'at al-Ahval-i Jahan Namah*, 392.

70. Bihbahani Kirmanshahi, *Mir'at al-Ahval-i Jahan Namah*, 392. For a different rendering see Cole, "Invisible Occidentalism," 11.

71. Bihbahani Kirmanshahi, *Mir'at al-Ahval-i Jahan Namah*, 392.

72. For instance, Kamran Arjomand claims that "in the second half of the nineteenth century there were serious efforts to defend traditional Islamic cosmology against modern European astronomy." See his "Emergence of Scientific Modernity," 10.

73. Mawlavi Abu al-Khayr (b. Mawlavi Ghiyas al-Din), *Majmu'ah-i Shamsi*. In the introduction Mawlavi Abu al-Khayr notes that his book is based on English-language sources, which he translated with the encouragement and assistance of Dr. William Hunter (1718–83). *Majmu'ah-i Shamsi* bears the following note in English: "A Concise View of the Copernican System of Astronomy. By Mouluwee Ubool Khuer. Under the superintendence of W. Hunter, M.D. Calcutta. Printed by T. Hubbard at the Hindoostanee Press, 1807." Writing about the status of modern science, particularly astronomy, in Iran, John Malcolm observed, "Efforts have recently been made to convey better information to them upon this important branch of human sciences. An abstract of the Copernican system, and the proofs which the labours of Newton have afforded of its truth, have been translated into Persian; and several individuals of that nation have laboured to acquire this noble but abstruse subject." He then added, "But it is not probable that these rays of light will soon dissipate the cloud of darkness in which a prejudiced and superstitious nation have been for centuries involved" (*The History of Persia*, 2:536–37).

74. Mawlavi Abu al-Khayr, *Majmu'ah-i Shamsi*, 2.

75. Hunter, "Some Account of the Astronomical Labours of Jaya Sinha."

76. Other texts on modern sciences, particularly astronomy, include Muhammad Isma'il Landani's *Tashil al-adrak fi sharh al-aflak*, available at Dar al-'Ulum Nadwat al-'Ulama, radif 3, no. 4; Muhammad Ayyub's *Risalah dar 'Ilm-i Nujum* (1216 H/1801), available at the Khuda Bakhsh Oriental Public Library, Acc. 334; Sayyid Ahmad 'Ali's *Muqaddamat-i 'Ilm-i Hay'at* (Calcutta: n.p., n.d.); and Rathan Singh Zakhmi Lakhnavi's *Hadayiq al-Nujum* (1838).

77. For a discussion of Martyn, see Ha'iri, *Nukhustin Ruyaruyiha-yi Andishahgaran-i Iran ba Du Ruyah-'i Tamaddun-i Burzhuvazi-i Gharb*, 507–45. For Martyn's Persian translation of the New Testament, see *Kitab al-Muqaddas va Huwa Kutub-i al-'Ahd-i al-Jadid-i Khudavand va Rahanandah-'i Ma 'Isa-'i Masih*.

78. Kokan, *Arabic and Persian in Carnatic*, 340–44, 345–48.

79. Discussing the "phenomenon of reification," Georg Lukács explained, "Its basis is that a relation between people takes on the character of a thing and thus acquires a 'phantom objectivity,' an autonomy that seems so strictly rational and all-embracing as *to conceal every trace of its fundamental nature*: the relation between people" (*History and Class Consciousness*, 83, emphasis added).

80. From the publisher's note appearing in the reprint edition of Jones's *A Grammar of the Persian Language.*

81. Cannon, *The Letters of Sir William Jones,* 2:798.

82. Jones to Charles Wilkins, 17 September 1785, in Cannon, *The Letters of Sir William Jones,* 2:683.

83. Mirza Itesa Modeen, *Shigurf Namah I Velaët,* 65–66.

84. Jones, "The Sixth Discourse"; Müller, *The Sacred Languages of the East,* 4:xx. Hans Aarsleff also views Jones as the founder of modern philology; see *The Study of Language in England,* 124.

85. The history of linguistics texts often opens with entries on William Jones. For instance, see Sebeok, *Portraits of Linguistics.* The first three essays in this volume are devoted to Jones.

86. According to Rehana Khatoon, "Khan-i Arzu is also the first scholar in both the East and the West who introduced the theory of similarities of two languages [*tavafuq al-lisanayn*], meaning that Sanskrit and Persian are sister languages. His ideas in this regard are contained in his monumental work being discussed here, i.e. the *Muthmir.* The work has not yet been thoroughly studied and made a subject of serious assessment; and this has prompted me to undertake and prepare a critical edition of the *Muthmir*" (introduction to Arzu, *Muthmir,* 43).

87. Müller, introduction to *The Sacred Books of the East,* iv–xx.

88. The term *tavafuq* literally means concordance or concurrence.

89. Arzu, *Muthmir,* 221.

90. Arzu offered a detailed definition of *tavafuq al-lisanayn* under the concept of *ang.* See his *Chiragh-i Hidayat,* 1017–18.

91. Arzu, *Chiragh-i Hidayat,* 1050, 1061, 1068, 1091, 1119, 1020–21, 1214.

92. Kristeva, *Language the Unknown,* 196.

93. This analysis is based on a comparison with Shiv Parshad's *Tarikh-i Fayz Bakhsh,* Bodleian Library, Oxford University, shelfmark Caps.Or.C.2.

94. Francklin, *The History of the Reign of Shah-Aulum.* The claim is based on Ghulam 'Ali Khan's *Ayi'in 'Alamshahi,* Bodleian Library, Oxford University, shelfmark Elliot 3.

95. Moorcroft, *Travels in the Himalayan provinces of Hindustan and the Panjab.*

96. See Mir 'Izzat Allah, *Ahval-i Safar-i Bukhara,* Bodleian Library, Oxford University, Bodl.OR.745.

97. Among English-language texts based on Persian works is Captain William Henry Sleeman, *Ramaseeana: or Vocabulary of the Peculiar Language Used by the Thugs,* which is based on *Mustalahat-i Thugan* of 'Ali Akbar.

98. On Sir Gore Ouseley's travel to Iran, see Wright, *The English amongst the Persians during the Qajar Period,* 12–17.

99. For a fraction of Mirza Salih's report, see Mirza Salih Shirazi, "Safar Namah-'i Isfahan, Kashan, Qum, Tihran." The official *mihmandar* of this delegation was Mirza Zaki Mustawfi-i Divan-i A'la. See 'Adb al-Razzaq Maftun Dunbuli, *Ma'asir-i Sultaniyah,* 247.

100. Price, *A Grammar of the Three Principal Oriental Languages.*

101. Price, *A Grammar of the Three Principal Oriental Languages,* vi. The text of Mirza Salih's "Persian Dialogues" appears on 142–88, followed by a French translation, "Dialogues Persians et Français," 190–238.

102. Ouseley, *Travels in Various Countries of the East,* 1:xvii.

103. The extract in Ouseley's *Travels in Various Countries of the East,* 1:xvii, is identical to the opening of Mirza Salih's text as it appeared in Price's *A Grammar of the Three Principal Oriental Languages,* 142–43.

104. The colophon of the manuscript, *Su'al va Javab,* held at the Bodleian Library, which belongs to the Ouseley Collection, notes that it was written for Sir William Ouseley (shelfmark Ouseley 390).

105. Price, *A Grammar of the Three Principal Oriental Languages,* vii. In a note Price remarked, "Since that period Mirza Saulih came to England with Col. Darsy, in order to learn the English Language, returned to Persia in 1819, and lately arrived on a special Mission from the King of Persia to his Majesty George the Fourth. On my presenting him with a copy of his own dialogues, he expressed himself much pleased, and promised to compose a new set" (vi).

106. Price, *Journal of the British Embassy to Persia.*

107. Ouseley, *Travels in Various Countries of the East,* 3:363, 2:16.

108. Wright, *The Persians amongst the English,* 73.

109. Binder and Wellbery, *The End of Rhetoric,* 16.

110. For instance see, 'Alavi, "Critical Writings on the Renewal of Iran," 253.

111. Writing on eighteenth-century Bengal, Richard Eaton has also observed, "Two stereotypes—one by students of Indian history, the other held by students of Islam—have conspired greatly to obscure our understanding of Islam in Bengal, and especially of the growth of a Muslim peasant community there. The first of these is the notion of eighteenth-century Mughal India as a period hopelessly mired in decline, disorder, chaos, and collapse" ("The Growth of Muslim Identity in Eighteenth-Century Bengal," 161).

References

Aarsleff, Hans. *The Study of Language in England, 1780–1860.* Minneapolis: University of Minnesota Press, 1983.

'Abbas Mazda. "Nufuz-i Sabk-i Urupa'i dar Naqashi-i Iran." *Payam-i Nau* 2, no. 10 (1325/1946), 59–72.

Abrahamian, Ervand. "European Feudalism and Middle Eastern Despotisms." *Science and Society* 39 (1975), 129–56.

———. *Iran between Two Revolutions.* Princeton, N.J.: Princeton University Press, 1982.

———. "Oriental Despotism: The Case of Qajar Iran." *International Journal of Middle Eastern Studies* 5 (1984), 3–31.

"An Account of the Life and Character of Tofuzel Hussein Khan, the Vakeel, or Ambassador, of the Nabob Vizier Assof-Ud-Dowlah, at Calcutta, during the Government of Marquis Cornwallis." *The Asiatic Annual Register, or, A View of the History of Hindustan, and of the Politics, Commerce and Literature of Asia for the Year 1803*. London: Cadell and Davies, 1804.

Adamiyat, Faraydun. *Andishah-'i Tarraqi va Hukumat-i Qanun: 'Asr-i Sipahsalar*. Tehran: Khwarazmi, 1351/1972.

'Adb al-Razzaq Maftun Dunbuli. *Ma'asir-i Sultaniyah*. 1241/1825. Tehran: Ibn Sina, 1351/1972.

Alamdari, Kazim. *Chira Iran aqab mand va gharb pish raft?* Tehran: Nashr-i Gam-i Naw: Tawsiah, 1379/2000.

'Alavi, Buzurg. "Critical Writings on the Renewal of Iran." *Qajar Iran: Political, Social, and Cultural Changes, 1800–1925*, ed. Edmond Bosworth and Carloe Hillenbrand. Costa Mesa, Calif.: Mazda Publishers, 1992.

Ardakani, Husayn Mahbubi. *Tarikh-i Mu'assasat-i tamadduni-i Jadid dar Iran*. Tehran: Anjuman-I Danishjuyan-I Danishgah-I Tihran, 1975–89.

Arjomand, Kamran. "The Emergence of Scientific Modernity in Iran: Controversies Surrounding Astrology and Modern Astronomy in Mid-nineteenth Century." *Iranian Studies* 30, nos. 1–2 (1997), 5–24.

Arzu, Siraj al-Din Khan. *Chiragh-i Hidayat*, and Ghiyas al-Din Rampuri, *Ghiyas al-Lughat*. Ed. Mansur Sirvat. Tehran: Amir Kabir, 1984.

———. *Muthmir (Musmir)*. Ed. Rehana Khatoon. Karachi: Institute of Central and West Asian Studies, 1991.

Ashraf, Ahmad. "Historical Obstacles to the Development of a Bourgeoisie in Iran." *Studies in the Economic History of the Middle East: From the Rise of Islam to the Present Day*, ed. M. A. Cook. New York: Oxford University Press, 1970.

———. *Mavani'-i Tarikhi-i Rushd-i Sarmayahdari dar Iran: Dawrah-'i Qajariyah*. Tehran: Zaminah, 1359/1980.

Bayat, Mangol. *Iran's First Revolution: Shi'ism and the Constitutional Revolution of 1905–1909*. Oxford: Oxford University Press, 1991.

Bernier, François. *Travels in the Mogul Empire, AD 1656–1668*. Trans. Archibald Constable. Revised by Vincent Smith. New Delhi: Atlantic Publishers and Distributers, 1989.

Bihbahani Kirmanshahi, Aqa Ahmad. *Mir'at al-Ahval-i Jahan Nama*. Ed. 'Ali Davani. Tehran: Intisharat-i Markaz-i Asnad-i Inqilab-i Islami, 1996.

Bihnam, Jamshid. *Iraniyan va Andishah-'i Tajaddud*. Tehran: Farzan Ruz, 1375/1996.

Binder, John, and David Wellbery, eds. *The End of Rhetoric: History, Theory, and Practice*. Stanford: Stanford University Press, 1990.

Binder, Leonard. "The Natural History of Development Theory, with a Discordant Note on the Middle East." *Islamic Liberalism: A Critique of Development Ideology*. Chicago: University of Chicago Press, 1988.

Burrows, Ruben. "A Proof that the Hindoos Had the Binomial Theorem." *Asiatic Researches* 2 (1790), 487–97.

Cannon, Garland, ed. *The Letters of Sir William Jones*. Oxford: Clarendon Press, 1970.

Chakrabarty, Dipesh. "Postcoloniality and the Artifice of History." *Representations* 37 (winter 1992), 1–26.

———. *Provincializing Europe: Postcolonial Thought and Historical Difference*. Princeton, N.J.: Princeton University Press, 2000.

Cole, Juan. "Invisible Occidentalism: Eighteenth-century Indo-Persian Construction of the West." *Iranian Studies* 25, nos. 3–4 (1992), 3–16.

Davari Ardakani, Riza. *Shimmah'i az Tarikh-i Gharbzadigi-i Ma: Vaz'-i Kununi-i Tafakkur dar Iran*. Tehran: Surush, 1363/1984.

Dussel, Enrique. "Eurocentrism and Modernity." *Boundary 2* 20, no. 3 (1993), 65–76.

———. *The Underside of Modernity: Apel, Ricoeur, Rorty, Taylor, and the Philosophy of Liberation*. Trans. Eduardo Mendieta. Atlantic Highlands, N.J.: Humanities Press, 1996.

Eaton, Richard. "The Growth of Muslim Identity in Eighteenth-Century Bengal." *Eighteenth-Century Renewal and Reform in Islam*, ed. Nehemiah Levtzion and John Voll. Syracuse, N.Y.: Syracuse University Press, 1987.

Fabian, Johannes. *Time and the Other: How Anthropology Makes Its Object*. New York: Columbia University Press, 1983.

Forbes, Eric. "The European Astronomical Tradition: Its Transmission into India, and Its Reception by Sawai Jai Singh II." *Indian Journal of History of Science* 17, no. 2 (1982), 234–43.

Foucault, Michel. "Of Other Spaces." *Diacritics* 16, no. 1 (1986), 22–27.

———. "What Is Enlightenment?" *Ethics: Subjectivity and Truth*, ed. Paul Rabinow. New York: New Press, 1994.

Francklin, William. *The History of the Reign of Shah-Aulum, the Present Emperor of Hindostaun*. London, 1798.

Gobineau, Arthur de. *Les religions et philosophies dans l'Asie centrale*. Paris: Didier et Cie, 1865.

———. *Toris ans en Asie, Voyage en Persian*. Paris: Librairie de L. Hachette et Cie, 1859.

Gurney, John D. "Pietro Della Valle: The Limits of Perception." *Bulletin of the School of Oriental and African Studies* (1986), 103–16.

Habermas, Jürgen. *The Philosophical Discourse of Modernity: Twelve Lectures*. Trans. Frederick G. Lawrence. Cambridge, Mass.: MIT Press, 1987.

Ha'iri, 'Abd al-Hadi. *Nukhustin Ruyaruyiha-yi Andishahgaran-i Iran ba Du Ruyah-'i Tamaddun-i Burzhuvazi-i Gharb*. Tehran: Amir Kabir, 1367/1998.

Hall, Stuart. "The West and the Rest: Discourse and Power." *Modernity: An Introduction to Modern Societies*, ed. Stuart Hall, David Held, Don Hubert, and Kenneth Thompson. Cambridge, Mass.: Blackwell, 1996.

Hegel, George W. F. *The Philosophy of History*. Trans. J. Sibree. Buffalo, N.Y.: Prometheus Books, 1991.

Hunter, William. "Some Account of the Astronomical Labours of Jaha Sinha, Raja of Ambhere, or Jayanagar." *Asiatic Society* 5 (1799), 177–210.

Huntington, Samuel. *The Clash of Civilizations?* Cambridge, Mass.: Harvard University, John M. Olin Institute for Strategic Studies, 1993.

I'tizad al-Saltanah, 'Ali Quli Mirza. *Falak al-Sa'adah*. Tehran: Dar al-Taba'ah-'i Aqa Mir Muhammad Tihrani, 1278/1861.

———. *Iksir al-Tavarikh*. Ed. Jamshid Kayanfarr. Tehran: Visman, 1997.

Ivanov, A. A. "Nadirah-'i Dawran Muhammad Zaman." *Davazdah Rukh: Yadnigari az Dawazdah Naqash-i Nadirahkar-i Iran*. Trans. Ya'qub Azhand. Tehran: Intisharat-i Mawla, 1377/1998.

Jones, William. *A Grammar of the Persian Language*. Menston: Scholar Press, 1969.

———. "The Sixth Discourse: On the Persians; Delivered 19 February 1789." *The Works of Sir William Jones in Six Volumes*, ed. Ann Maria Shipley-Jones. London: G. G. and J. Robinson, 1799.

Katouzian, Homa. "Arbitrary Rule: A Comparative Theory of State, Politics and Society in Iran." *British Society for Middle Eastern Studies* 24, no. 1 (1977), 49–73.

———. *The Political Economy of Modern Iran: Despotism and Pseudo-Modernism, 1926–1979*. New York: New York University Press, 1981.

Kaye, G. R. *The Astronomical Observatories of Jai Singh*. Calcutta: Archaeological Survey of India, 1918.

Keddie, Nikki. *Sayyid Jamal al-Din "al-Afghani": A Political Biography*. Berkeley: University of California Press, 1972.

Kedourie, Elie. *Afghani and Abduh: An Essay on Religious Unbelief and Political Activism in Modern Islam*. London: Cass, 1966.

Khan Ghori, S. A. "Development of Zîj Literature in India." *History of Astronomy in India*, ed. S. N. Sen and K. S. Shukla. New Delhi: Indian National Science Academy, 1985.

Kitab al-Muqaddas va Huwa Kutub-i al-'Ahd-i al-Jadid-i Khudavand va Rahanandah-'i Ma 'Isa-'i Masih (The New Testament of Our Lord Saviour Jesus Christ). Trans. Henry Martyn. London: British and Foreign Bible Society, 1876.

Kokan, Afzal-ul-Ulama Muhammad Yousuf. *Arabic and Persian in Carnatic, 1710–1960*. Madras: Hafiza House, 1974.

Kristeva, Julia. *Language the Unknown and Initiation into Linguistics*. Trans. Anne M. Menke. New York: Columbia University Press, 1989.

La Hire, Phillipe de. *Tabulae astronomicae . . .* 1687. Parisiis: Apud Joannem Boudot, 1702.

Lakhnavi, Rathan Singh Zakhmi. *Hadayiq al-Nujum*. Patna: Khuda Bakhsh Oriental Public Library, 1838.

Lewis, Bernard. "The Impact of the West." *The Emergence of Modern Turkey*, 2nd ed. London: Oxford University Press, 1961.

Lukács, Georg. *History and Class Consciousness: Studies in Marxist Dialectics*. Trans. Rodney Livingstone. Cambridge, Mass.: MIT Press, 1971.

Malcolm, John. *The History of Persia from the Most Early Period to the Present Time.* London: J. Murray, 1815.

———. *Tarikh-i Iran.* Trans. Mirza Isma'il Hayrat. Bombay: Matba'-i Datparsat, 1886.

Manafzadeh, Alireza. "Nukhustin Matn-i Falsafah-'i Jadid-i Gharbi bah Zaban-i Farsi." *Iran Nameh* 9, no. 1 (1991), 98–108.

Martin, François. *François Martin Mémoires: Travels to Africa, Persia and India.* Trans. Aniruddha Ray. Calcutta: Subarnarekha, 1990.

Mawlavi Abu al-Khayr. *Majmu'ah-i Shamsi: mushtamil-i bar masa'il-i 'ilm-i hay'at mutabiq-i tahqiqat-i 'ulama-yi muta'akhirin-i Farang.* Calcutta: Hindoostani Press, 1222/1807.

Menocal, Maria Rosa. *The Arabic Role in Medieval Literary History: A Forgotten Heritage.* Philadelphia: University of Pennsylvania Press, 1987.

Mercier, Raymond. "The Astronomical Tables of Rajah Jai Singh Sawa'i." *Indian Journal of History of Science* 19 (1984), 143–71.

Mirza Itesa Modeen. *Shigurf Namah I Velaët, or Excellent Intelligence Concerning Europe; Being the Travels of Mirza Itesa Modeen, in Great Britain and France.* Trans. James Edward Alexander. London: Parbury, Allen, 1827.

Mirza Salih Shirazi. "Safar Namah-'i Isfahan, Kashan, Qum, Tihran." *Majmu'ah-'i Safar namah-hayi Mirza Salih Shirazi.* Tehran: Nashr-i Tarikh-i Iran, 1364/1985.

Montesquieu. *Persian Letters.* 1721. New York: Penguin, 1973.

Moorcroft, William. *Travels in the Himalayan provinces of Hindustan and the Panjab; in Ladakh and Kashmir; in Peshawar, Kabul, Kunduz, and Bokhara . . . from 1819 to 1825.* London: J. Murray, 1841.

Müller, Max. *The Sacred Languages of the East.* New York: Pantheon Books, 1978.

Nashat, Guity. *The Origins of Modern Reform in Iran, 1870–80.* Urbana: University of Illinois Press, 1982.

Ouseley, William. *Travels in Various Countries of the East, More Particularly Persia.* London: Rodwell and Martil, 1819–23.

Price, William. *A Grammar of the Three Principal Oriental Languages, Hindoostani, Persian, and Arabic on a Plan Entirely New, and Perfectly Easy; to Which is Added, a Set of Persian Dialogues Composed for the Author, by Mirza Mohammed Saulih, of Shiraz; Accompanied with an English translation.* London: Kingsbury, Parbury, and Allen, 1823.

———. *Journal of the British Embassy to Persia; Embellished with Numerous Views Taken in India and Persia; Also, A Dissertation upon the Antiquities of Persepolis.* 2 vols. London: Thomas Thorpe, 1932.

Pulzer, Peter. *The Rise of Political Anti-Semitism in Germany and Austria.* New York: Wiley, 1964.

Rafi' al-Din Khan 'Umdat al-Mulk, Muhammad. *Rafi' al-Basar.* Calcutta: C. V. William Press, 1841.

Razaullah Ansari, S. M. "Introduction of Modern Western Astronomy in India during 18–19 Centuries." *History of Astronomy in India,* ed. S. N. Sen and K. S. Shukla. New Delhi: Indian National Science Academy, 1985.

Roberts, J. M. *The Triumph of the West*. London: British Broadcasting Corporation, 1985.
Sahlins, Marshall. *Culture and Practical Reason*. Chicago: University of Chicago Press, 1976.
Said, Edward. *Orientalism*. New York: Vintage, 1979.
Salis, Mahdi Akhavn. "Akhir-i Shahnamah." *Akhir-i Shahnamah*, 8th ed. Tehran: Intisharat-i Murvarid, 1363/1984.
Schwab, Raymond. *The Oriental Renaissance: Europe's Rediscovery of India and the East, 1680–1880*. Trans. Gene Patterson-Black and Victor Reinking. New York: Columbia University Press, 1984.
Sebeok, Thomas A. *Portraits of Linguistics: A Bibliographical Source Book for the History of Western Linguistics, 1746–1969*. Westport, Conn.: Greenwood Press, 1976.
Shahnavaz Khan, Samsam al-Dawlah. *Ma'asir al-Umara*. Ed. Mawlavi 'Abd al-Rahim and Mawlavi Mirza Ashraf 'Ali. Calcutta: Asiatic Society of Bengal, 1892.
Sharma, Virendra Nath. "Jai Singh, His European Astronomers and Its Copernican Revolution." *Indian Journal of History of Science* 18, no. 1 (1982), 333–44.
———. "Zīj Muhammad Shahi and the Tables of de La Hire." *Indian Journal of History of Science* 25, nos. 1–4 (1990), 36–41.
Shayegan, Dariush. *Cultural Schizophrenia: Islamic Societies Confronting the West*. Trans. John Howe. London: Saqi Books, 1992.
Shklar, Judith. *Montesquieu*. Oxford: Oxford University Press, 1987.
Shushtari, Mir 'Abd al-Latif. *Tuhfat al-'Alam va Zayl al-Tuhfah*. Ed. Samad Muvahhid. Tehran: Tahuri, 1363/1984.
Soudavar, Abolala. "European and Indian Influences." *Art of the Persian Courts: Selections from the Art and History Trust Collection*. New York: Rizzoli International, 1992.
Tavakoli-Targhi, Mohamad. "Going Public: Patriotic and Matriotic Homeland in Iranian Nationalist Discourses." *Strategies* 13, no. 2 (2000), 174–200.
Toulmin, Stephen. *Cosmopolis: The Hidden Agenda of Modernity*. 2nd ed. Chicago: University of Chicago Press, 1992.
Von Grunebaum, G. E. *Modern Islam: The Search for Cultural Identity*. Berkeley: University of California Press, 1962.
Weber, Max. *The Protestant Ethic and the Spirit of Capitalism*. New York: Scribner, 1958.
Wright, Denis. *The English amongst the Persians during the Qajar Period, 1787–1921*. London: Heinemann, 1977.

Part IV

Early Modernities of Tibetan Knowledge

CHAPTER ELEVEN

New Scholarship in Tibet, 1650–1700

KURTIS R. SCHAEFFER

The colophon to a minor medical work contained in several editions of the Tibetan Buddhist canon hints at a scene of great interest for the history of cultural exchange in late seventeenth-century South Asia. The year is 1664. The setting is the Potala Palace in Lhasa, then less than two decades old and not yet topped with the imposing Red Palace raised high above its white base. We are led to the courtyard in front of the palace, where a crowd has assembled. It is a gathering of various people from India, among whom Tibetan officials have heard that there is a man schooled in medicine. He is found, questioned about his background, found to be qualified, and for the next three years is employed by the Fifth Dalai Lama's court to procure and translate medical texts.

This physician was only one of several dozen South Asian intellectuals active at the Tibetan court in the latter decades of the seventeenth century. Their presence was made possible by the consolidation of power under the Fifth Dalai Lama, Ngawang Lobsang Gyatso (1617–82), who had been granted rule over Tibet in 1642 by the Mongol leader Gushri Khan following his decisive support of the Dalai Lama's faction in the central Tibetan civil war. These scholars wrote little, but their relevance for new intellectual developments in Lhasa is significant when viewed in light of new interests in medicine, grammar, astrology and divination, and poetics on the part of the Tibetan court. Nowhere is this interest displayed more often than in the writings of the last great regent of the Fifth Dalai Lama, Sangyé Gyatso (1653–1705), leader of the central Tibetan government from 1679 to 1702, who focused his writing efforts on the very forms of knowledge in which these Indian intellectuals specialized. When viewed together as two aspects of a broad complex of changes resulting from the formation of the new Tibetan court at Lhasa in 1642, Indians at the court and Sangyé Gyatso's literary output suggest how the prescription and practice of the five Buddhist classical forms of knowledge (*vidyāsthāna, rig gnas lnga*)—arts and crafts, medicine, grammar, logic, and Buddhist contemplative praxis—underwent significant changes in the late seventeenth century.

IIIIIIIIIIII

The rubric of the five forms of knowledge regularly employed by Tibetan writers was formalized by the fourth century CE at the latest, when it appears in Asaṅga's *Yogācārabhūmi* in a discussion of the "non-Buddhist" subjects to be studied by the bodhisattva. All knowledge to be acquired by the bodhisattva is subsumed under (1) interior or insider knowledge (*adhyātmavidyā, nang rig pa*), (2) logic (*hetuvidyā, gtan tshigs rig pa*), (3) language (*śabdavidyā, sgra rig pa*), (4) medicine (*cikitsāvidyā, gso ba rig pa*), and (5) arts and crafts (*karmasthānavidyā, bzo rig pa*).[1] Logic, language, and medicine have all been formalized into treatises according to Asaṅga, while the practical arts of the world, such as metalworking and jewelry making, should be learned from those who are skilled in those fields. This group was later codified (and reorganized in the commentarial literature) in a verse in the *Ornament for the Sūtras of the Great Way*:

> If [the bodhisattva] does not endeavor in the five forms of knowledge,
> Even a supreme noble will not attain omniscience.
> Therefore, in order to refute and to lead others,
> And to become omniscient oneself, endeavor in these.[2]

To these major forms, five "minor" forms are often added, making up ten forms of knowledge that are often prescribed for the bodhisattva: arts and crafts, medicine, language, logic, Buddhism, poetics (*kāvya, snyan ngag*), prosody (*chandas, sdeb sbyor*), synonyms (*abhidhāna, mngon brjod*), dramaturgy (*nāṭaka, zlos gar*), and astrology and divination (*gaṇita* or *jyotis, skar rtsis*).[3] This group of ten was the common formulation received and adapted in Tibet, though there were certainly others. Sangyé Gyatso provides an encyclopaedic survey of different lists found in classical Buddhist literature and select non-Buddhist treatises: ninety-eight arts studied by the Buddha in the *Lalitavistara Sūtra,* sixty-four in a commentary on the *Vinayavastu,* thirty practical arts, eighteen types of music, seven types of singing, and the nine dramatic airs (*rāsas*) from the ninth-century Tibetan-Sanskrit glossary known as the *Mahāvyutpatti,* sixty-four aspects of erotic practice according to the *Treatise on Desire,* and so forth.[4]

Despite this plethora of systems and lists, the rubric of five major forms of knowledge was the principal structure that guided Tibetan treatises on the arts. This way of organizing knowledge has been more important for Tibetans writing on language arts and medicine than for those writing about subjects classified as Buddhist. Since the fifteenth century logic has been considered a de facto Buddhist activity proper to monastic education, specifically the process of developing or gaining insight into the nature of reality, and thus has been differentiated from arts and crafts, medicine, the language arts, and astrology and divination. Sangyé Gyatso summarizes the categories' relevance in terms of the "five non-confusions" in regard to intellectual pursuits: "If you know poetics, you will not be confused in verbal ornamentation. If you know synonyms, you will not be confused regarding names. If you know dramaturgy, you will not be confused regarding languages. If you know divination you will not be confused regarding mathematics. Also, if you know poetics you will not be confused in verbal ornamentation."[5]

Tibetan translators inherited this rubric of the five forms of knowledge from the Sanskrit treatises of Indian Buddhism as early as the ninth century, though subsequent writers infused it with a life of its own in Tibet. In systematic treatises on the five forms of knowledge, Tibetan scholars not only describe individual arts, but also theorize the nature of cultural practice, artistic tradition, and religion. In such works we see how Tibetan intellectuals conceived the relationship between diverse practices—from astronomy and astrology to architecture, from logic to last rites, from medicine to meditation. Linking these practices was the Buddhist rhetoric of the bodhisattva—the ethically perfect and, one might say, fully civilized individual—for each form of knowledge was an essential component of his or her training. By the late seventeenth century scholars in turn linked the bodhisattva ideal, and thus the arts and sciences, to the ideology of the Tibetan central government, just as visiting scholars from India were infusing discussions of knowledge forms with new life.

THE AUTOBIOGRAPHY of the Fifth Dalai Lama is the principal source for information on Indian travelers in late seventeenth-century Tibet. This long work mentions nearly forty Indian travelers received at the Lhasa court between 1654 and 1681,[6] the majority of whom were heralded as experts either in medicine or language arts. The year 1677 was significant for Indian visitors at court; eighteen visitors, or almost half of the total number mentioned in the autobiography as a whole, arrived in that year. Indian scholars could spend considerable periods of time in Lhasa under the patronage of the Dalai Lama. Gokula, a Brahman scholar from Varanasi who is mentioned five times between 1654 and 1664, spent a decade or more either in residence in Lhasa or traveling between India and Lhasa. The scholar Kṛṣṇodaya is mentioned once in 1672 and again in 1677, suggesting either continuous residence or multiple return journeys, and the physician Manaha seems to have stayed for three years, from 1675 to 1677. Varanasi emerges as the most frequent point of origination in India, as fully ten of the scholars are said to be from that city. However, this does not necessarily mean that every figure was a native of the city. Although Gokula is said to be a scholar of Varanasi, a letter addressed to him from the Dalai Lama indicates that he was born in the region of Malaya.

A visit to the Dalai Lama was certainly an economic venture for Indian scholars, for the Dalai Lama bequeathed various items to his visitors, most usually gold, but also cotton, tea, clothing, silk, Chinese red satin, and provisions for the road. Indeed many of the entries record only that the Dalai Lama met a certain Indian and gave him certain goods, as two examples illustrate. In 1677 the Dalai Lama met with Paṇḍita Kṛṣṇodaya, a Brahman from the Kuru region of western India, and gave him four measures (*zho*) of gold, tea, cotton, clothing, silk, provisions for the road, and travel papers for his return home.[7] Again in 1677 he gave *sannyāsin*s from Mathurā named Kṣemagiri and Nīlakaṇṭha three measures of gold each, and to a Brahman of Varanasi named Sītādāsa two measures of gold. He also provided them with documents allowing them to travel as they wished between India and Tibet. The Dalai Lama routinely issued such travel documents to his Indian visitors. These were likely bilingual, for we read that in 1671 Dar Lotsawa translated a travel document into "an Indian vernacular, for the benefit of those going to India."[8] Occasionally the variety of items brought to Lhasa by travelers are mentioned, giving some sense of the trade that occurred along with intellectual pursuits. In 1677 the Brahman Devānanda brought tortoises and yellow orpiment, otherwise known as arsenic, to the court, for which he was given half a measure of gold, three loads of tea, seven lengths of cloth, red wool, a travel document, and other presents.[9] In 1672 Kṛṣṇodaya brought soil from Indian holy sites fashioned into pills as an offering, as well as leaves from the bodhi tree. He composed Sanskrit verses of homage to the Dalai Lama, which Darpa Lotsawa translated. Because his finely crafted verses delighted all those present, he was given numerous gifts of gold and other valuables.[10] Some visitors brought pots of water from the River Ganga to offer up to the Dalai Lama, as did two Brahmans from Varanasi named Haridāsa and Jayadāsa in 1675, for which they were given rice and flour.[11] A *sannyāsin* from the eastern Indian region of Rāṇadeva did the same in 1676 and received five measures of gold for his efforts.[12]

Scholars are the most prominent Indian visitors to the Potala, but merchants and messengers from regional rulers in India also came before the Dalai Lama, though they usually received no more than the merest mention in the autobiography. In 1668, for instance, the Dalai Lama met with a messenger of the king of Gaghaṭa in India.[13] In 1672 he met with representatives from Nepal, India, Jumla, and Ngari, all termed kingdoms at the border.[14] In a letter of 1663 we read that the Varanasi scholar Gokula has sent a merchant to China.[15] Often Indians are mentioned in passing within longer lists of people with whom the Dalai Lama met on a given day. In 1669, for instance, he met with the merchant of Paṇḍita Gokula, some people from Amdo, and other Tibetan dignitaries.[16]

Two of the five arts and sciences, language and medicine, were the focus of intellectual contact between these Indian visitors and the Tibetan scholars of the Potala court. Two prominent Tibetan members of the court led the way in these areas: Dar Lotsawa Ngawang Puntsok Lhundrup (b. 1633) in the case of grammatical literature and Darmo Menrampa Losang Drakpa in medicine. In 1654 Dar Lotsawa began his study of Sanskrit at the age of twenty-one under the guidance of Gokulamiśra (Gokulanātha), a Brahman from Varanasi. According to the Dalai Lama's autobiography, "Through applying effort and analysis, within ten months [Dar Lotsawa] became proficient in [Sanskrit], and subsequently his spoken language became very good."[17] When Gokula returned to India his elder brother, the scholar Balabhadra, was invited to visit "with a book of Pāṇinian grammar," and in 1658 Dar Lotsawa began studying Pāṇini with him.[18] By 1659 Dar Lotsawa had translated the text of Pāṇini's grammar and fashioned a summary chart.[19] In 1660 the Dalai Lama dispatched a letter so that the scholar Balabhadra and his brother might return safely to Varanasi.[20] On behalf of their efforts the two scholar brothers were each given a complete set of clothing, and together they were given 110 measures of gold and 200 half-ingots (*srang*) of gold, which "made them very happy."[21]

The same long-term interactions between Indians and Tibetans at the Dalai Lama's court occurred over medical scholarship. Recall the scene that opened this essay: the front of the Potala and the search for an Indian scholar well versed in medicine.[22] The person conducting the search was none other that the scholar of Sanskrit grammar, Dar Lotsawa, who continued to work at the Potala through the 1660s and 1670s. According to him, Tibetan medical knowledge was very likely missing out on Indian ayurvedic knowledge, and not for lack of opportunity. Indian experts in ayurveda had likely traveled to Tibet in large numbers in the past. Yet, as in the case of the Indian scholar Smṛtijñānakīrti, who traveled through central Tibet offering his scholarly services but was ignored because he could not speak Tibetan and no one who met him could speak his Indic dialect, so must Indian medical scholars have passed through Tibet unnoticed. Not wanting to miss another such opportunity, in 1664 he located a *sannyāsin* named Godararañja. Godara was born "in a pure Brahman family in the west-Indian region of Maharashtra." He became a renunciant, traveled "like water" throughout India and Nepal, and under the tutelage of four teachers became a scholar. Dar Lotsawa promptly employed Godara to study under the Tibetan physician Darmo Menrampa. This physician was another major figure at the court of the Dalai Lama, who also composed a work on medicinal substances with the assistance of Indian informants such as Godara.[23] The Tibetan showed him the many wondrous medical practices and charged him to return south in search of types of medicine not found in Tibet. It appears that Godara traveled only as far as the Kathmandu Valley and, with the assistance of the king and ministers of Kathmandu, was able to draw up a synthesis of medical systems. He returned to Lhasa with many types of medicine, and in the summer of 1677 Godara's summary of ayurvedic instructions was translated into Tibetan by Dar Lotswawa and edited by Darmo Menrampa. The Fifth Dalai Lama himself composed the dedicatory verses for the translation.

Godara's work is now contained along with five other medical works in the ever-popular "miscellaneous" section of later redactions of the Tibetan canon.[24] All six works were composed or redacted by Indian medical scholars at the Potala under the patronage of the Fifth Dalai Lama. Of the several medical scholars in this group, the name Manaha stands out. We first read of Manaha in an entry dated 1675 in the Dalai Lama's autobiography, where he is identified as an Indian physician learned in many aspects of the healing arts, particularly ophthalmology, who was then residing in Yorpo. He was entreated to come to Lhasa, where Darmo Menrampa began to study with him and thus became skilled in treating eye problems. For as long as this physician stayed in Lhasa the Fifth Dalai Lama provided for him and, upon his return, gave him gifts and a passport.[25] His four-folio treatise on ophthalmology, appropriately titled *The Eye-Opener Which Is Meaningful to Behold,* styles itself as "an arrangement in notes of the practical knowledge of Manaha, physician of the Indian Shah Jahan, from Baripur region."[26] Though short, it appears to have been of great relevance to the Fifth Dalai Lama. In 1677 Darmo Menrampa's work with Manaha on ophthalmology enabled him to successfully operate on cataracts developing in the Dalai Lama's eyes.[27] At first the Fifth Dalai Lama's left eye was losing clarity; then his right eye gradually was affected. Darmo applied a salve (*sman mar*) to his eyes, and for a few months they cleared up. When the condition flared up again Darmo employed techniques learned from two

Indian scholars, Gautal and Manaha, whereupon the Dalai Lama's eyelids began to sting and break out in a rash, but eventually his eyes cleared up.

IIIIIIIIIIII

I WILL SUGGEST later that it is not coincidental that the major writings of Sangyé Gyatso are dedicated to just those forms of knowledge in which the Indian visitors at the Potala were expert. But first we must gain a sense of the scope and contents of his work. The Fifth Dalai Lama composed only two works that fall clearly within the five forms of knowledge. The first was his commentary on Daṇḍin's *Mirror of Poetry*, written in 1656, a work that waxes eloquent on the five forms of knowledge in its introductory verses.[28] In the same year he composed a response to queries regarding select aspects of divination and the forms of knowledge. While the second half of this work provides responses to questions regarding the calculation of intercalary months and the origins of Chinese divination, it begins with an overview of the five major and five minor forms of knowledge.[29] The Dalai Lama's contribution to the new scholarship was not his writing, however, but his promotion and patronage. In addition to sponsoring Indian scholars at the court, he founded a medical college in 1653 at his home institution of Drepung Monastery near Lhasa and sponsored the printing of *Four Medical Tantras* in 1670 as well as the composition of new medical commentaries and biographies of the two iconic Tibetan physicians, the Elder and Younger Yutoks. (Yutok the Elder is held to have been active at the court of the imperial Tibetan ruler Trisong Detsen.)[30]

Nevertheless the Dalai Lama's efforts in no way matched those of his protégé, Sangyé Gyatso. Many Tibetan writers before and since Sangyé Gyatso composed works on the ten forms of knowledge, but no other Tibetan author expended so much effort addressing what are usually considered worldly (*'jig rten pa*) matters, topics largely judged by tradition not to merit extended exegesis over the course of one's career despite the prescriptions of *The Ornament for the Sutras of the Great Way*. Unlike most Tibetan Buddhist writers, whose collected works comprise primarily writings on Buddhist praxis, Sangyé Gyatso wrote on governance, history, biography, medicine, astrology, language arts, and the politics of ritual. Artistic techniques, funeral rites, astrological and calendrical theories, methods of healing, and rules for court servants—all of these find ample explication in his writings.

Nearly all of his works contain passages on the fivefold rubric of knowledge forms. Even his earliest work (completed in 1679, when he was made ruler of the Tibetan government at the age of twenty-six), *Guidelines Clarifying Regulations and Prohibitions* [*for Officials*], which offers in its introduction elements of a political theory, makes passing mention of the forms of knowledge in its opening verses.[31] Aside from his biographical writing on the Fifth Dalai Lama, Sangyé Gyatso's principal works include *The White Beryl* (completed 1685), on astrology and divination; *The Blue Beryl* (completed 1688), on medicine; the *Supplement* to the third section of *Four Medical Tantras* (1691); *Clarifications to the White Beryl* (1698), a lengthy response to criticisms of *The White Beryl*; *The Yellow Beryl* (1698), a history of the Gelukpa School of Buddhism; and *The Beryl Mirror* (1703), a history of medicine. The image of the beryl employed in the titles of five of the six works is drawn from the opening scenes of *Four Medical Tantras*, though each work uses the term differently: a necklace of white beryl in the case of his divination treatise, a blue beryl necklace in the medical work, and a yellow beryl mirror in the history if the Gelukpa School.[32]

Sangyé Gyatso's 654-folio treatise on astrology and divination, *The White Beryl*, was composed between 1683 and 1685, immediately following the Fifth Dalai Lama's death. It is heavily indebted to the work of Pukpa Lundrub Gyatso, *White Lotus Oral Instruction* (1447), which gave rise to the so-called Pukpa tradition of divination.[33] *The White Beryl*'s thirty-five chapters cover a vast range of mathematical, astronomical, astrological, divinatory, historical, and cultural topics, including chronologies of the Buddha's teachings (chapter 3), the calculation of solar and lunar eclipses (9), the history of Chinese divination (20) and its principles (21 and 22), marital, medical, and funerary divination (23, 25, and 29), geomancy (31 and 32), and the five forms of knowledge (35). Chapter 35 is Sangyé Gyatso's earliest treatment of the subject, taking up in turn the various groupings of the arts and sciences found in classical Buddhist literature, and then the history of language and literature, logic, arts and crafts, medicine, and Buddhism.[34] *The White Beryl* provoked controversy among central Tibetan intelligentsia to such a great extent that one writer, the otherwise anonymous Ngawang of Chongyé, submitted a lengthy criticism to Sangyé Gyatso consisting of 208 questions on specific topics in *The White Beryl*. More impressive is the fact that the regent felt compelled to respond to these criticisms by composing an entirely new work "in between writing periods for affairs of state and the composition of works on medicine" during the years 1687–88 of almost the same length as *The White Beryl* itself, *Clarifications to the White Beryl* (in 494 folios), and have these published along with Ngawang's queries. The response to queries regarding chapter 35 of *The White Beryl* represents not only Sangyé Gyatso's most detailed Tibetan presentation of the forms of knowledge and his lengthiest statement on Buddhism (treated fittingly as the last of the forms of knowledge and not as an isolated topic), but the most extended treatment of the five forms of knowledge in Tibetan literature as a whole.[35] In 1687 Sangyé Gyatso started researching medicine more seriously as he began to write a commentary on *Four Medical Tantras*, and in 1688 he completed his *The Blue Beryl*,[36] arguably the single most important commentary on the classic Tibetan medical treatise and the standard against which all later medical scholarship would have to be measured. Its four sections follow the order of the *Four Tantras* themselves. *The Root Tantra* provides mnemonic aids for diagnosis and treatment, overviews of human anatomy, and a summary of the contents of the work as a whole. *The Explanatory Tantra* sets forth medical theory under a series of topics, including pathology, conduct and diet, diagnosis, methods of treatment, and the qualifications and conduct of the physician. *The Instructional Tantra* discusses the etiology of various diseases and prescribes medication and treatment. This chapter is generally regarded as the most challenging in the book due to its detailed sections on medical plants and complex forms of treatment. Sangyé Gyatso was not wholly satisfied with his commentary on this chapter, and his *Supplement* was an attempt to present a synthetic treatment in verse of its practical instructions. *The Final Tantra* describes the techniques which underlie treatment, including sphygmology, urology, the preparation of medical compounds and purgatives, phlebotomy, and moxibustion.[37]

After almost a decade of writing biography, ritual encyclopaedias, and what we might think of as prescriptive place guides, such as his work on the circumambulation routes in Lhasa,[38] in 1698 Sangyé Gyatso turned to historiography in *The Yellow Beryl*, the largest history of the Gelukpa School up to that time. After a biography of the school's founder, Tsongkhapa, the work relates the history of the three major Gelukpa monasteries near Lhasa—Ganden, Sera, and Drepung—which collectively housed more than eight thousand monks at the time. It then provides a survey of nearly six hundred monasteries and temples either founded or converted by the Gelukpa throughout Tibet. The last quarter of the work is given over to the life and works of the Fifth Dalai Lama and the regent, ending with a brief account of the birth and education of the Sixth Dalai Lama, who was enthroned in Lhasa in 1697, only a year before the completion of *The Yellow Beryl*.

Sangyé Gyatso's last major work combined his interests in history and medicine. *The Beryl Mirror: The Interior Analysis of the Glorious Medical Art* largely consists of a history of medicine from ancient India up through imperial Tibet to Sangyé Gyatso's time. It also prescribes the place of medicine within the forms of knowledge and within the career of the bodhisattva. After introductory remarks on the definition of medicine (16–20) he narrates the spread of medical knowledge from the gods (20–40) to the human realm and ultimately to the Buddha himself (42–86) and beyond to Indian scholars (90–148) such as Vāgbhaṭa. Sangyé Gyatso begins Tibetan medical history with the international scene in the imperial Tibetan capital of Yarlung in central Tibet, where scholars from Persia, India, China, Nepal, and the Himalayan regions are said to have gathered at the request of Emperor Trisong Detsen (148–76). The central founding figure of Tibetan medicine, Yutok Yönten Gönpo, is the subject of a longer biographical passage (206–29). Before discussing the subsequent history of medicine in Tibet, Sangyé Gyatso presents a long schematic introduction to Buddhist textual traditions, dividing the literature into the Buddha's Word (229–44) and Hīnayāna and Mahāyāna literature (244–46, 246–53), concluding with remarks on the character and scope of Buddhist exegetical literature (*śāstra*, 253–74). These passages make no mention of medical literature and might be found in any Tibetan work on Buddhist literature, such as the first half of Büton's history of Buddhism. In Sangyé Gyatso's work this long section is, however, a lengthy preamble to his claim that *Four Medical Tantras* is the Word of the Buddha and not an indigenous Tibetan composition (274–84), a matter of heated debate for centuries before the regent's day. Returning to historical topics, he then relates the later history of Tibetan medical traditions, treating major schools such as the Drangti lineage at Sakya Monastery (291–96), the Jang school of western Tibet (306–11), and the Zur school of central Tibet (329–64), which was of great importance for Sangyé Gyatso's own formulation of medical scholarship and practice. The historical portion of the work ends with the contemporary activities of both the Fifth Dalai Lama (64–372) and Sangyé Gyatso himself (372–96) in promoting medicine, where we also find mention of the Indian physicians active at the court. There follows extensive treatment of three distinct yet related themes: the proper methods to study medicine (396–411, 537–47), the three Buddhist vows (monastic vows, the bodhisattva vow, and the tantric pledges, 411–508), and the characteristics of teachers and students of medicine (508–47). This last section digresses on the relation between medicine and the arts and crafts, which are in turn divided into the common threefold rubric of body, speech, and mind: work with the hands makes possible medical analysis, the physician's voice provides encouragement to the patient, and continual learning keeps the mind clear (513–26). *The Beryl Mirror* is also significant because it includes fragments of an intellectual autobiography. (It appears that since he was not a cleric or a recognized incarnation, he did not merit a full-fledged biography, and biographies do not appear until the late nineteenth century.)[39] Here he portrays himself as a young scholar-statesman completely dedicated to a life of learning.[40]

IIIIIIIIIIII

EVEN SO BRIEF a survey of Sangyé Gyatso's major writings illustrates the extent to which the forms of knowledge as a structured group served as a classical guideline for his work, providing him with a vision of culture and intellectual practice beyond the confines of Buddhist literature more narrowly defined. While he certainly drew heavily from the writings of previous scholars, his major works on medicine and divination are unprecedented in scope. It was customary for prolific Tibetan scholars to write something on these topics in addition to their voluminous efforts in philosophy, esoterica, and ritual, yet Sangyé Gyatso's nearly exclusive focus on the "non-Buddhist" forms of knowledge is unique. He concentrated precisely on those forms of knowledge that did not fit neatly into Buddhist scholarship, principally medicine and divination. If we review his corpus from the perspective of a single form of knowledge, such as Indian poetics and poetry, the pervasive influence of the forms of knowledge is also apparent. From his early commentary on *Four Medical Tantras* to his last work on the history of medicine, Sangyé Gyatso reveled in displaying a command of Indian poetics.[41] Daṇḍin's *Mirror of Poetry* frames his major works, either in the form of direct quotation or through introductory verses composed with poetic figures drawn from Daṇḍin's work.[42] Thirty-seven ornate *kāvya* verses begin *The Yellow Beryl*'s history of the Gelukpa School; 810 verses conclude *Treasury of Blessings*, his encyclopaedic catalogue of the Fifth Dalai Lama's reliquary and its attendant chapels in the Potala.[43] Sangyé Gyatso's three-volume continuation of the Fifth Dalai Lama's autobiography is composed in *campūkāvya*, or mixed prose and verse. (This is in fact a major difference between Sangyé Gyatso's addition and the Fifth Dalai Lama's autobiography proper, which contains no ornate verse.) All the verses moreover employ the poetic figures of Indian poetic theory, and each one is labeled with the name of a specific poetic figure that it exemplifies. In the mid-1690s Sangyé Gyatso extracted these verses to create a poetic biography of the Fifth Dalai Lama comprising nearly twenty-four hundred verses and concluding pyrotechnically with seven poems of praise in visually complex acrostic diagrams, based as well upon Daṇḍin's treatise. Often he cites verse 1.14 of *The Mirror of Poetry*, which introduces the characteristics of *mahākāvya*, or "great poetry,"[44] with the injunction that each work of great poetry must include benedictory verses. Lest the reader is not aware of this, Sangyé Gyatso typically includes this injunction at the close of his introductory verses, thereby insinuating that his compositions are indeed works of great poetry exhibiting the principles laid down in the classic Indian treatises.

If it is readily apparent that Sangyé Gyatso was directly influenced by classic formulations of knowledge in both his choice of topics and his literary style, it is more difficult to assess his relationship to Indian intellectuals at the court. The vast majority of Indians at the court were active during the 1670s, when Sangyé Gyatso would have been a young man in his late teens and early twenties engaged in a formal education at the court. We know that he studied with Tibetans who had worked with Indian scholars, but we have no evidence that he studied with Indians at this time. Nevertheless it is clear that he not only knew of these figures, but considered them to be important in the intellectual scene of his time, for only a decade after their visits he includes Balabhadra and Gokula in his historical survey of grammatical treatises and their Tibetan translations in *The White Beryl* in 1688.[45] The contributions of Indians at the court also figure into his history of medicine (1703), where they are noted in a discussion of the Fifth Dalai Lama's promotion of medical knowledge and practice.[46] Yet despite the fact that Manaha's work on ophthalmology provided some practical assistance to the Fifth Dalai Lama, Sangyé Gyatso does not reference it either in *The Blue Beryl* or his supplemental commentary on *The Instructional Tantra*. The Indian intellectuals active at the court in the decades immediately preceding his maturity were indeed part of Sangyé Gyatso's history of the present—a crucial aspect of his presentation of the Ganden government's achievements. They appear on the whole, however, to have had little direct impact on his medical writings.

Sangyé Gyatso did, however, work with Indians during his own scholarly career as he completed his commentary on *Four Medical Tantras*. At the same time he also worked on illustrations of the medicinal plants mentioned in *The Instructional Tantra*. He was fully aware that plants varied by region, and so contracted informants to provide descriptions of plants and verification of the illustrations. Some of these informants were from India, although Sangyé Gyatso disparages them as pseudo-scholars who were not valid sources of knowledge. Despite the fact that "the vast majority were fakes," he nevertheless made use of physicians from Magadha and Kashmir, East India and Nepal, as well as people from the Dakpo and Kongpo regions of southeastern Tibet who could not be called physicians (*sman pa*) but who possessed some expertise in plant identification.[47]

The collaborative scholarship on grammar and medicine between Indian and Tibetan intellectuals from 1650 through 1680 and the writings of Sangyé Gyatso on similar topics in the 1680s and 1690s coincide in part with new intellectual trends in certain regions of India during the seventeenth century. As Sheldon Pollock and others in this volume have shown, urban centers in northern India and throughout South Asia were increasingly centers for scholarly exchange during this period, serving as meeting points for mobile intellectual classes. "Varanasi in the seventeenth century," Pollock suggests, "witnessed a confluence of more or less free intellectuals . . . of a sort it had almost certainly never seen before, certainly not in such numbers."[48] What did Sangyé Gyatso, the Fifth Dalai Lama, and the other scholars of the Tibetan court know of these developments? This is a difficult question to answer with the materials at hand. While at least forty Indians are mentioned by name in the Fifth Dalai Lama's autobiography over a period of almost four decades, indications as to what these visitors and the Dalai Lama may have spoken about are largely absent, save in a few passages. In an entry dated 1677 he questions his guests on their skills and religious background. The two Brahmans from Varanasi, Jīvanti and Gaṇeśa, respond that they are learned in the science of mathematics and that they are followers of Vishnu. The passage ends too quickly, however, as we are told that for their visit they are rewarded with the usual measures of gold, a box of tea each, and woolen cloth.[49] When the Dalai Lama does offer descriptive remarks about his Indian guests, they are nothing but laudatory, as he praises both India and its scholars. In a letter dated 1663 he addresses the grammarian Gokula as a "son of the world's grandfather, Brahma, supreme in effort among those who speak of the Vedas." In a letter of 1670 to Gokula and his brother, Balabhadra, he praises Varanasi as "the great city where gather many scholars of vast intellect, skilled in all linguistic and philosophical topics." He bids the scholars farewell as they prepare to return south through the Himalayas, having illuminated the darkness of Tibet with the moonlight of Pāṇini's grammar. If we cannot make too much of these passages, they do at least suggest that Varanasi was renowned for its scholars and that being known as a scholar from Varanasi entailed a fair degree of prestige in Lhasa.

As I mentioned earlier, Tibetan scholars at the court explicitly compared the visits of Indian intellectuals to Lhasa with the golden age of imperial Tibet, when teams of Indian and Tibetan scholars collaborated to translate Buddhist sutras, tantras, and the treatises of the great Indian Buddhist masters. The new influx of Indian figures was certainly seen as something of a return to the imperial golden age under the Fifth Dalai Lama. Yet there was a crucial difference between the intellectual projects with contemporary scholars and those who had visited Tibet in the past or who had played host to Tibetans in India. This had nothing to do with Buddhism. The exchange was centered entirely on medicine, grammar, and language arts—in short, the major forms of knowledge save the "interior knowledge" of Buddhist praxis itself. This is not wholly an accident of the "post-Buddhist" phase, for to the activities of the court we may contrast a reconnaissance mission to Bodhgaya in 1752, the purpose of which was to seek out vestiges of Buddhist culture.[50] This was an attempt to expand the Tibetan repertoire of the forms of knowledge with new findings from India, an attempt which (though this cannot be fully argued here) is bound up with the Lhasa court's rhetorical efforts to portray itself as no less than a bodhisattva charged with protecting Tibet by all means available.

In broader geographic terms the significant presence of Indian figures at the Dalai Lama's court suggests that the position of Lhasa within the social networks of South Asian intellectual life during the seventeenth century be considered in greater detail than I have been able to offer here. Throughout the latter half of the century a more or less continuous exchange of goods, ideas, and practices took place between Lhasa and the cities of India, most prominently Varanasi. Yet one point that merits further consideration is the directionality of this exchange as portrayed in the Tibetan literature. None of the works considered here provide evidence that Indians transmitted anything of intellectual import back to their homelands along with the gold bequeathed to them. Nor do we hear of any Tibetan scholar making the journey south to Varanasi, though we know that Tibetans continued to travel at least to Bodhgaya well into the eighteenth century. The writings of the Fifth Dalai Lama and Sangyé Gyatso portray the relation between the Dalai Lama and visiting Indian intellectuals as that of patron and patronized, a relationship reminiscent of the Tibetan imperial patrons—figures of great importance for the Fifth Dalai Lama's leadership persona in other contexts—and their more famous Buddhist visitors from India. The Dalai Lama, Sangyé Gyatso, and the new Tibetan intellectuals sit above in the Potala while the Indians stand below in the courtyard, waiting to be called upon to provide their particular expertise. The anecdote at the beginning of this essay is rhetorically inclined upward, toward the Dalai Lama, a viewpoint that can yield only a certain perspective. For alternate perspectives on the new intellectual interactions between Indians and Tibetans in Lhasa at the end of the seventeenth century, we will have to look elsewhere.

Notes

1. Asaṅga, *Rnal 'byor spyod pa'i sa las byang chub sems dpa'i sa,* fol. 52a.2–5, 57a. See also Asaṅga, *Rnal 'byor spyod pa'i sa rnam par gtan la dbab pa bsdus pa,* fol. 24b.2–3. Schmithausen, *Ālayavyijñāna,* 1:13–14, 187–89, discusses the periodization of this literature.

2. *Mahāyānasūtrālaṃkāra* 12.60, *Sde dge Bstan 'gyur, Sems tsam,* vol. Phi, fol. 15b.4–5. See Griffiths, "Omniscience in the *Mahāyānasūtrālaṅkāra,*" 99–101.

3. Ruegg, *Ordre Spirituel et Ordre Temporel,* 101–32, offers a bibliographic overview of the forms of knowledge.

4. On the *Lalitavistara Sūtra,* see Bays, *Voice of the Buddha,* 234–35. On the *Vinayavastu,* see Tshul khrims bskyangs, *Lung phran tshegs kyi rnam par bshad pa, Sde dge Bstan 'gyur, 'Dul ba,* vol. Dzu, ff. 1–232a.5. On the *Treatise on Desire,* see Gedün Chöpel, *Tibetan Arts of Love,* 63. See also Sangs rgyas rgya mtsho, *Bstan bcos baiḍūr dkar po las dris lan 'khrul snang g.ya' sel,* fols. 258a.6–265a.6: response to question 195.

5. Sangs rgyas rgya mtsho, *Bstan bcos baiḍūr dkar po las dris lan 'khrul snang g.ya' sel,* fol. 261b.4.

6. Tucci, *Tibetan Painted Scrolls,* mentions some fifteen of the forty occurring in the Dalai Lama's autobiography (see 71, 74–75, 137). Meyer, "The Golden Century of Tibetan Medicine," has brought the translation of medical works at the Potala to our attention, as has Verhagen, *A History of Sanskrit Grammatical Literature in Tibet,* the translations of grammatical works.

7. Ngag dbang blo bzang rgya mtsho, *Du ku'u la'i gos bzang,* 3:81.5.

8. Ngag dbang blo bzang rgya mtsho, *Du ku'u la'i gos bzang,* 2:241.1.

9. Ngag dbang blo bzang rgya mtsho, *Du ku'u la'i gos bzang,* 3:95.16.

10. Ngag dbang blo bzang rgya mtsho, *Du ku'u la'i gos bzang,* 2:305.7.

11. Ngag dbang blo bzang rgya mtsho, *Du ku'u la'i gos bzang,* 2:476.5.

12. Ngag dbang blo bzang rgya mtsho, *Du ku'u la'i gos bzang,* 2:413.11. See also 3:82.8.

13. Ngag dbang blo bzang rgya mtsho, *Du ku'u la'i gos bzang,* 2:127.5.

14. Ngag dbang blo bzang rgya mtsho, *Du ku'u la'i gos bzang,* 2:273.17.

15. Ngag dbang blo bzang rgya mtsho, *Du ku'u la'i gos bzang,* 1:671.20.

16. Ngag dbang blo bzang rgya mtsho, *Du ku'u la'i gos bzang,* 2:163.18.

17. Ngag dbang blo bzang rgya mtsho, *Du ku'u la'i gos bzang,* 1:443.2.

18. Ngag dbang blo bzang rgya mtsho, *Du ku'u la'i gos bzang,* 1:525.8.

19. See Verhagen, *A History of Sanskrit Grammatical Literature in Tibet,* 138–40, 154–55, 304–20.

20. Ngag dbang blo bzang rgya mtsho, *Du ku'u la'i gos bzang,* 1:586.11.

21. Ngag dbang blo bzang rgya mtsho, *Du ku'u la'i gos bzang,* 1:534.17.

22. Godararañjo, *Tshe'i rig byed mtha' dag gi snying po bsdus pa* (Ui, *Catalogue-Index of the Tibetan Buddhist Canons,* no. 4438), colophon 666.4.

23. See Blo bzang chos grags, *Bshad pa'i rgyud.* Darmo Menrampa's efforts at pro-

moting medical literature are mentioned on a number of occasions: Ngag dbang blo bzang rgya mtsho, *Du ku'u la'i gos bzang,* 2:239 (1671). On his efforts in printing and promoting the *G.yu thog cha lag bco brgyad,* see 2:459, 3:19. On his composition of G.yu thog Yon tan mgon po's *rnam thar,* see 3:340 (1680), 3:384 (1680).

24. Ui, *Catalogue-Index of the Tibetan Buddhist Canons,* nos. 4438–43. Meulenbeld, *History of Indian Medical Literature,* lists all of these works, though he does not trace any to Indic originals. One wonders whether it is accurate to refer to them as translations at all; perhaps they are better thought of as works cocreated by Indian and Tibetan scholars.

25. Ngag dbang blo bzang rgya mtsho, *Du ku'u la'i gos bzang,* 2:460.8.

26. Manaha, *Mig 'byed mthong ba don ldan* (Ui, *Catalogue-Index of the Tibetan Buddhist Canons,* no. 4443), fol. 360b.5.

27. Ngag dbang blo bzang rgya mtsho, *Du ku'u la'i gos bzang,* 3:64.4. See Meyer, "Golden Century of Tibetan Medicine," 103–4.

28. Ngag dbang blo bzang rgya mtsho, *Snyan ngag me long gi dka' 'grel,* 8–9. Eppling, "Calculus of Creative Expression," 2:1482–501, briefly surveys the major *kāvya* works of seventeenth-century Tibet. See also Dimitrov, *Mārgavibhāga,* 25–62.

29. Ngag dbang blo bzang rgya mtsho, *Rtsis dkar nag las brtsams pa'i dris lan,* 573–84.

30. See Rechung Rinpoche, *Tibetan Medicine,* 141–327.

31. Sangs rgyas rgya mtsho, *Blang dor gsal bar ston pa'i drang thig dwangs shel me long,* 203.

32. Sangs rgyas rgya mtsho, *Gso ba rig pa'i bstan bcos,* 1:9.5–14; Sangs rgyas rgya mtsho, *Dpal mnyam med ri bo dga' ldan pa'i bstan pa,* 3.12–14.

33. Lhun grub rgya mtsho, *Legs par bshad pa'i pad ma dkar po'i zhal lung.* See Schuh, *Untersuchungen zur Geschichte der tibetischen Kalenderrechnung,* 22–46, for an overview of calendrical and astrological literature, and Dorje, *Tibetan Elemental Divination Painting,* for a translation and exquisite facsimile of an illustrated manuscript on divination based upon Sangs rgyas rdo rje's *White Beryl.*

34. Sangs rgyas rgya mtsho, *Phug lugs rtsis kyi legs bshad*: *rig gnas* (429–30), *sgra rig* (430–33), *gtan tshigs rig pa* (433–36), *bzo rig* (436–41), *gso rig* (441–43), *nang rig* (443–48).

35. Sangs rgyas rgya mtsho, *Bstan bcos baiḍūr dkar po las dris lan 'khrul snang g.ya' sel*: presentation of the forms of knowledge (Question 195) vol. 1, fol. 258a.6; minor arts (Q196) 265a.6; language (Q197) 266b.2; logic (tshad ma, Q198–99) 282a.2; arts and crafts (bzo, Q200–203) 290b.2; medicine (gso, Q204) vol. 2, fol. 30a.4; Buddhism (nang rig, Q205) 39b.2–226a.6. Sangs rgyas rgya mtsho is in constant conversation with Shes rab rin chen's *Rig gnas kun shes nas bdag med grub pa,* which he quotes (*Rig gnas kun shes,* 72) at the beginning of his general presentation of the forms of knowledge and to which he refers throughout.

36. Meyer, "The Golden Century of Tibetan Medicine," 104–11, surveys Sangyé Gyatso's medical scholarship.

37. See Dorje, "Structure and Contents of the Four Tantras," for a succinct description.

38. See Schaeffer, "Ritual, Festival, and Authority under the Fifth Dalai Lama," and Schaeffer, "Salt and the Sovereignty of the Dalai Lama."

39. Skal bzang legs bshad, *Rje btsun byams pa mthu stobs kun dga' rgyal mtshan gyi rnam thar*, 213.4–227.5.

40. Sangs rgyas rgya mtsho, *Dpal ldan gso ba rig pa'i khog 'bugs*, 372–96, 547–56.

41. On his commentary, see Sangs rgyas rgya mtsho, *Gso ba rig pa'i bstan bcos*, 1:5.1.

42. Sangs rgyas rgya mtsho, *Man ngag yon tan rgyud kyi lhan thabs*, 1–6.

43. Sangs rgyas rgya mtsho, *Dpal mnyam med ri bo dga' ldan pa'i bstan pa*, 5; Sangs rgyas rgya mtsho, *Mchod sdong 'dzam gling rgyan gcig*, 970–1055.

44. Eppling, "Calculus of Creative Expression," 1:100.

45. Sangs rgyas rgya mtsho, *Phug lugs rtsis kyi legs bshad*, 432.5–7.

46. Sangs rgyas rgya mtsho, *Dpal ldan gso ba rig pa'i khog 'bugs*, 371.2.

47. Sangs rgyas rgya mtsho, *Gso ba rig pa'i bstan bcos*, 2:1462.22–1463.11.

48. Pollock, "New Intellectuals in Seventeenth-Century India," 21.

49. Ngag dbang blo bzang rgya mtsho, *Du ku'u la'i gos bzang*, 3:89.17.

50. Bsod nams rabs rgyas, *'Phags pa'i yul dbus dpal rdo rje gdan*.

References

TIBETAN SOURCES

Blo bzang chos grags, Dar mo Sman ram pa. *Bshad pa'i rgyud kyi le'u nyi shu pa sman gyi nus pa bstan pa'i tshig gi don gyi 'grel pa mes po'i dgongs rgyan*. Xining: Kan su'u mi rigs dpe skrun khang, 1997.

Bsod nams rabs rgyas. *'Phags pa'i yul dbus dpal rdo rje gdan du garsha'i rnal 'byor pa bsod nams rab rgyas gyis legs par mjal ba'i lam yig dad pa'i snye ma*. Manuscript preserved at the Toyō Bunko Library, Tokyo. 7 folios. Composed 1752.

Lhun grub rgya mtsho, Phug pa (fifteenth century). *Legs par bshad pa pad ma dkar po'i zhal gyi lung. Rtsis gzhung pad dkar zhal lung*. Beijing: Mi rigs dpe skrun khang, 2002.

Ngag dbang. *Snga na med pa'i bstan bcos chen po bai ḍūrya dkar po las 'phros pa'i snyan sgron nyis brgya brgyad pa*. Vol. 2 of *The 18th Century Sde-dge Redaction of the Sde-srid Saṅs-rgyas-rgya-mtsho's Vai Ḍūrya Gya' sel with the Sñan sgron Ñis brgya brgyad pa*. 2 vols. Dehra Dun: Tau Pon Sakya Centre, 1976.

Ngag dbang blo bzang rgya mtsho, Dalai Lama V (1617–82). *Rgya bod hor sog gi mchog dman bar pa rnams la 'phrin yig snyan gnag tu bkod pa rab snyan rgyud mang*. Xining: Mtsho sngon mi rigs dpe skrun khang, 1993.

———. *Rtsis dkar nag las brtsams pa'i dris lan nyin byed dbang po'i snang ba*. Vol. 20 of *The Collected Works (Gsung-'bum) of the Vth Dalai Lama, Ngag-dbang blo-bzang rgya-mtsho*. Gangtok: Sikkim Research Institute of Tibetology, 1992. Composed 1656.

———. *Snyan ngag me long gi dka' 'grel dbyangs can dgyes pa'i glu dbyangs*. Vol. 20 of *The Collected Works (Gsung-'bum) of the Vth Dalai Lama, Ngag-dbang blo-bzang rgya-mtsho*. Gangtok: Sikkim Research Institute of Tibetology, 1992. Composed 1656.

———. *Za hor gyi ban de ngag dbang blo bzang rgya mtsho'i 'di snang 'khrul ba'i rol rtsed rtogs brjod kyi tshul du bkod pa du ku'u la'i gos bzang*. 3 vols. Beijing: Bod ljongs mi dmangs dpe skrun khang, 1989.

Sangs rgyas rgya mtsho, Sde srid (1653–1705). *Blang dor gsal bar ston pa'i drang thig dwangs shel me long nyer gcig. Bod kyi dus rabs rims byung gi khrims yig phyogs bsdus dwangs byed ke ta ka*. Tshe ring bde skyid, ed. Bod ljongs mi dmangs dpe skrun khang. Lhasa: 1987.

———. *Bstan bcos baiḍūr dkar po las dris lan 'khrul snang g.ya' sel don gyi bzhin ras ston byed. The 18th Century Sde-dge Redaction of the Sde-srid Saṅs-rgyas-rgya-mtsho's Vai Ḍūrya Gya' sel with the Sñan sgron Ñis brgya brgyad pa*. 2 vols. Dehra Dun: Tau Pon Sakya Centre, 1976.

———. *Dpal ldan gso ba rig pa'i khog 'bugs legs bshad bai ḍūrya'i me long drang srong dgyes pa'i dga' ston*. Lanzhou: Kan su'u mi rigs dpe skrun khang, 1982.

———. *Dpal mnyam med ri bo dga' ldan pa'i bstan pa zhwa ser cod pan 'chang ba'i ring lugs chos thams kyi rtsa ba gsal bar byed pa baiḍūrya ser po'i me long*. Beijing: Krung go bod kyi shes rig dpe skrun khang, 1989.

———. *Drin can rtsa ba'i bla ma ngag dbang blo bzang rgya mtsho'i thun mong phyi'i rnam thar du k'u la'i gos bzang gi don bsdur gyur pa dbyangs can 'phang 'gro'i rgyud las drangs pa'i rab snyan gzhan gsos*. 194 folios (incomplete: missing folios 1–12). Block-print. Copy at Tibetan Buddhist Rescource Center, New York.

———. *Gso ba rig pa'i bstan bcos sman bla'i dgongs rgyan rgyud bzhi'i gsal byed baiḍūr sngon po'i mallika*. 2 vols. Lhasa: Bod ljongs mi rigs dpe skrun khang, 1982.

———. *Man ngag yon tan rgyud kyi lhan thabs zug rngu'i tsha gdung sel ba'i katpū ra dus min 'chi zhags gcod pa'i ral gri*. Xining: Mtsho sngon mi rigs dpe skrun khang, 1991.

———. *Mchod sdong 'dzam gling rgyan gcig rten gtsug lag khang dang bcas pa'i dkar chag thar gling rgya mtshor bgrod pa'i gru rdzings byin rlabs kyi bang mdzod*. Lhasa: Bod ljongs mi dmangs dpe skrun khang, 1990.

———. *Phug lugs rtsis kyi legs bshad mkhas pa'i mgul rgyan baiḍūr dkar po'i do shal dpyod ldan snying nor*. 2 vols. Beijing: Krung go'i bod kyi shes rig dpe skrun khang, 1996. Also contained in *The Vaidūrya dkar po of Sde-srid saṅs-rgyas-rgya-mtsho*. 2 vols. New Delhi: T. Tsepal Taikhang, 1972.

———. *Snyan 'grel dbyangs can dgyes glu'i mchhan grel*. Xining: Mtsho sngon mi rigs dpe skrun khang, 1996.

Shes rab rin chen, Stag tshang Lo tsa'a ba (b. 1405). *Rig gnas kun shes nas bdag med grub pa*. Vol. Ka of *Stag tshang lo tsa'a ba shes rab rin chen gyi gsung skor*. Kathmandu: Sa skya rgyal yongs gsung rab slob gnyer khang, 2000.

Skal bzang legs bshad (nineteenth–twentieth century). *Rje btsun byams pa mthu stobs kun dga' rgyal mtshan gyi rnam thar*. Xining: Krung go'i bod kyi shes rig dpe skrun khang, 1994.

TIBETAN CANONICAL SOURCES

Asaṅga. *Rnal 'byor spyod pa'i sa las byang chub sems dpa'i sa*. Sde sdge Bstan 'gyur, Sems tsam, vol. Wi of *Sde dge Tibetan Tripiṭaka Bstan 'gyur: Preserved at the Faculty of Letters, University of Tokyo*. Tokyo: University of Tokyo, 1980.

———. *Rnal 'byor spyod pa'i sa rnam par gtan la dbab pa bsdu ba*. Sde sdge Bstan 'gyur, Sems tsam, vol. Shi of *Sde dge Tibetan Tripiṭaka Bstan 'gyur: Preserved at the Faculty of Letters, University of Tokyo*. Tokyo: University of Tokyo, 1980.

Godararañjo. *Tshe'i rig byed mtha' dag gi snying po bsdus pa. Sde dge Bstan 'gyur, Sna tshogs*, vol. No, fols. 299a.1–334b.6. Ui, *A Catalogue-Index of the Tibetan Buddhist Canons*, no. 4438.

Manaha. *Mig 'byed mthong ba don ldan. Sde dge Bstan 'gyur, Sna tshogs*, vol. No, fols. 360b.5–363a.7. Ui, *A Catalogue-Index of the Tibetan Buddhist Canons*, no. 4443.

OTHER SOURCES

Bays, Gwendolyn, trans. *The Voice of the Buddha*. Berkeley: Dharma Publishing, 1983.

Dimitrov, Dragomir. *Mārgavibhāga: Die Unterscheidung der Stilarten. Kritische Ausgabe des ersten Kapitels von Daṇḍins Poetik* Kāvyādarśa *und der tibetischen Übertragung* Sñan ṅag me loṅ *nebst einer deutschen Übersetzung des Sanskrittextes*. Marburg: Indica et Tibetica Verlag, 2002.

Dorje, Gyurme. "Structure and Contents of the Four Tantras." *Tibetan Medical Paintings: Illustrations to the Blue Beryl treatise of Sangyé Gyamtso (1653–1705)*, ed. Yuri Parfionovitch, Gyurme Dorje, and Fernand Meyer. New York: Harry N. Abrams, 1992.

———. *Tibetan Elemental Divination Paintings: Illuminated Manuscripts from the White Beryl of Sangs-rgyas rGya-mtsho with the Moonbeams Treatise of Lo-chen Dharmaśrī*. London: John Eskenazi in association with Sam Fogg, 2001.

Eppling, John F. "A Calculus of Creative Expression: The Central Chapter of Dandin's 'Kavyadarsa.'" 3 vols. Ph.D. dissertation, University of Wisconsin, 1989.

Gedün Chöpel (Dge 'dun chos 'phel). *Tibetan Arts of Love*. Trans. Jeffrey Hopkins with Dorje Yudon Yuthok. Ithaca, N.Y.: Snow Lion Publications, 1992.

Griffiths, Paul. "Omniscience in the *Mahāyānasūtrālaṃkāra* and Its Commentaries." *Indo-Iranian Journal* 33 (1990), 85–120.

Meulenbeld, G. Jan. *A History of Indian Medical Literature*. Groningen: Egbert Forsten, 1999.

Meyer, Fernand. "The Golden Century of Tibetan Medicine." *Lhasa in the Seventeenth Century: The Capital of the Dalai Lamas*, ed. Françoise Pommaret. Leiden: Brill, 2003.

Pollock, Sheldon. "New Intellectuals in Seventeenth-Century India." *Indian Economic and Social History Review* 38, no. 1 (2001), 3–31.

Rechung Rinpoche. *Tibetan Medicine*. Berkeley: University of California Press, 1976.

Ruegg, David Seyfort. *Ordre Spirituel et Ordre Temporel dans la Pensée Bouddhique de l'Inde et du Tibet*. Paris: Collège de France, 1995.

Schaeffer, Kurtis R. "The Fifth Dalai Lama Ngawang Lopsang Gyatso." *The Dalai Lamas: A Visual History*, ed. Martin Brauen. Chicago: Serindia Publications, 2005.

———. "Ritual, Festival, and Authority under the Fifth Dalai Lama." *Power, Politics, and the Reinvention of Tradition: Tibet in the Seventeenth and Eighteenth Centuries*, ed. Bryan J Cuevas and Kurtis R. Schaeffer. Leiden: Brill, forthcoming.

———. "Salt and the Sovereignty of the Dalai Lama, *circa* 1697." *Festschrift for Koichi Shinohara*, ed. Jinhua Chen. Unpublished manuscript.

———. "Textual Scholarship, Medical Tradition, and Mahayana Buddhist Ideals in Tibet." *Journal of Indian Philosophy* 31, nos. 5–6 (2003), 621–41.

Schmithausen, Lambert. *Ālayavyijñāna: On the Origin and the Early Development of a Central Concept of Yogācāra Philosophy*. Tokyo: International Institute for Buddhist Studies, 1987.

Schuh, Dieter. *Untersuchungen zur Geschichte der tibetischen Kalenderrechnung*. Wiesbaden: Franz Steiner, 1973.

Tucci, Giuseppe. *Tibetan Painted Scrolls*. Rome: Libreria Dello Stato, 1940.

Ui, Hakuji, et al. *A Catalogue-Index of the Tibetan Buddhist Canons*. Sendai: Tōhuko Imperial University, 1934.

Verhagen, Pieter C. *A History of Sanskrit Grammatical Literature in Tibet*: Vol. 1, *Transmission of the Canonical Literature*. Leiden: Brill, 1994.

CHAPTER TWELVE

Experience, Empiricism, and the Fortunes of Authority

Tibetan Medicine and Buddhism on the Eve of Modernity

JANET GYATSO

Sometime around 1670 Darmo Menrampa, one of an inner group of physicians close to the Fifth Dalai Lama, set up a laboratory in a park in Lhasa. He and his students proceeded to dissect four human corpses—two male, two female; two old, two young—in order to count their bones. He wrote briefly of the event in an anatomical treatise after surveying received tradition on how to count the bones in the body, which were said to add up to 360. The number 360, Darmo notes, is explained in the texts as entailing that one count the sections of the skull as four. He writes, "[But] I and my students based ourselves instead on there being nine sections of the skull, and thus [we counted] 365."[1]

We have to admire Darmo's deftness. The number of bones that the physician and his acolytes "determined with precision, and through naked illustration" confirms the canonical number of 360. He also hints at the possibility of variation, if, for example, one were to recognize a different number of sections in the skull. But then Darmo goes on to call into question the notion of a definitive count altogether, suggesting a horizon of undecidability. Such undecidability would ensue if, instead of counting, as he and his students did, only the bones that measure "the span between fingertip and elbow and the width of a finger," one were also to reckon "the numerous small bones that are merely the size of a roasted bean."

The idea that properties of the physical world cannot be represented definitively by canonical doctrine might remind us of issues germane to the birth of Western empiricism. Darmo's experiment may also strike us for its simultaneity with the public anatomy lessons of the Amsterdam Guild of Surgeons, especially memorable from a famous painting by Rembrandt. But Tibet had no part in the European Enlightenment. It saw no radical revolution in science, no salient notion of innovation, no widespread and publicly touted recourse to repeated dissection and further experimentation.[2] Still, Darmo's corpse dismemberment in the Tibetan capital encapsulates a climactic moment in the history of medicine in Tibet. Part and parcel of a series of momentous social changes, which culminated in the consolidation of a centralized Tibetan state under the rule of the Dalai Lamas, was that medicine came into its own as a system of knowledge distinct from mainstream Buddhism. In fact the process by which this coming into its own unfolded involved arguments and practices that we often associate with the birth of modernity. Participants in a growing network of physicians flourishing under the patronage of the Dalai Lama's court, medical theorists and historians like Darmo were caught up in debates about what constituted the authoritative sources for medical knowledge and how to construe the history of that knowledge. A notable part of these debates was a distinctively medical empiricism, with far-reaching implications for the prestige of medicine and its practitioners, as well as for the larger issue of how medicine relates to Buddhism.

To consider these developments in light of the question of modernity promises to enhance both our sense of that notion and our understanding of Tibetan history. To recognize features—and note that I am arguing that *some* features can be so identified, not *all* or exactly the same ones—of what is defined as modernity in a variety of historical contexts suggests that modernity might be a generalizable process. That process, though by no means universal, can be recognized in other times and places besides its emergence in full force as "the modern West." Such a realization stands apart from anything we might say about actual influence or interaction, although as our historiographical practice develops one can hope that we become better at discerning connections between European modernity or science and the medicine being practiced under the Ganden Podrang government and the surrounding cosmopolitan culture of seventeenth-century central Tibet. In any event, what follows is not meant to claim that seventeenth-century Tibetan society achieved full modernity. It is only to say that, again, *some* aspects of that society bear comparison with such moments around the world.

In this essay I examine attitudes and values associated with modernity—among them, a questioning of religious authority, a valuing of empirical evidence, a probative attitude to texts and practices, and a recognition of cultural difference—in the particular ways they developed in specifically Tibetan conditions. To recognize the variety of circumstances that can give rise to such attitudes is both to discover the broad descriptive power of the category of *modernity* and to appreciate the rich range of its instances. Furthermore, to take a category like modernity and use it heuristically to study a time and place where no such indigenous category is named is not necessarily to force our interests upon an incommensurate object. Rather, if the category is both apropos and general enough—*gender* would be another example; so would *culture* or *religion*—it can help us to recognize connections and identify patterns that we might not otherwise have seen.[3]

I focus on a few inflections of what I am identifying as empiricism in the shifting camps of Tibetan medical science from around the end of the fifteenth century through the momentous events of the seventeenth century and their legacy in the following years. In brief, I explore two salient clusters of ideas that contributed to such empiricism, both of which fell under the larger Tibetan rubric of *experience*.[4] One had to do with the special kind of knowledge that, contrary to book learning, is acquired only in practice, guided by a teacher, and involving daily immersion in the illnesses of individual patients. The other concerned the particular type of knowledge that comes from direct perception, that is, from contact between the sense organs of the researcher and something in the material world. These two senses of experience overlapped, but the second is more specific and pointed. Importantly it has a special authority of its own; in the polemical rhetoric of the medical writers under discussion it could trump what was predicted by ideology or doctrinal system.

Germane to this entire investigation is a troubling of the boundaries of science and religion in Tibet, that is, between medicine and mainstream Buddhist modes of writing and thinking, during the period under discussion. Buddhism and medicine grew up together in Tibet in a shared universe of institutions, conceptions, and modes of discourse. Buddhist texts certainly also were concerned with versions of both of the kinds of experience just mentioned. Still, there is a telling distinctiveness in the way that medicine came to construe the significance of direct perception and hands-on practice.

Background: Institutions and Literature

The emergence in Tibet of professional medicine was a gradual process that began with the kings of the Yarlung dynasty.[5] From everything we can tell from Tibetan historiography, the early Tibetan kings sponsored the visits of a stream of physicians from India, Nepal, Kashmir, China, Persia, and other areas of western Asia. The process started in earnest at the court of Songtsen Gampo (seventh century). An astonishing number of titles of medical works are recorded from this period that were either translated from other languages or were new works composed in Tibetan. By the time of Tridé Songtsen (late eighth century) there are reports of the title of court physician (*bla-sman*) in the royal court, a position that remained through the twentieth century, along with the granting of land and inherited rights to medical clan lineages, including the releasing of such clans from military duty.[6]

After the fall of the dynasty (ninth century), patronage for medical learning was taken up by the emerging Buddhist monastic centers. The premier translator of Indic Buddhist works in the "new" period, Rinchen Zangpo (eleventh century), also translated the medical work *Aṣṭāṅgahṛdaya*, whose presentation of ayurvedic tradition was especially influential for medicine in Tibet.[7] It was probably during the next century that Yutok Yönten Gönpo (1126–1202) and his students codified the work known as *The Four Treatises* (*rGyud bzhi*), although the text attributes its authorship to the Buddha.[8] *The Four Treatises* became the principal root medical text in Tibet. Already major Buddhist teachers and writers, such as Gampopa (1079–1153) and Jetsün Drakpa Gyeltsen (1147–1216), had been serving as physicians and composing medical works. Special schools for medical learning, sometimes conjoined with curricula in astrological calculation, began to be established at the major monasteries: at Sakya Mendrong during the twelfth century; at Zhalu, which specialized in the *Aṣṭāṅga* system, and Tsurpu, which saw much eclectic medical scholarship, in the fourteenth century; and at É Chödra at Bodong in the early fifteenth century. By the time the Chang line of physicians was consolidating at Chang Ngamring, the old Tibetan capital, in the fifteenth century, there were oral medical examinations and regimens of memorization. There was also much medical learning at Latok Zurkhar, which became the home of the other major line of Tibetan medicine, the Zur. The Drigung Kagyü developed its own medical tradition, branching off from the Zur.

By the early seventeenth century several key medical centers were thriving in the Lhasa area, at Drepung, Shika Samdruptsé, and Tsé Lhawangchok, where more formal methods of examination were established. These centers were the springboard for the further nurturing of medical learning under the reign of the Fifth Dalai Lama and the establishment by his regent, Desi Sanggyé Gyatso (1653–1705), of the Chakpori Medical College on a hill in the middle of Lhasa. Chakpori served as the center for the medical academy in Tibet until the Cultural Revolution. Other medical schools continued to appear at monastic centers, now in eastern Tibet, including at Degé, Pelpung, Katok, Kumbum, and Labrang Trashikhyil. Some of these institutions developed curricula that were at odds with medical orthodoxy at Chakpori, creating the conditions for debate and dissent.

In fact medical practice in Tibet was far from limited to monastic learning centers. Healing traditions also abounded in lay tantric circles, and oracle mediums were involved with healing as well. But even for the medicine fostered by the monastic schools we know little of the sociology of practice regarding, for example, what percentage of practicing physicians were actually trained in those schools, what the lay-monastic breakdown was, and to what extent physicians actually used medical writings, let alone all the questions one might raise about the economics and daily practice of medicine. We can only venture, for example, that the degree of professionalization of Tibetan medicine probably never approached that achieved in Chinese medicine due to the centralized bureaucratization there of qualifying examinations for physicians and, by the Song dynasty, regulations that physicians keep standardized case records.[9] On another note, we can also observe that familiarity with dissected bodies seems to have been unusually widespread among Tibetans due to the long-standing practice of dismembering human corpses to feed vultures; such charnel grounds were used to gather stray body parts for the making of certain ritual instruments, and they also served as sites for Buddhist meditations on death.

Lacking at this point detailed sociological knowledge for the period under discussion, I restrict my attention to insights we can glean from the scholastic medical literature. This provides, however, a huge and rich body of data, with many indications of tensions, debates, and changing mentalities and practices, as well as a heterogeneous array of medical traditions, from the ayurvedic physiology of the three humors, to Chinese-influenced pulse diagnostics and methods of treatment indebted to western Asian medicine. By the fourteenth century, writing a commentary on *The Four Treatises* had become a major way for medical scholars to demonstrate their learning and debate points of controversy. Other common literary genres included instructions that supplemented the root text, manuals for the preparation of medicines, manuals for medicinal plant recognition, and manuals of therapeutic techniques.[10] Another key genre that provides clear evidence of an effort to identify medicine as a separate tradition is the historical overview of medical personalities, literature, and institutions. Two very influential examples of this special medical history were written at the height of the consolidation of medical tradition, in the sixteenth and seventeenth centuries.[11]

Writing from Experience

One other distinctive medical genre that may be traced to the sixteenth century is the *nyams-yig*, a "writing from experience." This genre serves well as a flashpoint for the emerging empiricist dimensions of the medical mentality, and this is so not only because of its name. In fact the label only infrequently actually appears in a text's title.[12] One is immediately led to ask, then, how it is known that a text is a *nyams-yig*. Beyond the occasional specification that a given text was written on the basis of the author's experience,[13] being a *nyams-yig* appears to be a matter mostly of reader reception. Frequently the term is only retrospectively applied, as is the case for the famous but oddly titled *Bye ba ring srel* (Relic of Ten Million), a work describing medical practices written in the fifteenth century.[14] It may well be that the term was used only after the sixteenth century, with the explicitly labeled collection of one hundred *nyams-yig* by Gongmenpa.[15] The entire conception of *nyams-yig* as a prestigious genre written by an elite expert class may be the product of the very period we are looking at.

One of the main purposes of the *nyams-yig* seems to be to convey the special kind of knowledge that comes from hands-on practice, that is, the first sense of experience sketched above. In general the content of the *nyams-yig* seems to be construed as superior to the results of mere book learning.[16] The Desi even ventures to say in his own *nyams-yig* that *The Four Treatises* are of little use for actually treating patients, other than to provide information on recognizing medicinal plants, the basic structures of medical knowledge, and the location of the channels in the body. Even previous *nyams-yigs* didn't support actual treatment sufficiently, he claims, prescribing "one medicine for one hundred diseases" and failing to describe the course of an illness fully from beginning to end.[17] That is what the Desi implies his *nyams-yig* will provide. Thoroughness based on hands-on experience becomes the signal virtue of the *nyams-yig*. In the eighteenth century Degé Lamen maintains, "I wrote this *nyams-yig* from my own experience and what I have become familiar with. This would be equivalent to a vast textbook on what has been heard of the kind actions [of former teachers]."[18] The point is developed further in the nineteenth century, with the influential *nyams-yig* of Kongtrül, who chastises physicians who never had "oral teachings from an experienced teacher nor experience based on a long period of familiarization." Kongtrül stresses that merely checking the pulse and urine (the basic diagnostic tools of *The Four Treatises*) is not sufficient to diagnose disease. He insists instead upon asking a set of detailed, particular questions tailored to the individual patient and listening carefully to the response.[19]

The upshot of the increasing demand for more reliance on experience-based training was that *The Four Treatises* fell out of clinical use, even if the medical colleges still compelled students to memorize it.[20] At least by the nineteenth century authors of *nyams-yigs* could speak directly of the limitations of *The Four Treatises.* Kongtrül describes the many sources to which he had to resort for information that was lacking in the root text, and Mipam can distinguish the way he read pulses based on his own experience from an array of authoritative precedents in Tibet and China.[21]

Innovation if not actual deviation from the authoritative was always a risky business in Tibetan literary culture. In the face of such risks, what could be gained from bringing forth one's own experience to surpass the root text can be seen in the rhetoric of prestige surrounding the fact that so-and-so wrote a *nyams-yig.*[22] Equally so, the information that a *nyams-yig* conveyed became valuable property. In the sixteenth-century collection of one hundred, each *nyams-yig* was dedicated to one of the author's students, whose name and often clan or toponyms were specified. Here the *nyams-yig* would be a kind of patrimony, a possession to be guarded against competitors.

The Weight of Experience

Both to the extent that the *nyams-yigs* described what was learned idiosyncratically in the clinic and that they recorded information not known in *The Four Treatises,* the genre reflected an important direction in which medical practice was moving by the sixteenth century. The possibility of newness and innovation had definitely become an open question in learned medical circles.

There can be no question that in tandem with this move was a recurring insistence on the value of textual study and the learning of system.[23] Certainly such a value is evident in the perduring popularity of writing commentaries to *The Four Treatises,* and that reminds us once again of the shared universe of values between medicine and Buddhist scholasticism. We often find three principal sources of valid knowledge—scriptural authority, logic, and experience—valued in the medical commentaries of the period.[24] The pair "empirical examination and critical analysis" is also invoked.[25] But the medical writers very frequently emphasized experience in particular, and denigrated the barrenness and even dangerousness of a physician who had book learning alone.[26]

A critique of book learning alone and a stress on the importance of experience are also encountered in Buddhist rhetoric, especially with regard to meditative practice and spiritual advancement. But differences can be detected. Tibetan Buddhist writers display a decided ambivalence about experience as a valid source of knowledge, unless it has been thoroughly informed by the "right view" on key points of doctrine. The issues involved might be compared to debates about empiricism and the relation between mind, matter, and divine design in early modern Europe.[27] In Tibet neither Buddhist nor medical writers considered experience ever to be entirely free of ideational content. Hence their presumption that it is necessary to educate experience in the right way; left uneducated, experience is subject to emotional prejudice and error. But while medical writers also worried about physicians who practiced only on the basis of their own experience, we find far less suspicion of experience as a category as such than we do in Buddhist epistemology and meditation theory.[28] Experience appears to have been unambiguously a good thing in medical learning, even if it could not suffice on its own.

I suggest there was a fundamental disparity in basic orientation that overdetermined this difference. The goal of medical practice, the orienting horizon of what constituted success, was the patient's recovery. The ascertainability of this telos—its empirical demonstrability—is of an entirely different order than that of the Buddhist summum bonum of enlightenment. The success of the latter was determined by a far more socially complex set of criteria than the pretty indisputable fact of whether or not a patient died.[29]

I suspect that the very different ways these goals were ascertained affected the mentalities of their respective traditions in far-reaching ways. But we can also note more generally a greater respect for the realities of the physical world in medicine than in Buddhism, particularly in the substantial medical treatises that were being written by the sixteenth century and whose debates I consider below. One sometimes finds an openness to the physical world revealing itself in ways that no discursive knowledge can fully anticipate. One detects a confidence that there is something out there that has its own integrity—the number of bones in the body, say—that stands fully apart from what any text might say, and that one can consult and that can reveal new information. This becomes particularly clear in the second, more specific notion of experience at play in medical tradition by this point, namely the authority of direct observation. In the following section I consider arguments that sometimes turned on "what can be seen in actuality" or what is known "through direct perception" as a way to prove someone else's theory wrong.[30] While we might find some analogue in classical Buddhist epistemology regarding the role of direct perception in moments of meditative breakthrough, that is a very different matter, since what is perceived then is not everyday reality. And on those few occasions when Buddhist polemicists did indeed invoke some obvious fact in the everyday world, it was more a rhetorical ploy than a precise argument. For Buddhist theory what appears to direct perception in the conventional world will on close analysis prove to be but an illusion.[31] Again the contrast is instructive. On no account will we find the medical tradition arguing that the physical death of a patient is an illusion.

With their ultimate goals so disparate (and despite the frequent characterization of the Buddha himself as healer par excellence), we find medicine in Tibet struggling with classical Buddhist doctrine throughout its history. Some of these struggles issue simply from a discrepancy of system: Is illness to be understood as bad karma, or physically, as an imbalance in the humors of the body? Although these very different kinds of etiology coexist in *The Four Treatises*, they do so sometimes in an uneasy mix. What's more, in actual medical practice a truly weighty implication emerged in those moments when the physical world could be perceived as having a reality of its own, as when Darmo resorted to the body itself to determine the number of its bones. We begin to recognize in some quarters of the emerging medical academy both the suspicion, and then an unease with that suspicion, that the "word of the Buddha" itself could be subject to correction.

When Systems Collide: The Channels of the Body

Empirical evidence in its most overtly physicalistic sense posed on several occasions an estimable challenge to Buddhist revelation. How this challenge was mounted, and then fielded, is illustrated well by a debate on the anatomy of the channels.

The Four Treatises describes four kinds of channels that transmit substances and energies through the body: (1) initial "growth channels," which give rise to the fetus's body; (2) "channels of being," which are matrices of channels at the brain, heart, navel, and genitals responsible for perception, memory, and reproduction; (3) "connecting channels," which consist of two main sets of "soul channels," one white and one black, that control the nervous and cardiovascular systems; and (4) "life channels" through which a life force moves around the body. There was debate in the commentaries about what these categories actually refer to and exactly where in the body they are. But the discussion took a new turn entirely when, in the fifteenth century, a medical writer casually remarked that the life channels follow along the path of *lalanā* and *rasanā*.[32]

This writer was referring to the tantric conception that describes a "central channel," a straight tube that runs between the crown of the head and the genitals, and two others, *lalanā* and *rasanā*, that run along its sides. This system is extremely well known in Indic and Tibetan Buddhist tantras as the basis for yogic cultivation. But starting in the fifteenth century there developed in Tibet a sustained effort to locate this tantric system within the medical system of channels in *The Four Treatises*.

A generation or so later Sönam Yeshé Gyeltsen identified the tantric channels not only with the life channels, which in any case had always been viewed as derivative of the tantric system, but also with the more properly medical growth channels, and especially with the connecting channels.[33] He called the medical white soul channel, often understood to be the spinal column, the "outer" *lalanā*, and he labeled the black soul channel, which is something like the vena cava, the "outer" *rasanā*. This seems to be the first time a medical writer equated the medical nervous and cardiovascular systems with these two tantric channels. But we already see in that effort a device to make the equation palatable, a distinction between an "outer" and an "inner" version. This device mirrors a larger tendency to distinguish the average human body—in this case, the medical body—from the body of the meditator or indeed a Buddha. And it serves to avoid a very large problem: the central channel, *lalanā*, and *rasanā* are simply not visible in the body in the way they are described in the tantras. Clearly people had been looking inside corpses to find these channels, and the discrepancy had already been noted several centuries before.[34] What is new is the medical community's increasing attention to the problem. And yet the problem has not really been solved: in Sönam Yeshé Gyeltsen's solution, the real, or correct—the inner—*lalanā* and *rasanā* are still invisible.

Different writers tried different schemes. The brilliant Kyempa was most interested in locating the most important tantric channel, the central one, which, controversially, he identified with the spinal column itself.[35] This would seem promising, since like the central channel, the spinal column is a single straight channel running from the top of the torso to the bottom. It took the master commentator Zurkharwa Lodrö Gyelpo (b. 1509) to disqualify this attractive solution, largely because it fails to respect the signal characteristics of the central channel as described in the tantras.[36] The deft solution he proposes instead is emblematic of where medicine was headed by the latter part of the sixteenth century.

Like other participants in this fray, Zurkharwa let the embryonic growth channels be the tantric channels.[37] In a way this is a safe solution; those initial channels disappear once the fetus's body is fully formed. So, although he does not say this, there is no chance to disprove their existence later by investigation. But apparently, finding the tantric channels in the first weeks of life did not satisfy the quest to locate them in the adult body. So Zurkharwa also had recourse to the earlier suggestion that lined up the two side tantric channels with the white soul channel (the spinal column) and the black soul channel (the vena cava) in the adult body. However, a close reading of his language reveals that he doesn't actually equate them. At one point he says that *rasanā* "gives rise" to the black soul channel.[38] But this seems to have only the general sense that, as he says elsewhere, all of the wind channels in the body are the central channel, all of the blood channels in the body are *rasanā*, and all of the liquid channels are *lalanā*.[39] As for the central channel itself, he indicates his agreement with a variety of tantric passages that describe it, but he does so in the context of the growth channels. The upshot is that he actually locates the tantric central channel only in the general channel (*srog-rtsa*) that grows in the embryo's body.[40] So he rejects the views of those who argue that the central channel "always exists," that is, in the adult body.[41] He certainly never physically pinpoints any of the tantric channels the way he does the medical channels, whose location with respect to the spine, for example, he can specify by digit.

Here and elsewhere Zurkharwa is quite conscious of a larger question of incommensurability.[42] But when he raises the possibility that it could be inappropriate to introduce ideas into medicine from what is clearly another system, that of the tantras, he doesn't jump at the chance to disqualify the medical effort to find the tantric channels in the physical body. He almost sounds like a modern historian when he argues instead that in fact it *is* appropriate to bring in tantric ideas into medical description, for the medical system has always had multiple sources, which included the Vedas and disparate Buddhist sources such as the Vinaya, the *Suvarnaprabhāsottama*, and the *Kālacakra*. But the signature of his complex polemics is all too evident when he maintains that while *The Four Treatises* system "roughly accords" with the tantric system, it is important to separate the terminology, for *The Four Treatises* is not talking about the same thing as the tantras do, which is the fruits of meditation. Whatever is tantric about the anatomy of *The Four Treatises* is "hidden."[43] Note that in the process Zurkharwa has sustained his allegiance to tantric truths.

Indeed Zurkharwa rejects as "invalid" the argument of previous writers that the tantric channels are merely matters of meditation, existing in the imagination but not present in the average body such that they would be perceptible in the dismembered corpse.[44] Others came up with the theory that the tantric channels do exist concretely in the body but evaporate at death, just as the mind does. Zurkharwa's reason for rejecting all of these views is telling: if the tantric channels were only a matter of the imagination, the fruits of tantric yoga would not be obtained. The next major *Four Treatises* commentator, the Desi Sanggyé Gyatso, makes a similar argument for the physicality of the tantric channels: yogis can attain immortality by holding the winds in the central channel.[45]

In arguing for the concrete efficacy of exercises involving the tantric channels, Zurkharwa and the Desi would probably say (if they were reading this essay) that they were having recourse to another kind of empirical truth: the evident efficacy (from their perspective) of tantric practice. It is significant that they were committed to this efficacy being physically based. That already says a lot about the ambitions of the period to establish tantric ideology in some sort of physical reality, and I will return to this below. Note for now that while the influential Desi repeats most of Zurkharwa's solution verbatim, the subtlety of the ambiguity is such that the question continues to dog the medical commentators, and we find major nineteenth- and twentieth-century writers still at pains to demonstrate that what Zurkharwa and the Desi established was the validity of the tantric system.[46] But that perduring ambiguity also meant that the door had been opened for dissent. Lingmen Trashi (b. 1726), a close student of the polymath Situ Chökyi Jungné at the outlying medical center at Pelpung, might have been far enough from the dominion of the Desi's Chakpori to offer a more rigorously empirical account. He is willing to concede that the soul channel is sometimes called an outer central channel; this distinction lines up with his general approach to treat the human body differently from that of a Buddha.[47] But regarding that human body, Lingmen has little patience for anything that is not directly observable. He can declare categorically that the tantric channels are out of court in an anatomy of the medical, that is, "material," body. The tantric system is meant solely as a map for meditation.[48]

The Seventeenth Century, Medicine, and the Word of the Buddha

The move to separate tantric anatomy from the medical eventuated in writers such as Lingmen setting aside tantric anatomy altogether as valid for medical knowledge. In another debate that similarly pits scripture against empirical evidence, Lingmen will explicitly invoke what he has seen with his own eyes in medical examinations and autopsies in order to disprove anatomical statements in *The Four Treatises*.[49] Still, the fact that influential writers like the Desi and Zurkharwa were motivated to find some place for a separate functioning of tantric anatomy indicates that it was hard to displace entirely.

There can be no question that the medical writers were grappling with a knotty set of issues about authority. While we can find examples of commentators overtly correcting statements in *The Four Treatises,* these tend to be cautious and minor. More commonly *The Four Treatises* was upheld in displays of loyalty that are more important for what that says socially than what such statements actually meant for the practice and theory of medicine. A twentieth-century commentator's warning probably reflects a long-standing sentiment: "If [the channels are] explained other than this, that is, if one makes a claim that goes against . . . the tantras that explain the natural condition of the body's channels, that would establish a position that invalidates scripture. [Therefore] it is best to abandon personal arrogance and follow the experts [as represented in scripture]."[50]

Both the display of compliance to system and the urge to distance medicine from it make eminent sense for the tumultuous period leading to the centralization of the Tibetan state in the seventeenth century. The patrons of medical writers like Zurkharwa and Sönam Yeshé Gyeltsen—warring factions such as the Rinpung clan lords and the Karmapa lamas, jockeying for power during the sixteenth century—made for much political insecurity. The final hegemony of the Fifth Dalai Lama's government in the Potala in the seventeenth century coincided with the aspirations of the Ch'ing dynasty in Tibet, which meant serious vulnerability for renegade monasteries and intellectuals.[51] We are only beginning to appreciate the impact of cultural contact between Tibetans and the Chinese empire during this period.[52] But one undoubted effect of the consolidation of a centralized government in Tibet was a consolidation of the fortunes of the medical academy.

The combination of increased bureaucratization of the Tibetan state and the highly rationalized apparatus of the Chinese empire created a climate that conferred status on public accountability and empirical testing.[53] A principal agent in creating that status was the Great Fifth himself, who actively sought out medical experts from abroad, for his own well-being but also clearly with a view to broadening the profession's repertoire of diagnostic, therapeutic, surgical, and pharmacological tools. Again the message is that *The Four Treatises* do not contain everything one needs to know to practice medicine successfully. The Dalai Lama's search for supplements, both foreign and local, to medical knowledge in Tibet was wide-ranging. He brought an Indian physician to his court in 1675 for that physician's expertise in cataract operation, a technique that he induced Darmo to master and which the latter performed successfully on the Dalai Lama himself.[54] The Dalai Lama also invited a Chinese expert in eye treatments, along with other physicians from South Asia for whom he sent emissaries to India. He put his own court physicians in contact with these foreign experts, encouraging them to study the new techniques; new medical works were also translated into Tibetan. His catholic vision extended to the past as well. Old works of Tibetan medicine were sought in archives, recognized as important, edited, and carved for block printing; biographies of the founders of medical tradition in Tibet were codified; several scholars attempted to codify a definitive edition of *The Four Treatises*.[55]

We also read repeatedly of the granting of lands and income to monastic medical schools. State support of medical learning reached its climax with the Dalai Lama's commission in 1694 of his key administrator, the Desi, to set up a new medical school in the wake of the deterioration of the school at Drepung. Perhaps for the first time in medical academia the new school would admit both monks and lay students.[56] The granting of "tax monks" for assuring student enrollment, the degree of trans-Tibetan enrollment, and the standardization of medical examinations, degrees, and sites for plant collection in central Tibet all reached new specificity. The Desi's other great accomplishment for medicine was overseeing the production of a spectacular set of medical paintings illustrating anatomy, pharmacology, therapeutics, and other vignettes of medical theory, practice, and learning.[57] In striking ratification of our thesis about empiricism during this period, the Desi records that his artists executed the anatomical drawings by looking at real corpses.[58] It is also reported that he employed artists to draw plants based on their local knowledge of particular specimens. The Desi is aware that drawing from life produced new knowledge which surpassed that in *The Four Treatises*.[59] The set's precise images constituted "unprecedented paintings which provide direct instruction that can introduce [medical learning] as if pointing to it with the finger."[60]

Yet the revelatory visions that lay at the base of the Desi's agenda—seeing Chakpori Mountain as the heaven of the medicine Buddha, glorifying the tomb of the Great Fifth and its place in Lhasa[61]—show that we must also remain alert to other dimensions of this moment. In their actions with regard to state, religion, and medicine alike, agents like the Great Fifth and the Desi were creating what is now loosely called the Tibetan "theocracy," a state whose very essence was founded upon the entire edifice of the Tibetan Buddhist universe of imagination, tantric and otherwise.[62]

It is not too much to say that the stridently authoritative political climate infected intellectual life in the monastery as well as the practical sciences. Ironically the political climate fostered conservatism at the very time of the sweeping moves toward rationalization. This conservatism is not the same as censorship, however; rather, when we see prominent medical scholars of the period protecting the root text or looking for the tantric channels in the empirical body, we understand that they are simply part and parcel of the same worldview that was being built into the whole political and cultural basis of Tibet. Tantric meditation is efficacious in the real world; the words of the Buddha are true.

The upshot is a dual movement. The same medical colleges that are being given fiscal autonomy and experimental license are also ordered to conduct prayer ceremonies every day for the health of government officials. Nowhere do we see this twofold urge more clearly than in a long-running debate in Tibet about which Zurkharwa quips, "Whenever three or more people gather, it gets discussed."[63] The debate is none other than the very grave question of whether or not *The Four Treatises* is the "word of the Buddha." If it is, it is a sacred revelation on par with the other canonical teachings of the Buddha, a work that was translated into Tibetan from the holy language of Sanskrit. If not, it would have been composed by Tibetans in Tibetan.[64] This is a familiar issue in Tibet; demonstrating a text to be an authentic translation from an Indic original had long been the dividing line between a true Buddhist teaching and a debased apocrypha by an impudent upstart. If the root text was originally Tibetan rather than Indic its authority could be in jeopardy.

Evidence for the Tibetan composition of *The Four Treatises* had already been noticed as early as the thirteenth century,[65] but it was Zurkharwa who worked it out most explicitly. He drew on arguments devised for another, parallel debate, also at issue by the thirteenth century, regarding the authorship of the so-called Treasure scriptures and the nature of the word of the Buddha.[66] Yet again the contrast is stark: where the advocates of the Buddhist Treasure scriptures always concealed their Tibetan composition and constructed elaborate ways to keep authorship attributed to the Buddha, the medical scholar Zurkharwa is driven in one of his essays to finally blurt it out: "If [*The Four Treatises*] were not made to appear as if it were Buddha-word, Tibetans—wise, dumb, and middling alike—would have a hard time believing it."[67]

Zurkharwa makes *The Four Treatises* instead a *śāstra*, a genre that in Buddhism denotes highly regarded writings but not the work of the Buddha himself.[68] In one essay he builds three positions: on the outside *The Four Treatises* is Buddha-word, on the inside it is a *śāstra* written by an Indian scholar, and secretly (this always connotes the deepest truth) it is a *śāstra* written by a Tibetan. Then he shows the first two to be untenable. *The Four Treatises* mentions tea, not typically mentioned in Indian *śāstras*; it describes diagnostic methods using pulse and urine, not known in Indian medicine: these and other facts show the text must be a Tibetan composition. This kind of evidence, noted even prior to Zurkharwa, is emblematic of the empiricist mentality and historical sense that the medical writers could muster.[69] And once again Lingmen, writing in eastern Tibet in the eighteenth century, illustrates this mentality even more clearly: Indian medical texts such as the *Aṣṭāṅgahṛdaya* consider goat meat to be good, he declares, while *The Four Treatises* considers it to be one of the worst kinds of meat; this shows the Tibetan character of *The Four Treatises,* a work that accords with the cold climate of Tibet.[70]

What illustrates the tension surrounding such empiricism is the striking fact that the Desi himself accedes to so many of the same arguments about the Tibetan character of *The Four Treatises* but still makes it Buddha-word. The medicine Buddha, he avers, granted the text to the Tibetan compiler Yutok, who then fixed it in a few spots for the Tibetan context.[71] We cannot read this position without thinking about the Desi's political investment in the power of the word of the Buddha. But the risks in making the medical root text anything but the word of the Buddha are already evident a century earlier in Zurkharwa's obvious caution: how he saves his most radical statement for a separate essay; how in his more widely known commentary he is oblique, avoiding overt discussion of Tibetanness and focusing instead on the status of kinds of *sastra*; and how in a third work he bemoans the delicacy of the question, insisting that in the end it is undecidable and what is really important is to realize that *The Four Treatises* is highly valuable and should be respected *as if* it were Buddha-word.[72]

The Alternate Space of Medicine

The high stakes of empiricist leanings in medicine have as their background larger issues in Tibetan intellectual circles regarding the value of individual direct experience and the possibility of its dissonance with doctrinal system. We have seen how that larger question had specific valences for medicine. Personal experience was sometimes construed as more valid than system and as capable of producing innovation. But personal experience also serves to flesh out system, and even owes its existence to system, as when foreign physicians, invited by the Dalai Lama, teach new concepts and techniques that Tibetan physicians then put into practice. It has been instructive to compare the salience of experience in medicine with its status in scholastic Buddhist discussions, a comparison which has suggested, for one thing, that medicine had particular tolerance for the innovation that direct experience fosters.

Much of that difference is doubtless attributable to the higher investment that the Tibetan state had in defending the traditional authority of Buddhism—its ethics, its institutions, its imagery, its ritual regalia, the status of its scholastic doctrine—than did the more circumscribed world of academic medical theory and practice. But once we take stock of how fundamental the entire universe of Buddhism had become to the society that the Tibetan government represented, we confront the limitations of any heuristic distinction between Buddhism and medicine. There was much mutual indebtedness, and both participated in the larger mix that constituted the texture of Tibetan life, especially in the very cosmopolitan Tibetan capital during the centuries we are considering. Medicine had some very special things, symbolic and otherwise, to offer to that mix. But we would probably still do well to consider that mix under the general banner of Buddhism in this more expansive sense, even if medicine sometimes represented a dissent from Buddhist scriptural authority.

In essence medicine provided a special spin to experience, even while it shared with mainstream Buddhist scholasticism an interest in experience. For both kinds of experience identified in this essay, the medical versions were distinctively all about the physical. The thick world of practice that informed *nyams-yig* writing, for example, had for the physician everything to do with the idiosyncrasy of material existence: the texture of daily routine; the cultivation of habits of clinical technique; the development of dexterity, an often cited virtue; familiarity with plants; and most of all the variability and indeed unpredictability of the course of illness itself. Even more so, the force of the second, more specific sense we have identified for experience—direct contact with the empirical—was actually about the authority of the physical world. Here especially the evidence that polemicists like Zurkharwa and Lingmen could cite to decide a dispute was there to be seen by the eye; unpredictable again, it had a reality of its own, of a different order beyond the grasp of system.

The case par excellence of the recalcitrant physical fact that exceeds system is the death of the patient. I speculated earlier that the possibility of death defined the field of medicine in fundamental ways. I would add now that this special allegiance to materiality conjoined, in the particular historical circumstances of seventeenth-century Tibet, with a recognition on the part of the Dalai Lama's government that medicine provided a special service in a Buddhist society. The externality of the physical world became, as it were, the rare case of a reality standing outside system, and it therefore could serve as a checkpoint, an independent confirmation of what often seemed otherwise to be an all-encompassing Buddhist universe. I suspect that these are the stakes in the channel debate: medicine offered a vehicle by which to confirm the truth of the systems of tantric yoga, in turn a very key ingredient of the reincarnation system that founded the office of the Dalai Lama. Even if empirical evidence was not actually forthcoming to prove the physical existence of the tantric channels, the medical thinkers almost made it so. But however the debate came to be decided (and it was decided differently by each of the commentators) at least we must note that the in/visibility of the channels, along with the larger question of what can be seen, or proven, through dissection, had been put on the table, forever to be reckoned with.

The Lhasa government created a special institutional space for medical learning, a place apart, to a degree not seen before. But medicine had already created a place apart for itself by virtue of its distinctive kinds of arguments and practices. Still, in the volatile case of the authorship of *The Four Treatises* the limits and risks of such a separation meant that an argument against the Buddha's authorship could be advanced only with extreme caution and some amount of dissimulation. But it would be wrong to conceive of these medical thinkers as thoroughgoing empiricists who only had to stay out of the punishing way of the "Church." Buddhist ideology and tantric truths were as basic to the medical writers' worldviews as was their interest in saving patients from death. We can even note cases where the dynamic also went in the other direction, whereby something like tantric thinking could also serve to legitimize the medical. As one example, tantric theorizations of subtle matter sometimes helped medical theory to talk about imperceptible functions in the body.[73]

If we find an uneasy tension remaining between the claims of perceptibility and the claims of soteriological transformation, the tension may turn out to be mainly in our eyes. In the particular circumstance of seventeenth-century Buddhist Tibet, medicine helped provide the government with the grounds for an episteme in which the ideals and images of religion could coexist with the everyday practices of governance and power, together to display a coherent universe.

Notes

I am grateful to Charles Hallisey for conversations that were critical in the conceptualization of this essay, and to Yang Ga of the Tibetan Medical College in Lhasa and Thupten Phuntsok of the Central University for Nationalities in Beijing for much information and insight on some of the historical issues discussed herein.

1. This and the following quote from Dar-mo, "Rus pa'i dum bu," 22–24, excerpted from his unpublished *gSer mchan rnam bkra glegs bam gan mdzod*.

2. The degree of the revolution in European science in the early modern period is itself subject to question: see Shapin, *The Scientific Revolution*.

3. I took a similar approach to the issues of individualized selfhood, the writing of autobiography in Tibet, and modernity in Gyatso, *Apparitions*.

4. The common Tibetan terms are *myong-ba* and *nyams*, or some combination thereof. See Gyatso, "Healing Burns."

5. The following survey of the history of Tibetan medicine is based largely on Zur-mkhar-ba, *sMan pa rnams*, 287–321, and modern surveys such as dKon-mchog-rin-chen, *Bod kyi gso rig*, and Khro-ru Tshe-rnam, "Bod lugs gso rig." For more on the early period, see Beckwith, "The Introduction of Greek Medicine."

6. Zur-mkhar-ba, *sMan pa rnams*, 291–99.

7. See Vogel, *Vāgbhaṭa's Aṣṭāṅgahṛdayasaṃhitā*; Vāgbhaṭa, *Vāgbhaṭa's Aṣṭāṅgahṛdayasaṃhitā*; Emmerick, "Sources of the *Four Tantras*."

8. [gYu-thog], *Four Treatises*. Yang Ga is currently completing a doctoral thesis in the Committee for Inner Asian and Altaic Studies at Harvard on the varied medical traditions and texts that contributed to the composition of this work.

9. See Cullen, "*Yi'an*."

10. The respective Tibetan genre labels are *lhan-thabs, sman-sbyor, 'khrungs-dpe*, and *lag-len*.

11. Zur-mkhar-ba, *sMan pa rnams*; sDe-srid *Khog 'bugs*.

12. Sometimes a work self-identifies as a *nyams-yig*, e.g., Kong-sprul, *'Tsho byed*, 2.

13. For example, sDe-dge-bla-sman, *Nad sman*, 417–18.

14. See sDe-srid, *Techniques*, 566, 568–69.

15. Gong-sman, *Nyams yig rgya rtsa*. Compare the sources listed by sDe-srid, *Techniques*, 566–69.

16. Such a sentiment is not necessarily at odds with the prestige accorded medical scholarship in other contexts, regarding which see Schaeffer, "Textual Scholarship."

17. sDe-srid, *Techniques*, 566–67.

18. sDe-dge-bla-sman, *Nad sman*, 401.

19. Kong-sprul, *'Tsho byed*, 1–3.

20. Zur-mkhar-ba already notes that *The Four Tantras*'s *bShad brgyud* is rarely read: *Mes po'i*, 1:95. Compare sDe-srid, *Techniques*, 566. Certainly by the twentieth century physicians almost always used recent *nyams-yigs* as their actual handbooks, usually Kong-sprul, *'Tsho byed*, or one of the ones by mKhyen-rab-nor-bu (1883–1962).

21. Kong-sprul, *'Tsho byed*, 32–33; 'Ju Mi-pham, *bDud rtsi*, 260–62.

22. See dKon-mchog-rin-chen, *Bod kyi gso rig*, 187, 188.

23. Schaeffer, "Textual Scholarship." Compare sDe-srid, *Techniques*, 569, emphasizing his own recourse to authoritative works (*lung*) and reasoning (*rigs*), even after arguing that experience is most essential.

24. That is, *lung* (Skt. *āgama*), *rigs* (Skt. *yukti*), and *myong-ba* (Skt. not standard). See, e.g., Zur-mkhar-ba, *Mes po'i*, 3; sDe-srid, *Techniques*, 569.5.

25. *[brTag]- dpyad* (Skt. *vicāra*) is sometimes a synonym for *myong-ba* in the trio cited in n.24: Byams-pa-'phrin-las (quoting sDe-srid), "sDe srid sangs rgyas," 415.

26. For example, sDe-srid, *Techniques*, 566, railing against "nyams-len byed-mi mi-'dug" (those who are not people who practice), even while also insisting on the need for learning. Compare sMin-gling Ngag-dbang-sangs-rgyas-dpal-bzang's criticism of Darmo and others who are "attached to the words of the Great Tantra (*Four Treatises*) but fail to do practice; leaving behind clinical examination, they murder patients" (quoted in Byams-pa-'phrin-las, *Gangs ljongs*, 318).

27. Israel, *Radical Enlightenment*, 252–56, 477–85, 535–40.

28. Gyatso, "Healing Burns."

29. For an early example of death as a key issue in medical ethics, see Sum-ston-pa "'Grel ba 'bum chung," 297.

30. Tib. *dngos-su-mthong-zhing* or *mngon-sum-du*.

31. Compare, for example, Candrakirti's arcane analysis of the status of things that are directly perceived: Sprung, *Lucid Exposition*, 60–63.

32. Byang-pa rNam-rgyal-dpal-bzang, *bShad rgyud kyi 'grel chen*, 90.

33. bSod-nams-ye-shes-rgyal-mtshan, *dPal ldan*, 63a–82a. His position is likely based on an innovation of his father's, the famed scholar Byang-pa bKra-shis-dpal-bzang, whose commentary on this part of the *rGyud bzhi* is not available at present. In a previous work I mistakenly identified the author of *dPal ldan* as bKra-shis-dpal-bzang: Gyatso, "The Authority of Empiricism."

34. Yan-dgon-pa, *rDo rje lus*, 434–35.

35. sKyem-pa, *rGyud bzhi'i rnam bshad*, 129.

36. Zur-mkhar-ba, *Mes po'i*, 133, 152, 159.

37. Zur-mkhar-ba, *Mes po'i*, 152.

38. Tib. *bskyed cing*. Zur-mkhar-ba, *Mes po'i*, 162. He also says that *lalanā* and *rasanā* are the basis (*gzhi*) of the white and black soul channels (133), or exist "in connection" (*dang 'brel ba*) with them (165).

39. Zur-mkhar-ba, *Mes po'i*, 155.

40. Zur-mkhar-ba, *Mes po'i*, 133.10–12. This *srog-rtsa* is different from the black and white *srog-rtsa* specified under the heading of the "connecting channels" in the mature body. See also 166.20–22. Zur-mkhar-ba and others also consider *The Four Treatises'* fourth kind of channel, the life channel, as a place to juxtapose the tantric system of channels, but these discussions largely quote tantric sources and avoid specific medical anatomy (166–74). I examine the interesting details of this debate more fully in a

forthcoming book titled *The Way of Humans in a Buddhist World: Contributions to an Intellectual History of Tibetan Medicine.*

41. Zur-mkhar-ba, *Mes po'i,* 133.

42. Zur-mkhar-ba, *Mes po'i,* 154.

43. Zur-mkhar-ba, *Mes po'i,* 133, 154.

44. *Mi-'thad-pa.* Compare sDe-srid, *Blue Beryl,* 2:152. He calls this view "stupid."

45. sDe-srid, *Blue Beryl,* 2:173–74.

46. Compare sDe-srid, *Blue Beryl,* 2:151; Zur-mkhar-ba, *Mes po'i,* 153. See, e.g., Kong-sprul, "rNal 'byor." This is reportedly also the position of a *Four Treatises* commentary by 'Ju Mi-pham. For a recent example by the twentieth-century scholar Tsültrim Gyeltsen, see Garrett and Adams, "The Three Channels."

47. Gling-sman, *gSo ba rig pa'i gzhung,* 44–45.

48. Gling-sman, *gSo ba rig pa'i gzhung,* 46; "material" = *gdos bcas.*

49. On the debate, see Gyatso, "The Authority of Empiricism," 89–90. Gling-sman, *gSo ba rig pa'i gzhung,* 478: "bDag gis ni pho mo mang po'i ro bshas pa mthong / Rang gis kyang gri snying blangs pas pho mo thams cad snying rtse cung zad g.yon phyogs brang ngos la bsten pa mthong."

50. Translation adapted from Garrett and Adams, "Three Channels," 19.

51. For an overview of the period, see Shakabpa, *A Political History;* Petech, *China and Tibet.*

52. An admirable beginning is Pommaret, *Lhasa.*

53. The archival sources for this period are substantial, but our information depends for the moment largely on the secondary work of contemporary Tibetan scholars who are systematically combing the lengthy biographies and autobiographies of the key figures in the period; examples include dKon-mchog Rin-chen, *Bod kyi gso rig,* and some of the essays in Byams-pa-'phrin-las, *Byams pa 'phrin las kyi gsung rtsom.*

54. The physician is styled "(R)manaho." dKon-mchog-rin-chen, *Bod kyi gso rig,* 99; Byams-pa-'phrin-las, *Gangs ljongs,* 315, quoting the Fifth's autobiography, *Dukula'i gos bzang.*

55. dKon-mchog-rin-chen, *Bod kyi gso rig,* 100–104; Byams-pa-'phrin-las, "sDe srid," 414–17.

56. Byams-pa-'phrin-las, "sDe srid," 417.

57. Parfionovitch, Dorje, and Meyer, *Tibetan Medical Paintings.* For a history of the set, see Byams-pa-'phrin-las, "Bod kyi gso rig," 370–81.

58. Byams-pa-'phrin-las, "sDe srid," 425–26, quoting *Blue Beryl.*

59. For example, he notes that when you look at an actual body you see that, in contrast to what *The Four Treatises* claims, the heart tip faces left in both male and female (Byams-pa-'phrin-las, "sDe srid," 425–26).

60. "Dmar khrid mdzub tshugs su ngo sprod sngon med kyi bris cha." Byams-pa-'phrin-las, "sDe srid sangs," 424, quoting sDe-srid's *mChod sdong 'dzam gling rgyan gcig rten gtsug lag khang dang bcas pa'i dkar chag,* f. 281.

61. Byams-pa-'phrin-las, "sDe srid," 419, quoting sDe-srid's *Baidurya Ser po.* sDe-

srid's *mChod sdong* is also a key source here: it has been studied by Schaeffer in "Controlling Time."

62. For striking images of the rituals of the Tibetan state by the mid-twentieth century, see Richardson, *Ceremonies*.

63. Zur-mkhar-ba, "rGyud bzhi bka'," 64. I treat this debate in detail in my forthcoming work *The Way of Humans*.

64. See also Karmay, "Four Tibetan Medical Treatises."

65. Karmay, "Four Tibetan Medical Treatises," 230, n.15. Others who argued for Yutok's authorship were Bo-dong Phyogs-las-rnam-rgyal (1376–1451) and sTag-tshang Shes-rab rin-chen (b. 1405).

66. See Gyatso, "The Logic of Legitimation."

67. Zur-mkhar-ba, "rGyud bzhi bka'," 70.

68. Compare Zur-mkhar-ba's detailed discussion in *Mes po'i*, 4.

69. See the list in Karmay, "Four Tibetan Medical Treatises," 234–37.

70. See Gling-sman, *gSo ba rig pa'i gzhung*, 4–8.

71. sDe-srid, *Khog 'bugs*, 274–76.

72. Zur-mkhar-ba, *sMan pa rnams*, 311–13. For Zurkharwa's most radical statement, see Zur-mkhar-ba, "rGyud bzhi bka'." For his commentary on kinds of *sastra*, see *Mes po'i*, 21–22. Here he vacillates between calling *The Four Treatises sastra* and another category of scriptures composed by figures other than the Buddha but which in some sense were inspired by the Buddha (*rjes-su-gnang-ba'i bka'*) and which therefore still count as canonical. I am saving the details of this intricate argument for my forthcoming book.

73. An example would be the invocation of the invisibility of the tantric central channel as a model for the imperceptibility of certain fine channels connected to the liver: 'Bri-gung dKon-mchog-'gro-phan-dbang-po, "gSo ba rig pa'i gzhung lugs," 134–38.

References

Beckwith, Christopher. "The Introduction of Greek Medicine into Tibet in the Seventh and Eighth Centuries." *Journal of the American Oriental Society* 99, no. 2 (1979), 297–313.

'Bri-gung dKon-mchog-'gro-phan-dbang-po (b. 1631). "gSo ba rig pa'i gzhung lugs chen po dpal ldan rgyud bzhi'i dka' gnad dogs sel gyi zin bris mdo." *'Bri gung gso rig gces bsdus*, ed. 'Bri-gung Chos-grags et al. Beijing: Mi-rigs-dpe-skrun-khang, 1999.

bSod-nams-ye-she-rgyal-mtshan (fifteenth–sixteenth century). *dPal ldan bshad pa'i rgyud kyi 'grel pa bklag pa don tham chad grub pa*. Photocopy of manuscript held in Lhasa.

Byams-pa-'phrin-las (b. 1928). "Bod kyi gso rig rgyud bzhi'i nang don mtshon pa'i sman thang bris cha'i skor la rags tsam dpyad pa." *Byams pa 'phrin las kyi gsung rtsom phyogs bsgrigs*. Beijing: Krung-go'i Bod-kyi-shes-rig-dpe-skrun-khang, 1997.

———. *Gangs ljongs gso rig bstan pa'i nyin byed rim byon gyi rnam thar phyogs bsgrigs.* Beijing: Mi-rigs-dpe-skrun-khang, 2000.

———. "sDe srid sangs rgyas rgya mtsho'i 'khrungs rabs dang mdzad rjes dad brgya'i padma rnam par bzhad pa'i phreng ba." *Byams pa 'phrin las kyi gsung rtsom phyogs bsgrigs.* Beijing: Krung-go'i Bod-kyi-shes-rig-dpe-skrun-khang, 1997.

Byang-pa rNam-rgyal-dpal-bzang (1395–1475). *bShad rgyud kyi 'grel chen bdud rtsi'i chu rgyun.* Chengdu: Si-khron Mi-rigs-dpe-skrun-khang, 2001.

Cullen, Christopher. "*Yi'an*: The Origins of a Genre of Chinese Medical Literature." *Innovation in Chinese Medicine*, ed. Elizabeth Hsu. Cambridge: Cambridge University Press, 2001.

Dar-mo sMan-rams-pa Blo-bzang-chos-grags (b. 1638). "Rus pa'i dum bu sum brgya drug cu'i skor bshad pa." *Bod lugs gso rig sman rtsis ched rtsom phyogs bsdus.* Lhasa: sMan-rtsis Khang, 1996.

dKon-mchog-rin-chen (b. twentieth century). *Bod kyi gso rig chos 'byung baidurya'i 'phreng ba.* Lanzhou: Kansu'u Mi-rigs-dpe-skrun-khang, 1992.

Emmerick, R. E. "Sources of the *Four Tantras*." *Zeitschrift der Deutschen Morgenländischen Gesellschaft* (Wiesbaden) 3, no. 2 (1977), 1135–42.

Garrett, Francis, and Vincanne Adams. "The Three Channels in Tibetan Medicine, with a Translation of Tsultrim Gyaltsen's 'A Clear Explanation of the Principal Structure and Location of the Circulatory Channels as Illustrated in the Medical Paintings.'" *Traditional South Asian Medicine* 8 (2008), 86–115.

Gling-sman bKra-shis (b. 1726). *gSo ba rig pa'i gzhung rgyud bzhi'i dka' 'grel.* Chengdu: Si-khron Mi-rigs-dpe-skrun-khang, 1988.

Gong-sman dKon-mchog-phan-dar (1511–77). *Nyams yig rgya rtsa: The Smallest Collection of Gong-sman Dkon-mchog-phan-dar's Medical Instructions to the Students.* Leh: Lharje Tashi Yangphel Tashigang, 1969.

Gyatso, Janet. *Apparitions of the Self: The Secret Autobiographies of a Tibetan Visionary.* Princeton, N.J.: Princeton University Press, 1998.

———. "The Authority of Empiricism and the Empiricism of Authority: Medicine and Buddhism in Tibet on the Eve of Modernity." *Comparative Studies of South Asia, Africa, and the Middle East* 24, no. 2 (2004), 83–96.

———. "Healing Burns with Fire: The Facilitations of Experience in Tibetan Buddhism." *Journal of the American Academy of Religion* 67, no. 1 (1999), 113–47.

———. "The Logic of Legitimation in the Tibetan Treasure Tradition." *History of Religions* 33, no. 1 (1993), 97–134.

gYu-thog Yon-tan mgon-po (1126–1202). *Four Treatises* = *bDud rtsi snying po yan lag brgyad pa gsang ba man ngag gi rgyud.* Lhasa: Bod-ljongs Mi-dmangs-dpe-skrun-khang, 1992.

Israel, Jonathan I. *Radical Enlightenment: Philosophy and the Making of Modernity 1650–1750.* Oxford: Oxford University Press, 2001.

'Ju Mi-pham (1846–1912). *bDud rtsi snying po'i rgyud kyi 'grel pa drang srong zhal lung las dum bu bzhi pa phyi ma rgyud kyi rtsa mdo chu mdo'i tika. gSo rig skor gyi rgyun mkho gal che ba bdam sgrigs,* ed. Yon-tan-rgya-mtsho et al. Beijing: Mi-rigs-dpe-skrun-khang, 1988.

Karmay, Samten. "The Four Tibetan Medical Treatises and Their Critics." *The Arrow and the Spindle: Studies in History, Myths, Rituals and Beliefs in Tibet.* Kathmandu: Mandala Book Point, 1998.

Khro-ru Tshe-rnam (1928–2004). "Bod lugs gso rig slob grva rim byung gi lo rgyus gsal ba'i gtam dngul dkar me long." *Bod sman slob gso dang zhib 'jug* 1 (1996), 1–11.

Kong-sprul Blo-gros-mtha-yas (1813–99). "rNal 'byor bla na med pa'i rgyud sde rgya mtsho'i snying po bsdus pa zab mo nang di don nyung ngu'i tshig gis rnam par 'grol ba zab don snang byed." *Zab mo nang gi don zhes bya ba'i gzhung gi rtsa 'grel,* ed. Karma Rang-byung-rdo-rje and Kong-sprul Yon-tan-rgya-mtsho. Xining: mTsho-sngon Bod-lugs-gso-rig-slob-grva-chen-mo, 1999.

———. *'Tsho byed las dang po la nye bar mkhor ba'i zin tig gces par btus pa bdud rtsi'i thigs pa. gSo rig skor gyi rgyun mkho gal che ba bdam sgrigs,* ed. Yon-tan-rgya-mtsho et al. Beijing: Mi-rigs-dpe-skrun-khang, 1988.

Meyer, Fernand. *Gso-ba Rig-pa: Le système médical tibétain.* Paris: Centre National de la Recherche Scientifique, 1981.

Parfionovitch, Yuri, Gyurme Dorje, and Fernand Meyer, eds. *Tibetan Medical Paintings: Illustrations to the Blue Beryl Treatise of Sangye Gyamtso (1653–1705).* 2 vols. London: Serindia, 1992.

Petech, Luciano. *China and Tibet in the Early XVIIIth Century: History of the Establishment of Chinese Protectorate in Tibet.* Leiden: Brill, 1972.

Pommaret, Françoise, ed. *Lhasa in the Seventeenth Century: The Capital of the Dalai Lamas.* Leiden: Brill, 2003.

Richardson, Hugh. *Ceremonies of the Lhasa Year.* Ed. Michael Aris. London: Serindia, 1993.

Schaeffer, Kurtis R. "Controlling Time and Space in Lhasa." Paper delivered to the annual meeting of the American Academy of Religion, Toronto, November 2002.

———. "Textual Scholarship, Medical Tradition, and Mahayana Buddhist Ideals in Tibet." *Journal of Indian Philosophy* 31 (2003), 621–41.

sDe-dge-bla-sman Chos-grags-rgya-mtsho (eighteenth century). *Nad sman sprod pa'i nyams yig. gSo rig skor gyi rgyun mkho gal che ba bdam sgrigs,* ed. Yon-tan-rgya-mtsho et al. Beijing: Mi-rigs-dpe-skrun-khang, 1988.

sDe-srid Sangs-rgyas-rgya-mtsho (1653–1705). *Blue Beryl = gSo ba rig pa'i bstan bcos sman bla'i dgongs rgyan rgyud bzhi'i gsal byed bai dur sngon po'i ma lli ka.* 2 vols. Leh: D. L. Tashigang, 1981.

———. *Khog 'bugs = dPal ldan gso ba rig pa'i khog 'bugs legs bshad baidurya'i me long drang srong dgyes pa'i dga' ston.* Lanzhou: Kansu'u Mi-rigs-dpe-skrun-khang, 1982.

———. *Techniques of Lamaist Medical Practice, Being the Text of Man ngag yon tan rgyud kyi lhan thabs zug rngu'i tsha gdung sel ba 'i katpu ra dus min 'chi zhags gcod pa'i ral gri.* Leh: S. W. Tashigangpa, 1970.

Shakabpa, Tsipon *A Political History of Tibet.* New Haven, Conn.: Yale University Press, 1967.

Shapin, Steven. *The Scientific Revolution*. Chicago: University of Chicago Press, 1996.

sKyem-pa Tshe-dbang (fifteenth century). *rGyud bzhi'i rnam bshad*. Xining: mTsho-sngon Mi-rigs-dpe-skrun-khang, 2000.

Sprung, Mervyn, trans. *Lucid Exposition of the Middle Way*. London: Routledge and Kegan Paul, 1979.

Sum-ston-pa Ye-shes-gzungs (twelfth century). "'Grel ba 'bum chung gsal sgron nor bu'i 'phreng mdzes." *gYu thog cha lag bco brgyad*, vol. 1. Lanzhou: Kansu'u Mi-rigs–dpe-skrun-khang, 1999.

Vāgbhaṭa. *Vāgbhaṭa's Aṣṭāṅgahṛdayasaṃhitā*. Ed. Rahul Peter Das and Ronald Eric Emmerick. Groningen: Forsten, 1998.

Vogel, Claus ed. *Vāgbhaṭa's Aṣṭāṅgahṛdayasaṃhitā: The First Five Chapters of Its Tibetan Version*. Wiesbaden: F. Steiner, 1965.

Yan-dgon-pa rGyal-mtshan-dpal (1213–58). *rDo rje lus kyi sbas bshad. The Collected Works (Gsun 'bum) of Yan-dgon-pa Rgyal-mtshan-dpal*, vol. 2. Thimphu: Kunsang Topgey, 1976.

Zur-mkhar-ba Blo-gros rGyal-po (b. 1509). *Mes-po'i = rGyud bzhi'i 'grel pa mes po'i zhal lung*. 2 vols. Beijing: Krung-go'i Bod-kyi-shes-rig-dpe-skrun-khang, 1989.

———. "rGyud bzhi bka' dang bstan bcos rnam par dbye ba mun sel sgron me." *Bod kyi sman rtsis ched rtsom phyogs bsdus*, ed. Bod Rang-skyong-ljongs sMan-rtsis-khang. Lhasa: Bod-ljongs Mi-dmangs-dpe-skrun-khang, 1986.

———. *sMan pa rnams kyis mi shes su mi rung ba'i shes bya spyi'i khog dbubs*. Chengdu: Si-khron Mi-rigs-dpe-skrun-khang, 2001.

CHAPTER THIRTEEN

Just Where on Jambudvīpa Are We?

New Geographical Knowledge and Old Cosmological Schemes in Eighteenth-century Tibet

MATTHEW T. KAPSTEIN

South Asia's traditional ways of knowledge belong not to South Asia alone. Beginning in antiquity and throughout the medieval period the ongoing transmission of Indian learning, religion, and art across Central, Southeast, and East Asia ensured that thought and practice stemming from the subcontinent would have a deep and enduring legacy, in many cases down to the present day. To inquire into the transformations of South Asian knowledge systems at the cusp of modernity, therefore, one must sometimes reach beyond South Asia itself. In the case considered here, we shall indeed be far removed from the region as it is ordinarily understood: our subject, the learned monk Sumpa Khenpo Yeshé Peljor (1704–88), was born to a family of high status among the Oirat Mongols. The monastic center with which he was affiliated, however, lay within the domains of another Inner Asian ethnic group, the Monguor, whose territory lies to the east of the Kokonor Lake. As inhabitants of a part of the area traditionally known among Tibetans as Amdo, the Monguor were definitively incorporated into the expanding Manchu Empire in 1723 with the inclusion of Qinghai, corresponding to the contemporary Chinese province of that name, into the administrative region of Gansu. The language of learning in which Sumpa Khenpo was educated and in which he wrote, however, was classical Tibetan, so that he was a participant in the generally Indianized cultural framework that was normative for the Buddhist clergy of Tibet.[1]

Sumpa Khenpo's education was exemplary. By the age of twenty-three he had advanced through the clerical ranks to become for a time an abbot of Gomang, one of the major colleges belonging to what would become Tibet's, and the world's, largest monastic community, Drepung, in the suburbs of Lhasa.[2] As such he was intimately familiar with a wide range of writings on

Indian Buddhist learning, including such matters as the logic of Dharmakīrti, Candrakīrti's dialectics, and Guṇaprabha's code of monastic discipline, together with other Indian authors whose works in translation dominated Tibetan religious thought.[3] Like many learned Tibetans, however, Sumpa Khenpo was also interested in and wrote extensively on history, a topic that was not primarily inspired by Indian models and that therefore, because it had no place in the formal curriculum, was sometimes regarded as a frivolous distraction.[4] Another unofficial subject that similarly aroused his curiosity was geography.

TO THE EXTENT that geography was known and taught in Tibet's Buddhist academies, it formed a part of the *abhidharma*, the "metadoctrine" of classical Indian Buddhism that was generally studied through the writings of Vasubandhu.[5] In the "teaching of worlds" (*lokanirdeśa*), the third chapter of that master's *Treasury of Abhidharma* (*Abhidharmakośa*), one found a thorough description of the *axis mundi*, Mt. Meru, symmetrically surrounded by four continents, each with two subcontinents, and ringed by successive oceans and mountain ranges. India was identified with the southern continent, the sawtooth-shaped Jambudvīpa (Tib. *'dzam-bu-gling*), so-called owing to the presence there of the *jambu* tree, often identified as the "rose-apple."[6] Places like Tibet, China, and Mongolia, which ordinary humans could in principle reach by traveling from India, were assumed to be part of Jambudvīpa; the blue color of the cloudless daytime sky in all known places, a reflection of the lapis southern slopes of the world mountain, assured you of that. The three remaining continents, though present in the imagination, were for all practical purposes off the map. The essentials of the scheme were drilled into the heads of all young monks through the daily recitation of the *maṇḍala*-offering, in which Mt. Meru and the world surrounding it are symbolically dedicated to the Buddha in the form of a pile of grain as one recites the names of the objects offered: Mt. Meru, the four continents, the eight subcontinents, and so on (figure 1).[7]

This ancient Indian cosmology was not, to be sure, the only one known in Tibet. Toward the end of the first millennium the autochthonous Bön religion had devised its own *abhidharma*, including a chart of the sacred land of Ölmolungring, located somewhere to the west of Tibet (figure 2). Speculation about the geography depicted in this diagram has been rife: a distinguished Russian geographer and Tibetanist concluded that what was represented was in fact the Iranian world of about the second century BCE. They believed that,

Figure 1. A nineteenth-century Tibetan scroll painting of the Mt. Meru world-system as a "*maṇḍala*-offering." Courtesy of Rubin Museum of Art, New York, C2006.66.558 (HAR 1038). Photograph by Bruce M. White.

among sixty-four named locations, they could identify more than forty, including the Persian capital of Parsogard recorded at the center as Bar-po-so-brgyad, Palestine as Mu-le-stong, Jerusalem as Lang-ling, and Alexandria as Ne-seng-dra-ba. The same authors also concluded, on the basis of a Tibetan version of the Mt. Meru system, that the eastern continent of Videha represented there could only be the Americas, leading them to speculate that the ancient Indians had crossed the Pacific on at least three occasions. "For the Tibetans," they enthused, "as for the medieval Arabs, geography was not only a practical but also a theoretical science." They added, "These geographers essentially appear to have been familiar with the entire surface of the earth, with the exception of Australia and Antarctica."[8] Though their results have been generally greeted with skepticism, their hypothesis that elements of ancient Iranian geography reached Tibet cannot be dismissed out of hand; although many of their suggestions seem extravagant it would be unwarranted to exclude categorically the possibility that some of the names in question, such as those identified with Parsogard and Alexandria, were indeed derived from foreign sources.[9]

A third geographical scheme that was widely invoked in traditional Tibet appears to be both purely Tibetan and quite practical in its origins. Under the old Tibetan Empire (seventh–ninth centuries CE) the Tibetans came to think of their land as strategically placed in the central square of a checkerboard of powers, the most important of which, after Tibet itself, were China, India, Tazik (the Iranian world), and Gesar. The last, specifically designated Phrom Gesar, "Caesar of Rome," has sometimes been identified with Byzantium, though it may also be connected with the Turkic rulers of what is today Afghanistan, who sometimes made use of the western royal title. Moreover notes found among the old Tibetan documents discovered at Dunhuang suggest, as would have been required by the interests of empire, that geographical information about specific locations was being regularly gathered; one manuscript contains a report concerning the region of Besh-baliq, in the empire of the Uighur Turks. Some knowledge of the Tibetan imperial geographical traditions was preserved in later times through historical writings, as well as in general compendiums of knowledge.[10] An intriguing fragment of old Tibetan geography may also be found in the puzzling chart copied by the Japanese monk Zenkaku in 1194 on the basis of an original said to have come from Chang'an that had been brought to Japan during the ninth century by the celebrated priest of the Tendai order, Enchin (814–91).[11]

The main literary sources of information on local geography outside of Tibet were to be found in the journals and autobiographical accounts of Ti-

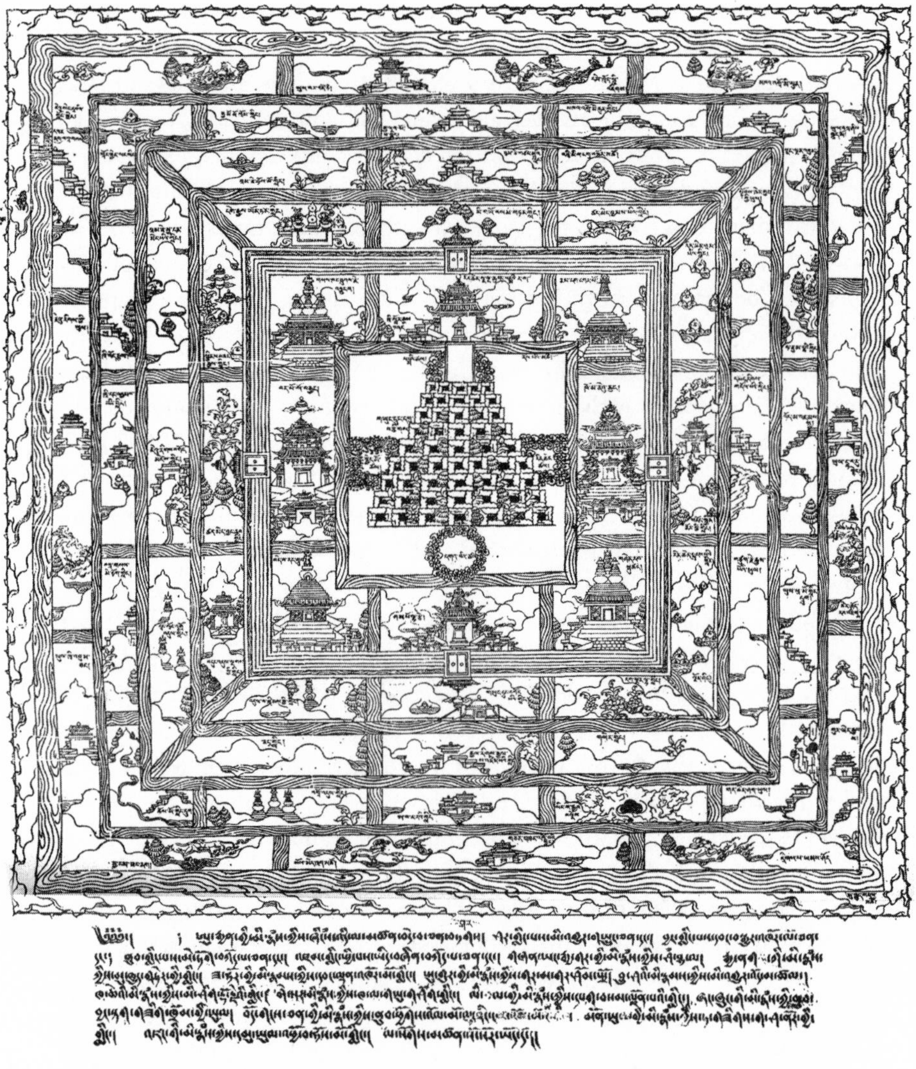

Figure 2. A version of the famous Bönpo map of Ölmolungring, "The Nine-Stage Swastika Mountain," from *The Nine Ways of Bon* by David Snellgrove (1967). Bar-po-so-brgyad, identified by Gumilev and Kuznetsov with Parsogard, is located at nine o'clock in the first belt surrounding the center. Ne-bo-seng-dra-ba, which they consider to be Alexandria, is in the upper left corner. Reprinted by permission of Oxford University Press and the School of Oriental and African Studies.

betans who had traveled, for instance, the famous records of the translator Chak Chöjé-pel (1197–1263) and of the adept Urgyenpa (1229–1309) of their peregrinations in India.[12] For local geography within the Tibetan world, bureaucratic requirements for records of taxes, labor services, and the like dictated that, under most regimes, considerable information was set down, sometimes supplemented with local maps. These generally took the form of more or less detailed sketches or paintings of landscapes with key features carefully labeled.[13] In addition biographical and historical writings, together with pilgrims' guides (*gnas-yig*), provided more or less thorough accounts of itineraries, toponyms, and famous sites. In short there was plenty of geographical information, derived from a broad range of sources, that was available in traditional Tibet. What tended to be lacking was any clear framework with which to organize and assess it as a coherent whole.

This was the result principally of want of interest, and not of deficient means. Joseph Schwartzberg, in his study of Tibetan charts and maps, takes his subject to embrace all types of graphic representations of space, not just terrestrial maps as we usually conceive of them. His approach makes clear that Tibetans knew very well how to employ geometric methods to achieve mathematically precise results; the intricate art of the *maṇḍala* emphatically demonstrates this.[14] In principle, therefore, it would not have involved too great a conceptual step for the Tibetans to have applied mathematics to cartography. If they never did so it was perhaps because they never felt the need for mathematical precision in this area. They were of course not a maritime nation, and their religion did not require (as does Islam, for instance) an exact orientation toward a particular sacred center. The approximations and reckoning by prominent landmarks that are suitable for landlubbers traveling by foot or on horseback generally served them just fine, and, for purposes of agricultural production, fields were thought to be better measured by the volume of seed they could receive than by an otherwise meaningless quantification of surface area.[15]

Within these broad parameters of Tibetan geographical interest the ancient Indian concept of Jambudvīpa served as a malleable template that could be adopted to serve the needs at hand. Chögyel Pakpa (1235–80), the imperial preceptor of Khubilai Khan, to note one example, duly adds his Mongol patrons to the register of the great kings of Jambudvīpa, but he seems little interested in the question of whether their conquests, which brought even Europeans into the court circles with which he was intimately familiar, should have any major implications for his conception of Jambudvīpa as our world. In a textbook written for Khubilai's eldest son, Jiṅ-gim, he did not waste words: "Three thousand two hundred and fifty years after the Buddha's Nirvāṇa, Jiṅ-gir (Genghis) became King up North in Hor (Mongolia). . . . He brought many countries of different languages and races under his power, and by his strength he became like a [Wheel-turning] King."[16] Just where the "many countries" were supposed to be situated was left unsaid.[17]

Despite this, the affirmation that one did indeed inhabit Jambudvīpa was religiously important. It meant that, temporally and geographically, one belonged to the very world-system in which Śākyamuni had appeared and in which therefore, with the right measure of good karma and personal endeavor, one might make some progress on the path oneself. For this reason Tibetan biography and autobiography often began by setting up their story with a sort of narrative zoom-in, as we find in the opening paragraphs of the early nineteenth-century lama Zhabkar's account of his life:

> Within the realm tamed by the peerless Buddha Shakyamuni, north of the Diamond Throne of India, the center of the southern continent of Jambudvipa, lies the Golden Valley of Rekong where Jetsun Kalden Gyatso . . . benefited countless beings. To the west lie the Pure Realms of U and Tsang where the Buddhas Amitabha and Padmapani emanated [as the Panchen and Dalai Lamas]. To the north, in Domey, stands the mountain Tsongkha Kyeri, the birthplace of the Second Buddha, the great Tsongkhapa, who reigns supreme over the three worlds. . . .
>
> In this region, ten villages of various sizes lie scattered in all directions. Among these is Nyengya, a village at the foot of the local god Jadrön's mountain abode. This is my homeland, the place of my birth.[18]

Just where we are on Jambudvīpa, therefore, assuredly does matter. By Zhabkar's time, however, one's sense of place had begun to be relativized by the appearance of new information, arriving, so it may have seemed to well-informed persons, from all directions at once.

Part of this story has already been told. The late Michael Aris, in a superb essay on the record of India and the West set down by the famed mystic Jikmé Lingpa in 1788, has shown that in some circles Tibetans were eager for new knowledge of the world. Though the master himself never traveled beyond Tibet and Nepal, a Bhutanese disciple who visited British Bengal served as his informant, supplying reports that Jikmé Lingpa recorded in notes on the Dutch spice trade, British shipbuilding and music boxes, Mughal-Ottoman relations, and more. As Aris astutely recognized, however, such interest in novelty was counterpoised against powerfully conservative trends. While he describes George Bogle and Samuel Turner, who visited Tibet in 1774 and

1783, respectively, as having been "deluged with questions by the lamas and officials they met on their journeys," he mentions too a contemporaneous Bhutanese biography that is "redolent with suspicion and prejudice at the fascination for European mechanical devices aroused at the Bhutanese court by Turner's scientific instruments and his gifts to the local dignitaries."[19] The debate between Tibetan partisans of old and new learning would continue well into the twentieth century and prove in the end profoundly debilitating for the integrity of the nation overall.[20]

IIIIIIIIIIII

SUMPA KHENPO's *General Description of Jambudvīpa* (*'Dzam-gling spyi-bshad*) was written in 1777, when he was already seventy-three years of age. Besides his Tibetan education, he had cultivated throughout his long life a broad range of Mongol contacts, particularly among the Zunghar, and had lived and traveled in China for extended periods as well. He was one of a generation of clergy from Amdo who, in becoming the interlocutors between the Tibetan, Mongol, and Manchu Chinese worlds, cultivated a brand of Inner Asian cosmopolitanism that was in many respects distinctive and new. Their milieu was characterized above all by the combination of political allegiance to the Manchu court (though Sumpa Khenpo had earlier been a partisan of the Zunghar), with a religious commitment to Tibetan Buddhism, finding in the latter a suitable international culture through which to mediate relations among the various linguistic and ethnic groups concerned. The fact that the Monguor were a tiny minority, distinct from and yet at the same time engaged with their more prominent neighbors, may have also contributed to their assuming this interlocutory role.[21]

The *General Description of Jambudvīpa* opens without promise. Its first pages summarize Buddhist canonical descriptions of Jambudvīpa, focusing on the regions of India and scriptural lists of peoples and lands.[22] The major tantric places of pilgrimage (*pīṭha*) are also detailed,[23] followed by remarks on the "barbarians" (*mleccha*) who live beyond India. Here Sumpa Khenpo follows primarily the tradition of *Tantra of the Wheel of Time* (*Kālacakratantra*), with its concealed reference to the prophet Mohammed as "Honey Mind" (Madhumati).[24] But once we escape the Indian sphere a transition begins as we journey to Khotan, Nepal, and then Tibet, where the author's path is forged no longer exclusively by scripture, but by varying degrees of familiarity with the terrain.[25] A brief detour to the north takes us to mythical Shambhala, with its twelve million towns, before we swing east to Manchuria, Korea, and Japan.[26] Returning by stages (including a pass through the land of dog-headed men) to China, we find a list of provinces and notes on mountains, rivers, lakes, towns, riches, diet, religion, and above all the wonders of Wutaishan, the bodhisattva Mañjuśrī's terrestrial abode.[27] Leaving China we enter the Mongol domains, and from these, at last, our author takes his readers on an unprecedented adventure into the unknown:[28]

> Beyond Khalkha, to the north, is Russia, the yellow expanse.[29] Concerning that, on the banks of the Caspian Sea,[30] which is at the frontier of Torghut and Russia, there is Russia's citadel of Astrakhan. On travelling past many citadels beyond that for a month and a half, in the center of Russia, there is [the place] called Moscow, the citadel of a girl-king called the White Khan[31] who belongs to the lineage of Chinggis Khan. It has three or nine concentric stages [i.e., walls] of which the outermost is said to be crystal. The palace in the center, with its spire, has pillars and so forth whose surfaces are gilded and decorated with gems. From the sound of the bells at the sides, one knows the changes of time, as well as good or bad tidings. At intervals there are splendid mansions, many-storied houses, cool houses, theatres and many small bridges.[32]
>
> The men of that land have measureless power and wealth, and their realm is exceedingly great. With silver coin, they trade in such goods as wigs for the bald, the flesh and blood of sea monsters, and serge.[33] They are nourished by the flesh of fish and fowl, and various grains. There is a fortress of magnetic iron.
>
> On the banks of the northern ocean that lies beyond is Sweden,[34] whose people dwell in small settlements. They are very skilled at the manufacture of various mechanical devices such as clocks.[35] In that country, there are black and red foxes, marten,[36] serge, weapons of fine steel, and utensils made of gold, silver, and crystal. They consume poultry, eggs, fish, tree-milk,[37] and various grains.
>
> In Kamchatka and Yakutia[38] in that country [Russia], as well as in [the region of] the Kem Kemchik [River],[39] which belongs to the Torghut territory, and elsewhere, at the height of the summertime, when the sun sets, the red remains in the sky as it rises again and the dawn begins. When the full moon of the fifteenth is still at its height in the west, the sun of the sixteenth rises, and when the sun sets on the twentieth, the moon rises at the same time. In summer and fall, the Swedes and the Russians cross back and forth over the sea in boats, but in winter and spring they travel on the frozen ice, in so-called *tshan* (from Russ. сани via Mong. *tsana*, "sled") dragged by dogs.

Again, if you travel via Khur lug beyond Kokonor and cross the Great Desert, then, by stages, there are the fortresses of the white-headed Muslims: Hami and Barkul, then Yarkand, Kashgar, the land where lambs are born when they plant the sheep-kidney worm, and Badakhshan.[40] To the left are A tsa yan (= prob. Andijan) and Namangan, Marghilan and Tashkent.[41] Then, there are Bukhara and Samarkand, where there is great wealth, with plentiful corals and pearls, carpets woven of silver and gold, and more. There are also the Aqtaghliq (the "White Mountain" order of the Naqshbandi sufis), who wear white hats. . . .[42] To the left [i.e., north] are the Kazakh, and to the west there is the great valley of Ferghana.[43]

Among them, although neither the Kazakh nor the Russians have a proper religion, the latter wear matted locks.[44] Moreover, the Muslims never sport matted locks or the queue.[45] As for the white-headed Muslims, all the others were wearing a long red headdress, but because the teacher called Abdullah made them wrap their skulls in a white turban, or however it was, all the Muslims in this land are called white-heads.[46] For the most part, they adhere to various theistic systems and have different sects, and they affirm a future life and [the distinctions of] virtue and sin. In accord with their own tradition, they delight very much in maintaining their vows, liturgical recitation, and ablutions. For this reason, they are ten times better than the two groups without religion described above, as well as the other frontierfolk who have no religion at all, whether of the *tīrthika*s or [other] outsiders, and who are thus the worst of the barbarians in that the future life and virtue and sin have not even entered their ears. The Muslims wear gold and silver brocade, and other silks, serge and linen, and hair ornaments,[47] etc. They enjoy animal meats except pork but including fish, poultry, beef and mutton, as well as various grains and fruits.

In the region to the east of the Oirat Zunghar, to west of the gate in the Chinese wall called Jiayuguan,[48] is the so-called Mongol left banner, along the course of the two rivers, Ili and Ir chid (Irtysh). Beginning with the great river called Thor gwos (Torgai), northwest of there, and along the streams of the E jel (Volga) and Jal (Sal),[49] the Torghut territory runs east-west for a distance that takes five or six months to traverse, though it is not so extensive from south to north. Beyond the Zunghar's river, the Er chid (= Ir chid, the Irtysh), is Russia, the yellow expanse. In the area east of the Russian border, as demarcated with a wooden fence, and beginning from the Chinese population of So lon, there is the so-called western Khu yur,[50] which is about equal to the Torghut in area. Thus, the aforementioned three—Turkestan,[51] Oirat and Russia—are arranged [in a series] south to north. From the Torghut to Shambhala are various lands and very many principalities. In most of those countries, they do not value gold or silver highly,[52] and, because they are rich, thieves and beggars exist not even in name.

To the southwest of Russia is the citadel of Kha zal ba shi [Kızılbaş, Qizilbash],[53] where there are great Khu ril [bronze?] fortresses of the regions of both Khurasan and Fars, as well as *lab tu'i dgus 'khor* and the northwest.[54] There, there are elephants, swamp lions, and such birds as the *shang-shang*.[55] The wealth of those lands rivals that of Russia, repeatedly as it were. When their kings set out on the highway, they scatter golden and silver coins that are transported in a cart.

Again, to the north of Russia is the ocean called the White Sea,[56] whose northern limit is not perceptible and whose extent is exceedingly grand. On its bank is the land of the great group called Na mi si, where there are white bear and other varied carnivores, as well as a lizard the size of a dog.[57] The men and horses of that country are larger than others, and they have silks of more than a hundred spans, a single [bolt of] serge that is sufficient for blankets for thirty men, and amazing contraptions of crystal, gold, silver and iron that tell the time and the date. Their royal palace is called Ne me shing. The troops who protect their palace and borders have one officer for each ten soldiers and so on by tens until the supreme commander of ten million who is called Ra ma rtsad.[58] To ward off military attack, and to march into offensive battle, whatever, their surface army dons *phe tshem* and takes up various wooden arrows, small firearms, medium and large artillery, as well as lassos, spears, iron hooks, many bladed javelins, and all sorts of swords.[59] Their subterranean army takes up hundreds and thousands of torches of wool [drenched in] fat, while others grasp shovels, hoes and picks, and they go forward digging the armies' path and then ignite a great fire beneath the fortress [of the enemy], incinerating it, whereupon they emerge on the earth's surface and strike. Their army that travels by the sea and great rivers wears the skins of seabirds and moves on by stealth in order to attack. In these and other ways they have amazing methods beyond the imagination. It is a sign that their land is exceedingly vast.

To the west of the E jel (Volga) river, after ten days' march there is the region of the Ottoman king, who in former times in Tibet was known as "Sunbeam."[60] That group has the name Turk, or, as corrupted, Turuṣka. Their frontiers are enclosed by a stone wall with 360 gates, between any two of which there are 360 *li* according to their own measure, or 720 Chinese *li*. Within, there are said to be four great mountains, various lakes and woods, and different sorts of grain. The royal citadel has twenty-four gates, and the central palace has a gilded roof. In the four cardinal directions around it are palaces of brick with porcelain enamel in blue, green, etc. The flaps of the red serge men's hats, the cuffs of their sleeves, and their belts, etc., are of silk brocade with jewel ornaments, with buckles made of lapis, etc.[61] They wear, too, gem ornaments, bracelets, and very fine gold and silver brocade, serge, and cotton. It is well known in that land that their lord is the king of Jambudvīpa overall,[62] and that in a single palace the thrones of thirty-three kings of Jambudvīpa are arranged. In that land, every one or two years, there arrives the fear of a great enemy wind. Because, when this occurs, it is a worry, in order to look out for it there is a great building in the form of a vase that is very tall with a narrow summit, atop which there must be a watchman, while in the middle of the other citadels there is a so-called minaret, like a Chinese pestle, whose height is that of five or six long ropes.[63] In the quarters and frontiers there, there are twelve ethnic groups speaking twelve dissimilar languages. The mouths of their wells and their water vessels are made of gold and silver and are bejewelled. They have elephants, walruses, hippopotami,[64] and also horses, cattle, sheep, etc., as well as many sorts of precious steel weapons and varied grains and fruits. When the king goes out for recreation, a distribution is made from many carts bearing gold and silver coins to onlookers and those who come to meet [the king]. In this and other ways, they rival the gods in riches. I heard this myself from a Kashmiri Muslim merchant,[65] who really heard it from a native of that land.

To the west from there, in the sea called Thing thing gi se [Mediterranean?],[66] there are [creatures with] human bodies and horse heads, as well as sea monsters with antlers, etc. There, in glass boats they go to the surface and depths of the sea and so retrieve various gems.[67]

To the south of the Ottomans, at Mecca, the *tīrthikas*' shrine, is a blue stone of about a meter, or four or five *mtho* around, [held] in space at about the height that can be reached by the hand of a rider on horseback, as I heard from a Chinese Muslim who had seen it.[68] Nearby, in a hollow at a distance of five or six arrow shots, there is a *tīrthika* temple within which there is neither pillar nor beam and there is a so-called self-manifestation that is gilded and bejewelled. About a league from that there is their shrine, a white stone with the sign of Īśvara, pierced and touching nothing, above or below, before or behind. So they say.[69]

Here and there, in those parts, there are many great regional groups whose felt tents have a double roof, and such like, and where there are varied sorts of houses and citadels. Moreover, the cities, towns, villages, lands, fortresses, and the jewelled rivers and lakes are very many.

Once again, to the west, successively, there are those with human bodies and horse heads and dog heads, the one-legged *tsu ta*, those with male and female sex organs in their legs, those with big ears who can sleep using one as a pillow and the other as a cover, and the so-called "shoulder-eyed bodies," who have no head, eyes in their shoulders and mouths at their waists.[70]

AT FIRST GLANCE Sumpa Khenpo's account recalls Western medieval tales of marvelous itineraries, such as we find in the travels of John Mandeville.[71] We should note carefully, though, that this is not merely a record of prodigies. In fact the amount of information here that reflects factually valid sources is impressive: Moscow's walls and bells, for instance, were widely famed and much remarked by travelers close to Sumpa Khenpo's time. Jacques-Henri Bernardin de Saint-Pierre, for instance, who visited Russia during the years 1762–64, commented, "Nothing is so magnificent as the appearance of this city, where nearly twelve hundred bell-towers have been raised, of which some are gilded, their pointed roofs culminating in crescents surmounted by the cross. At the center of the city is the Kremlin, the former residence of the sovereign. It is surrounded by a triple belt of crenellated walls, and flanked by towers."[72] Russia moreover was indeed ruled at the time by a "girl-king," Catherine II (r. 1762–96), called Catherine the Great. Though she may not have claimed Chingisid antecedents for herself, it is possible that Sumpa Khenpo reflects a belief about the tsars' ancestry that was current among Mongol populations. At least Ivan IV (r. 1547–84), known as Ivan the Terrible, had nurtured such associations.[73]

Dog sleds, accurately named in the text, were of course employed in the frozen winter, and the peculiar behavior of the summer sun in the higher latitudes, where white bears may also be found, is duly noted. Sweden's eighteenth-century clockmakers, kvass drinkers, and furs ring true. Numbers of Central Asian locations are correctly designated, including major rivers as far west as the Volga and the Sal. The famed winds of Istanbul appear to be mentioned, and perhaps even the Galata tower, which served as a weather observatory during the reign of the Ottoman sultan Murad III (r. 1574–95).[74] Sectarian rivalries among the Muslims of Inner Asia seem accurately reflected. Though some of the details regarding the *hajj* appear confused, the Black Stone is nevertheless unmistakable. Moreover certain of the legendary material we find in Sumpa's account may be ultimately due to Western sources as well; the tale of the glass diving bell, for instance, has a long history going back to the Romance of Alexander the Great, which was distributed throughout much of the medieval world, in Europe and the Middle East and as far as India.[75] Even certain of the monstrosities mentioned were well known to European conventions of illustration and may reflect shared Eurasian tropes (figure 3).[76] The most striking element of sheer fantasy is the account of the otherwise unidentified Na mi si of the far north. Let us recall, however, that even the great Erasmian humanist and cartographer Gerardus Mercator (1512–94) allowed his rational principles to slide into mythomania when it came to the north polar lands.[77]

Besides its many interesting details, our text is notable for its overall outlook. Though Sumpa Khenpo was by no means ready to abandon the authority of the traditional book learning to which he was heir, he clearly recognized the need to supplement the texts by turning to eyewitness accounts. His interviews with Muslim travelers are exemplary in this regard. What Anthony Grafton has written of the tension between novelty and intellectual conservatism in the work of the cosmographer Sebastian Münster, who was active some two centuries prior to Sumpa Khenpo, may be applied *mutatis mutandis* to the latter as well: "Münster added more new details and modern maps to the original text [of Ptolemy] than anyone else had. . . . New information did not modify or cancel the old, but piled up alongside it like fresh coal beside clinkers. New facts about the world . . . did not modify the old structure."[78] Just as Ptolemy's world lived on in such works as Münster's *Cosmographia* long after the data had surpassed it, so too did Jambudvīpa in the writings of Sumpa Khenpo and his successors.

Sumpa Khenpo concluded his work in a strikingly modern vein, however, calling for the improvement of geographical knowledge in Tibet. "In general," he writes, "if you have not seen, have not heard and do not know at least roughly the features of the continent on which you were born, and in particular the greatness of the king, the specificities of the terrain, and the species of animal of your own birthplace, then, no matter how much you have heard, seen or come to know of religious and worldly things . . . you are not better than, as the proverb states, 'the astrologer who knew the paths of the moon and stars, but failed to perceive the infidelity and carryings-on of his own wife.'"[79] This local knowledge, however, is for Sumpa Khenpo only the beginning, for he adds that the knowledge of the world beyond can free one from the conceit that one's homeland is the center of the earth and that its people

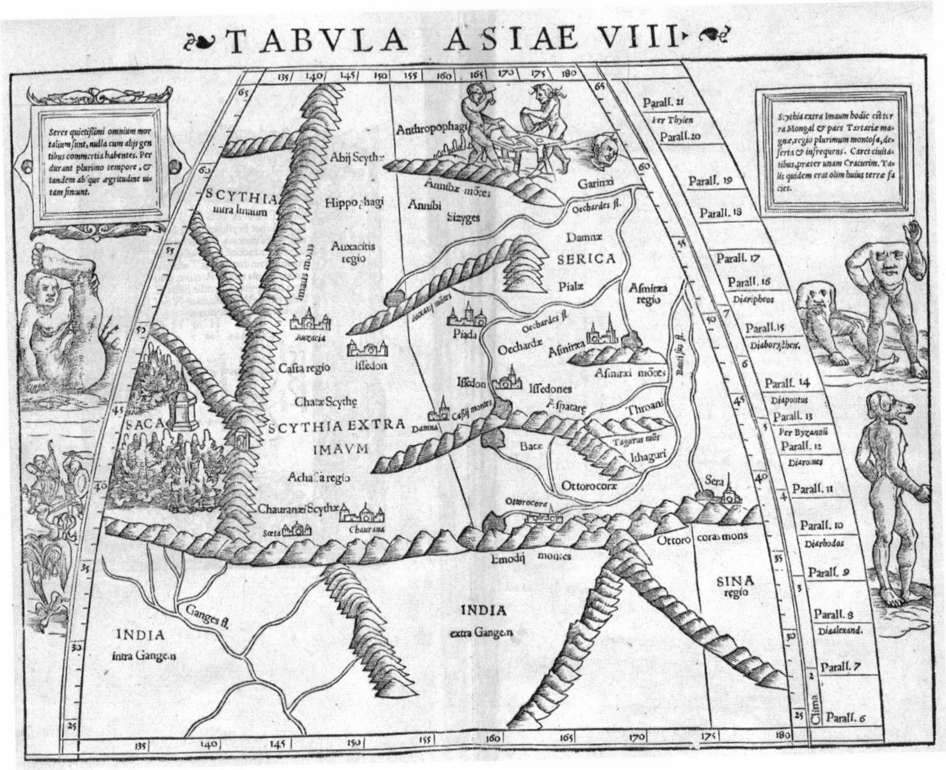

Figure 3. Monstrosities of Central Asia, as illustrated in Sebastian Münster's 1542 Basel edition of Ptolemy's *Geography*: Tabula Asiae, *Geographia Vniversalis, Vetus et Nova, Complectencs Clavdii Ptolemaei*, plate VIII. Courtesy of the Rare Books Division, New York Public Library, Astor, Lenox and Tilden Foundations.

and gods, one's own kith and kin, can be regarded as without peer elsewhere. Humility, therefore, is in the end geography's most essential lesson.

The project thus modestly initiated by Sumpa Khenpo found its greatest succession in *A Full Exposition of Jambudvīpa* (*'Dzam-gling rgyas-bshad*), completed in 1830 by the fourth Tsenpo incarnation, Tendzin Trinlé (1789–1838), whose chosen title clearly refers to his predecessor's work. Here we find ample discussions of Tibet, India, China, and many other lands, with such details in its account of Europe as a carefully transcribed list of the departments of France and, continuing west, even mention of Louisiana and Cuba.[80] It is perhaps an irony that this remarkable work, in which Jambudvīpa clearly begins to merge with the world as we know it in modern geography, was written just five years before the most celebrated slap at traditional Indian cosmology, that given in the conclusion of T. B. Macaulay's "Minute on Education" of 2 February 1835:

> The question now before us is simply whether, when it is in our power to teach this language [English], we shall teach languages in which by universal confession there are no books on any subject which deserve to be compared with our own; whether when we can teach European science, we shall teach systems which by universal confession whenever they differ from those of Europe differ for the worse; and whether, when we can patronise sound philosophy and true history, we shall countenance at the public expense medical doctrines which would disgrace an English farrier, astronomy which would move laughter in girls at an English boarding-school, history abounding with kings thirty feet high and reigns 30,000 years long, and geography made up of seas of treacle and seas of butter.[81]

Although Tendzin Trinlé demonstrated the capacity of a traditional Indic knowledge system to absorb considerable elements of Western learning on its own terms, his efforts were largely stillborn. With Tibet's nineteenth-century retreat into itself, geographical science progressed no more. In the early twentieth century modernists in the circle of the Thirteenth Dalai Lama sought to make up for lost time and embraced the anglophone education that Macaulay had urged on India. They were, however, resisted and ultimately defeated by monastic conservatives.[82] It was only in 1938 that the controversial artist and scholar Gendün Chöpel published an article in the *Mirror*, a Tibetan-language newspaper printed in Kalimpong, West Bengal, that included a sketch-map of the world with an accompanying text that invited his readers to pose the question whether the world was flat or spherical.[83] In this way he may be thought to have initiated a conversation that has become something of a commonplace in recent years, concerning the relations and conflicts holding between Tibetan Buddhism and contemporary science.[84] Perhaps, however, we can find a first hint of these queries in an eighteenth-century master's reflections on Jambudvīpa, with its white bears, dog sleds, and odd movements of the sun and moon in the summer arctic sky.

Notes

I thank my former research assistant, Ms. Rachel Lindner, and the staff of the Regenstein Library of the University of Chicago for their help in locating certain materials consulted in the course of work on this essay. In connection with the often puzzling text passage from Sumpa Khenpo studied herein, I am grateful to several colleagues for generous suggestions significantly contributing to the translation and annotation given here: Christopher Atwood (Indiana University), Alex Cherniak (Jerusalem), Johan Elverskog (Southern Methodist University), Todd Gibson, Vladimir Korobov (Vilnius University), Joseph Schwartzberg (University of Minnesota), E. Gene Smith (Tibetan Buddhist Resource Center, New York), Jan-Ulrich Sobisch (University of Copenhagen), Daniel Waugh (University of Washington, Seattle), Vesna Wallace (Oxford University and the University of California, Santa Barbara, when this work was in progress), and Dominik Wujastyk (University of Vienna and the Wellcome Trust Centre for the History of Medicine, when this work was in progress). On occasion I have referred precisely to their remarks in the notes below, using initials for purposes of attribution. This collaborative dimension of the project was one of its major pleasures, without which my results would have been much poorer indeed. Of course I alone am responsible for the final decisions about which interpretations to adopt and hence for whatever errors and omissions remain.

1. On the Monguor, the classic study remains Schram, *Monguors*. At present the Monguor are frequently designated by their Chinese ethnonym, Tu(zu). On Sum-pa mkhan-po, see Petech, preface, and my *Tibetan Assimilation*, chap. 7. The Indianized milieu of Tibetan learning is surveyed in Ruegg, *Ordre spirituel*, and my "The Indian Literary Identity."

2. For an introduction to 'Bras-spungs monastery, refer to Goldstein, "The Revival of Monastic Life."

3. Tibetan monastic education is studied in depth in Dreyfus, *The Sound of Two Hands*.

4. Sum-pa mkhan-po's major historical contributions are found in Das, *Pag Sam Jon Zang*; Lokesh Chandra, *Dpag-bsam-ljon-bzaṅ*; Yang, *The Annals*.

5. Refer to Dietz, "Cosmogony." Geographical annotations accompanying Tibetan commentaries on the *Abhidharmakośa* merit separate study. For some preliminary remarks on this, see Martin, "Tibet at the Center."

6. Although the *jambu* has been often identified as *Eugenia jambos*, the "rose-apple," Wujastyk, "*Jambudvīpa*," shows that it is probably in fact *Eugenia jambolana*, the "black plum." I thank DW for calling his work on this to my attention.

7. The "*maṇḍala*-offering" is detailed in many works on the essentials of Tibetan Buddhist practice, e.g., Padmakara, *The Words*, chap. 10.

8. Gumilev and Kuznetsov, "Two Traditions," 570, 578.

9. Martin, "'Ol-mo-lung-ring," dismisses the conclusions of Gumilev and Kuznetsov, while Schwartzberg, "Maps of Greater Tibet," 638–42, tentatively accepts them

(though he also calls for "reexamination and independent confirmation or refutation" [673]). While the former is certainly correct in his remarks on the arbitrariness of many of Gumilev and Kuznetsov's speculations, he nevertheless errs, I think, in his assumption that Bon legendary traditions should be expected to refer to the Middle East with respect to the places mentioned in the map. The legends, like medieval European paintings of the Holy Land, depict exotic locations through familiar conventions. It therefore remains possible, in my view, that a small number of the posited identifications are valid. The entirely implausible reading of Lang-ling as "Jerusalem," it may be noted, has taken on a life of its own: in 1995 I had the pleasure of hearing the outstanding scholar of Islamic art, Oleg Grabar, bring the "Tibetan world map showing Jerusalem" to the attention of his audience in a lecture devoted to the architectural history of the holy city. His remarks no doubt reflected the widespread publication of Gumilev and Kuznetsov's findings in Israel. We may note in passing too that if there is any plausibility to the proposed identification of Ne-seng-dra-ba with Alexandria, then the reference is no doubt to Alexandria on the Oxus or Alexandria of the Caucasus, and not at all to Alexandria in Egypt, as Gumilev and Kuznetsov believed.

10. On Phrom Gesar, see now Humbach, "New Coins." The eighth-century (or perhaps ninth) report on Besh-baliq is studied in Bacot, "Reconnaissance." Martin, "'Ol-mo-lung-ring," treats variant Bon-po geographical schemes.

11. Nakamura, "Old Chinese World Maps," 19–22.

12. Roerich, *Biography of Dharmasvāmin*; Tucci, *Travels of Tibetan Pilgrims.*

13. Schwartzberg, "Maps of Greater Tibet," 648–81, offers a valuable compendium of such maps. For a detailed study of a single example, see Huber, "A Tibetan Map of lHo-kha."

14. Schwartzberg, "Maps of Greater Tibet," 615–17.

15. The basic Tibetan units of land measurement, the *rkang* and the *'don,* were both variable and defined in terms of the number of seed loads that could be planted.

16. Hoog, *Prince Jiṅ-gim's Textbook,* 42.

17. The same text does, however, offer a short list of Buddhist lands on p. 41.

18. Ricard et al., *The Life of Shabkar,* 15.

19. Aris, "India and the British," 8. See also Aris, *'Jigs-med-gling-pa's "Discourse on India."*

20. Goldstein, *A History of Modern Tibet,* supplies plentiful evidence of the political entailments of these debates.

21. It is certainly no accident that three of the eighteenth century's most distinguished Tibetan Buddhist multiculturalists were closely associated with the Monguor: Sum-pa mkhan-po, Lcang-skya Rol-pa'i rdo-rje (1717–86), and Thu'u-bkwan Chos-kyi-rdo-rje (1737–1802).

22. Sum-pa mkhan-po, *'Dzam gling spyi bshad* (1975), plates 944–52; *'Dzam gling spyi bshad* (1986). In the notes that follow I refer to both the editions of Sum-pa mkhan-po's *'Dzam gling spyi bshad* that have been available to me, though I have not found any substantial differences between the texts.

23. Sum-pa mkhan-po 1975, plates 952–56, 959–60 (note that plates 957–58 are out of sequence and should be placed just before plate 963); 1986.

24. Hoffman, "Kālacakra Studies"; Orofino, "Apropos of Some Foreign Elements."

25. Sum-pa mkhan-po 1975, plates 960–62; 1986.

26. Sum-pa mkhan-po 1975, plates 962, 957–58; 1986.

27. Sum-pa mkhan-po 1975, plate 958; 1986.

28. Sum-pa mkhan-po 1975, plates 958, 963–67; 1986.

29. The term *rgya-ser,* applied to lands north of Tibet, parallels *rgya-nag,* the "black expanse," i.e., China, and *rgya-gar,* i.e., *rgya-dkar,* the "white expanse," India.

30. *Mtsho-chen thing-gir-srid,* the "great lake *thing-gir-srid.*" Here *thing-gir-srid* is perhaps a Tibetanized transcription of the Mongolian *tengis,* "sea," usually transcribed in Tibetan (as it regularly is below) as *teng-gi-si/se.* As *si* and *srid* are near or exact homonyms in many Tibetan dialects, the unusual orthography may be due to a scribe's taking dictation. In any event, given the clear mention of Astrakhan, the Caspian must surely be the reference.

31. *Cha gwan han* transcribes Mong. *čaγan qan,* "White Khan," used by Mongol writers as a title of the tsar from the late seventeenth century on. The origins of the title, however, can be traced back to the fifteenth-century use by the Tatar khans of the term *belyi tsar'* (white tsar), a convention that was adopted in referring to Ivan IV. The designation was thought to relate the tsar to the Qipchak Khanate. Refer to Ostrowski, *Muscovy and the Mongols,* 181–82.

32. "Cool houses" (*bsil khang*) perhaps refers to dachas (VK).

33. "Silver coin," *dngul-gyi dong-cha.* Later in the text we find *dong-tshe,* clearly equivalent to *dong-tse,* "coin." The Mongolian translation supports this interpretation (JE). *Sgra-rigs,* "dialects," makes no sense at all here, but later in the text we find *skra-rigs,* apparently meaning "hair ornaments." Assuming that *go-dmar* is defective for *mgo-dmar,* "bald," the reference would seem to be to hairpieces. Why these should have aroused such interest escapes me, though the eighteenth-century European fashion for wigs may well have struck Central Asians as exotic. It is not clear whether the sea monsters refer to sturgeon and caviar from the Caspian (JE) or to whale, walrus, and so on, from the northern seas. In an eighteenth-century text given in Claude de Grève, *Le Voyage en Russie,* we read, "L'océan septentrional est plus remarquable par la quantité et la singularité de ses poissons. On y trouve le cheval marin, animal monstrueux, dont la gueule est armée de dents longues et tranchants; sa peau, de plus d'un pouce d'épaisseur, est à l'épreuve du fusil; l'ours blanc, sorte d'amphibie qu'il ne faut pas confondre avec l'ours des forêts" (994–95). Caviar is also mentioned later in the same passage.

34. *Zhi-yang.* This interpretation is by no means certain but seems probable given the exile of thousands of Swedes to Siberia following the battle of Poltava in 1709. Some of these exiles encountered aspects of Tibetan Buddhism, and their scientific and technical skills were employed in developing the regions to which they were sent. See Norwick, "The First *Tsha-tsha.*" It is possible that Sum-pa mkhan-po's Zunghar

interlocutors therefore had some passing knowledge of Sweden, which they conveyed to him. A nineteenth-century Mongol manuscript, citing this passage, reads *se-ven,* which supports this identification (JE). At the same time, *zhi-yang* may be simply a transcription of Ch. *xiyang,* "western ocean" (CA). In either case the Nordic reference seems certain.

35. Though Sweden is not today associated with watch making, this was not the case during the eighteenth century. Gottschalk, MacKinney, and Pritchard, *The Foundations,* remark that "extraordinarily well gauged cogwheels for clocks were produced in series by the Swedish technological genius Christopher Polhem" (921).

36. *Pu la gwan,* Mong. *bulaγan.*

37. JUS writes, "This seems to be the sap that was quite commonly drawn in Sweden (and Russia) from the birch trees during the rise of the sap. It was consumed either fresh or after fermentation (which gave it a low alcoholic content). It is called variously bjørkkvass, bjørkvin, bjørklake, bjørksirup and more generally 'kvass.'"

38. The text appears to read *ya gong,* but *ya god* is certainly to be preferred. In the eighteenth century "Yakutia" (or "Iakuti") was sometimes used to designate much of northern Siberia. See, for instance, the map by Gerhard Friedrich Müller published by the Imperial Academy of Science, Saint Petersburg, in 1754, and reproduced in Nebenzahl, *Mapping the Silk Road,* as map 5.5B.

39. *Khem khem chi* in the Torghut territory is the River Kem-Kemchik, in the Tuvan Republic of Russia, and a tributary of the Yenisey. If the designation is being extended here to the Yenisey itself, that would explain the association with phenomena characteristic of the far north.

40. Khur lug is unidentified, though it is possible that "Karluk" (or Qarluq) was still, anachronistically, being used by Tibetan scholars to name this region. Unless otherwise noted, the remaining place-names in this paragraph are quite transparent. "White-headed Muslims," *kho tan,* from Mong. *qotang,* "Muslim," is used throughout this passage to refer to Central Asian Muslims generally, though on some occasions it serves as a toponym meaning roughly "Turkestan." The term derives of course from Khotan. On the designation "white headed," JE comments, "*Mgo dkar* is a translation of the Mongolian *caγan malaγai,* 'White Turbans.' It derives from the Chinese *chantou* (Pelliot, "Le Hoja et le Sayyid Husain de l'Histoire des Ming," *T'oung pao* 38 [1948], 130–32), and is also a term used to identify the Muslims of Inner Asia. These are the people who live in Hami, Kashgar, etc." "Sheep-kidney worm," *'bu lug gi mkhal ma btab na lug skye pa'i yul,* eludes me altogether.

41. The perspective is that of one facing east from Badakhshan, so that left is north.

42. *Zhwa dkar gyon pa'i e ges lig u lan the men zhes bu rad.* My interpretation is pure speculation: *e ges lig* is at best a very imperfect transcription of Aqtaghliq. The entire last half of the phrase, *u lan . . . bu rad,* remains a mystery to me. Despite these linguistic difficulties, the interpretation is historically attractive. One of the contenders for the Zunghar throne was Amursana (1722?–57), who submitted to Manchu authority in 1755 but soon rebelled. During this period he allied himself with the Aqtagh-

liq Khwajas against their rivals, the Qarataghliq ("Black Mountain") Naqshbandiya. Sum-pa mkhan-po likely would have been familiar with these events.

43. Uncertain. The term found here, *khud-khur,* might be emended to read *khud-khung,* which could more plausibly represent Kokand.

44. A reference perhaps to the Orthodox priesthood.

45. *Ra tshugs 'phru dmar.* My interpretation is a guess, but I can't imagine what else it might be. As the Qing's Chinese subjects were required to wear the queue, and Tibetan laymen generally kept their hair long and braided, a close-cropped haircut may have been notable.

46. This no doubt refers to the conversion of the "red-turbaned" Kızılbaş ("red heads," using here the modern Turkish orthography, or Qizilbash, after the Persian transcription; see below) to more orthodox brands of Islam. The origin of the name, as understood in later times, is explained by Garthwaite, *The Persians*: "Later tradition depicts Haydar [d. 1488, the father of the Safavid leader Isma'il] as Shi'a and as the originator of the distinctive red turban with its twelve folds—emblematic of the twelve Shi'a imams—worn by later Safavid supporters, which gave them the name *Qizilbash,* or red-headed ones" (160). On the waning of the Qizilbash in Safavid Iran, see Babayan, *Mystics, Monarchs, and Messiahs,* chap. 10. As she notes, "Within the religious domain, [Safavi centralization] served to tame the Qizilbash in an attempt to convert them to Imami Shi'ism" (361).

47. *Skra rigs.* The exact meaning here is uncertain. See also n. 33 above.

48. *Kya ye gon.* This fortress marks the western extreme of the Great Wall in Gansu.

49. The designations of these rivers represent the Mongol terminology throughout.

50. As CA comments, "The Solon Ewenkis were mostly distributed in the Hulun Buir area in northwest Manchuria. . . . Khu yur is thus a copyist's error for Khu [lun Bu] yur. . . . Since the bannermen of the Solons and Daurs of Hulun Buir were resettled in N. Xinjiang as garrison soldiers by the Qing, evidently this area of Tarbaghatai was called 'western Hulun Buir.'"

51. *Kho tan.*

52. Sumpa Khenpo here no doubt refers to the custom, much observed in Muslim lands, of gold traders just throwing nets over their stalls when leaving the market for meals and other brief business.

53. "Kızılbaş" was often used to designate Safavid Iran, particularly by the Russians. On the basis for this, see Efendiev, "Le rôle des tribus de langue Turque"; Babayan, *Mystics, Monarchs, and Messiahs.*

54. *U ru su'i nub lhor kha zal ba shi'i mkhar se bar zhag gnyis kyi 'khor dang lab tu'i dgus 'khor dang / nub byang du khu ril gyi mkhar chen re yod.* The interpretation of this sentence remains quite problematic. If *'khor* is taken just to mean "the vicinity of," and the first occurrence of *mkhar* in fact belongs to the transcription, then the first two of the toponyms specified would be *mkhar se bar zhag gnyis,* perhaps "both Khurasan (*mkhar se*) and Fars (*bar zhag*)." The third, *lab tu'i dgus 'khor,* however, eludes me completely.

55. Traditionally the *shang shang* is considered to be a mythical bird with a human head. Might this be an allusion to the various human-headed winged creatures that adorn pre-Islamic Persian and Mesopotamian edifices?

56. *Cha kwan thing gi si*, from Mong. *čaγan tengis*, "white sea." Though there is indeed a body of water whose proper name is the White Sea, where the port of Arkhangelsk is located, it seems equally likely that the reference here is to the Arctic Ocean in general. On eighteenth-century maps this was often designated the "glacial sea" (*mer glaciale*).

57. Na mi si, and Ne me shing, below, recall the ethnonym Naiman, though the Naiman in fact never moved so far north. One is reminded nevertheless of the tales of the Hyperboreans and the kingdom of Prester John, the latter having been sometimes identified with the Naiman, who were also famed for wizardry in warfare. Is it possible that we have here a western legend of the Naiman, recirculated in Asia? CA suggests, however, that there is a distant possibility that Na mi si represents Russ. немецкий, "German." The most plausible hypothesis received to date is from AC, who proposes that the name might be from Russ. Ненцы, the plural form of the ethnonym of the Nenets peoples of Russia's far north, who in the past were often the subject of marvelous legend. A search on the Internet readily produced a short article that begins:

> The first information about Khanty and Nenets peoples appeared in Russian chronicles in the 11th century. The tradesmen from Novgorod visited the northern Trans-Urals regions long before. They named this land Jugra. At first the Novgorodian's ideas of Jugra, its people and wealth, were to a considerable extent fantastic. The travelers told about squirrels and reindeer coming to the ground like a rain from the clouds. This "original menagerie" and "furs depository" attracted bands of Novgorod warriors and merchants. . . . The intensive trade in furs, mammoth tusks, fish glue, birds' down and feathers, birch fungi, boats, fur coats and other goods expanded over the northern territories. (www.yamal.ru/oi_ht_e.htm)

"A lizard the size of a dog" no doubt refers to the seal.

58. Or Ra-ma-rtsang. In the spirit of Gumilev and Kuznetzov, one might perhaps suggest Ramses, but in the Arctic? The decimal organization of Mongol armies is widely documented. Refer to Atwood, *Encyclopedia*, 139–40.

59. *Phe tshem* is unidentified, but is evidently a type of armor or mail.

60. The Ottomans are designated as *khung 'ur/-khur* in Sum-pa's text. On this usage, JE comments:

> Khung 'ur is a confusion with the name of Chagatai's son Könggür, who in Mongol historiography is said to have founded the Ottomans. In Gombojab's *Flow of the Ganges* (*γangā-a-yin urusqal*) we find: "Qinwang Chaghatai became khan in the White Turban Khoichi nation and founded his

> capital in Yirkel. Of his five children, the eldest Abdulan Khan ruled his father's territory. The second son Immakhuli became khan of the Sartuul, and made Samarkand the capital. Adarmamad became khan of India and lived in the city of Balsha. The fourth son, Könggür, was khan of Rum and lived in Istanbul. The fifth son Temür became khan of the Red Hat Orongga nation and lived in Bukhari. They all inherited these lands at the time of Taizong Khan [Ögedei]" (GU 55–57).

"Sunbeam," *nyi ma'i 'od zer*. This occurs as the name of a well-known form of Padmasambhava: Dudjom, *The Nyingma School,* 1:471.

61. "Buckles," *thob che*. Meaning uncertain.

62. This is no doubt because the Ottoman ruler was the recognized caliph.

63. "Watchman": the text reads *tsho ba,* which makes no sense. I emend to *so pa,* "spy, scout," though with some hesitation about so doing. At any rate, the Galata Tower indeed fits the description and was used as a weather observatory briefly during the late sixteenth century. See n. 74 below. "Chinese pestle," *rgya'i tun gyi 'tsab mu nar*. Uncertain, but *mu nar* seems likely to represent *minar* ("minaret") in any case, and these are indeed particularly impressive in Istanbul. I have taken *tun* as var. for *gtun,* "pestle," though perhaps it is a shortening of Ch. 墩臺 *duntai,* "beacon-tower." In Chinese *dun* alone would mean a "heap" or "mound."

64. The animals mentioned have no relation to anything one is likely to find in Istanbul except at the zoo. Except for the first, the interpretations are conjectural. The *chu glang,* "water bull (or elephant)," is elsewhere said to have tusks suitable to replace ivory and so may be either a walrus or hippopotamus. The *chu rta,* "water horse," has only an etymological connection with the hippopotamus (or perhaps hippocampus?). Note too that walrus etymologically means "horse-whale" and was referred to as the *cheval marin* in eighteenth-century French (n. 33 above). DW suggests that exotic placements of animals may sometimes be due to the misreading of old European maps, which were typically decorated with exotica of various kinds. Sum-pa mkhan-po, or his Mongol (esp. Zunghar) and Manchu informants, may well have been familiar with such documents.

65. "Kashmiri," *kha che*. The term is sometimes applied to other Central Asian Muslims, e.g., Yarkandis, besides Kashmiris.

66. The interpretation of the initial *thing* is uncertain. Only the context suggests that the Mediterranean might be the referent, though JS notes that it may be more accurately the Aegean.

67. This recalls the famous and widely distributed tale of Alexander descending in a glass diving bell. For examples, see (from the Greek) Stoneman, *The Greek Alexander Romance,* 118–19; (Latin) Kratz, *The Romances of Alexander,* 75–76; (Hebrew) van Bekkum, *A Hebrew Alexander Romance,* 188–91; and (Ethiopian) Budge, *The Life and Exploits,* 2:282–86. The story is well known too in the Iranian world through the *Khamsa* of Nizāmi (1141–1209), which was subsequently recast by Amīr Khusrau Dih-

lawī (1253–1325), in which form it was widely diffused in India. For an illuminated version of the last-mentioned work, prepared for the Mughal emperor Akbar (1542–1605) and including a full-page illustration of Alexander in his diving bell, see Seyller, *Pearls of the Parrot of India.*

68. The "blue stone" is, of course, the Black Stone of the Kaʿba. The measurements given here (a *mtho* is the distance from the tip of the outstretched thumb to that of the middle finger) are roughly accurate: the length of the Black Stone is variously described as two or three cubits (Peters, *The Hajj,* 63), and it is built into the Kaʿba at a height of about four feet (14). "A Chinese Muslim," *hu tsi,* from Ch. *huizu.*

69. The state of the text of this paragraph is problematic. Could the white stone of Īśvara be the pillars at Mina representing Satan?

70. This is a typical catalogue of Inner Asian monstrosities (here, however, placed in the "West," i.e., Europe and North Africa), such as appeared frequently in European illuminations of Marco Polo (who doesn't mention such things; see Larner, *Marco Polo,* plate 4) and John Mandeville (who does), as well as on early modern maps. Noteworthy is the mention of the *tsu ta,* as this also occurs in 'Jigs-med-gling-pa's 1788 *rgya gar gtam;* see Aris, "India and the British," 1:7–15 and appendix 3–11. As a Tibetan borrowing from Sanskrit *cūta,* the word literally means "mango tree."

71. Moseley, *The Travels of Sir John Mandeville.*

72. Jacques-Henri Bernardin de Saint-Pierre, *Observations sur la Russie* (1818), as given in Grève, *Le Voyage en Russie,* 385–86, my translation.

73. Heller, *Histoire de la Russie*: "Les mourzas de Nogaï considèrent en outre Ivan IV le Terrible comme 'un des leurs', en plus noble: le tsar de Moscou est en effet perçu comme l'héritier de Gengis Khan. Dans la correspondance qu'Ivan entretient avec les mourzas, il ne réfute pas cette glorieuse ascendance" (200). The antecedents of Ivan's policy in this respect may be found, as notes Ostrowski, *Muscovy and the Mongols,* 137–38, in what he refers to as the Rus' "secular virtual past," i.e., their status as tributaries of the Chingisid Qipchak Khanate. This relationship, in turn, was derived from the thirteenth-century relations of Byzantium with the Mongol khans, relations that were sealed in marital alliances between the ruling households.

74. "The prevailing northeast wind, or *poyraz,* comes from the Black Sea, giving way at times during the winter to an icy blast from the Balkans known as the *karayel,* or 'black veil,' capable of freezing the Golden Horn and even the Bosporus. The *lodos,* or southwest wind, can raise storms on the Sea of Marmara" ("Istanbul," *Encyclopædia Britannica,* 2005, online). If I am correct in proposing that Istanbul is indeed the reference, then the only observatory about which Sum-pa mkhan-po can be writing is the Galata Tower, even though it was used for this purpose for only a short while long before his time. As Philip Mansel, *Constantinople,* explains, "In 1580 the Mufti incited a mob to destroy a large, ultra-modern observatory completed three years before in Galata for Sultan Murad III: the Mufti considered it a bad omen and a source of calamity for the empire. The next Ottoman observatory did not open until 1868" (46).

75. Refer to n. 67 above.

76. On this, see White, *Myths of the Dog-Man.*

77. Crane, *Mercator,* 208–11, 242–44, 247. See especially plate 4, "The Arctic Lands," between 242 and 243. One of Mercator's Arctic captions reads, "Here live pygmies whose length in all is 4 feet, as are also those who are called Screlingers in Greenland" (210).

78. Grafton, *New Worlds,* 105–6.

79. Sum-pa mkhan-po 1975, plates 968–69; 1986.

80. Wylie, "Dating the Tibetan Geography." Wylie believes that the work may have been completed in 1820. EGS believes that it is in fact a question of two separate editions, one issued by the author in 1820 and the other ten years later. For the sections dealing with Tibet and Nepal, respectively, see Wylie, *The Geography of Tibet* and *A Tibetan Religious Geography.*

81. Muir, *The Making of British India,* 301.

82. Goldstein, *A History of Modern Tibet,* 421–26.

83. Lopez, *The Madman's Middle Way,* 15–18.

84. On this topic, see Lopez, *Buddhism and Science.* Although referring to many aspects of the dialogue between Buddhism and Western science, Lopez focuses particularly upon the role of Tibetan Buddhism in the conversation.

References

Adle, Chahryar, and Irfran Habib, eds. *History of Civilizations of Central Asia.* Vol. 5, *Development in Contrast: From the Sixteenth to the Mid-nineteenth Century.* Paris: UNESCO Publishing, 2003.

Aris, Michael. "India and the British According to a Tibetan Text of the Later Eighteenth Century." *Tibetan Studies,* ed. Per Kvaerne. Oslo: Institute for Comparative Research in Human Culture, 1994.

———. *'Jigs-med-gling-pa's "Discourse on India" of 1789.* Tokyo: International Institute for Buddhist Studies, 1995.

Atwood, Christopher. *Encyclopedia of Mongolia and the Mongol Empire.* New York: Facts on File, 2004.

Aziz, Barbara Nimri, and Matthew Kapstein, eds. *Soundings in Tibetan Civilization.* Delhi: Manohar, 1985.

Babayan, Kathryn. *Mystics, Monarchs, and Messiahs: Cultural Landscapes of Early Modern Iran.* Cambridge, Mass.: Harvard University Press, 2002.

Bacot, Jacques. "Reconnaissance en Haute Asie septentrionale par cinq envoyés ouigours au VIIIe siècle." *Journal Asiatique* 144 (1956), 137–53.

Budge, E. A. Wallis. *The Life and Exploits of Alexander the Great.* Vol. 2. London, 1896.

Crane, Nicholas. *Mercator.* New York: Henry Holt, 2002.

Das, Sarat Chandra, ed. *Pag Sam Jon Zang.* 1908. Kyoto: Rinsen, 1984.

Dietz, Sieglinde. "Cosmogony as Presented in the Tibetan Historical Literature and Its Sources." *Tibetan Studies*, vol. 1, ed. Z. Yamaguchi et al. Narita: Naritasan Shinshoji, 1992.

Dreyfus, Georges B. J. *The Sound of Two Hands Clapping: The Education of a Tibetan Buddhist Monk*. Berkeley: University of California Press, 2003.

Dudjom Rinpoche, Jikdrel Yeshe Dorje. *The Nyingma School of Tibetan Buddhism: Its Fundamentals and History*. Annotated translation by Gyurme Dorje and Matthew Kapstein. 2 vols. London: Wisdom Publications, 2002.

Efendiev, Oktaj. "Le rôle des tribus de langue Turque dans la création de l'état Safavide." *Turcica* 6 (1975), 24–33.

Enterline, James Robert. *Erikson, Eskimos, and Columbus: Medieval European Knowledge of America*. Baltimore: Johns Hopkins University Press, 2002.

Garthwaite, Gene R. *The Persians*. Oxford: Blackwell, 2005.

Goldstein, Melvyn C. *A History of Modern Tibet: The Demise of the Lamaist State*. Berkeley: University of California Press, 1989.

———. "The Revival of Monastic Life in Drepung Monastery." *Buddhism in Contemporary Tibet: Religious Revival and Cultural Identity*, ed. Melvyn C. Goldstein and Matthew T. Kapstein. Berkeley: University of California Press, 1998.

Gottschalk, Louis, L. C. MacKinney, and E. H. Pritchard, eds. *History of Mankind, Cultural and Scientific Development*. Vol. 4, *The Foundations of the Modern World, 1300–1775*. New York: Harper and Row, 1969.

Grafton, Anthony, with April Shelford and Nancy Siraisi. *New Worlds, Ancient Texts: The Power of Tradition and the Shock of Discovery*. Cambridge, Mass.: Belknap Press, 1992.

Grève, Claude de, ed. *Le Voyage en Russie: Anthologie des voyageurs français aux XVIIIe et XIXe siècles*. Paris: Robert Laffont, 1990.

Gumilev, L. N., and B. I. Kuznetsov. "Two Traditions of Ancient Tibetan Cartography." *Soviet Geography: Review and Translation* 11, no. 7 (1970), 565–79.

Harley, J. B., and David Woodward, eds. *The History of Cartography*. Vol. 2, book 2, *Cartography in the Traditional East and Southeast Asian Societies*. Chicago: University of Chicago Press, 1994.

Heller, Michel. *Histoire de la Russie et de son empire*. Trans. Anne Coldefy-Faucard. Paris: Flammarion, 1997.

Hoffman, Helmut. "Kālacakra Studies 1. Manichaeism, Christianity, and Islam in the Kālacakra Tantra." *Central Asiatic Journal* 13 (1969), 52–73, and 15 (1972), 298–301.

Hoog, Constance, trans. *Prince Jiṅ-gim's Textbook of Tibetan Buddhism: The Śes-bya rab-gsal (Jñeya-prakāśa) by 'Phags-pa Blo-gros rgyal-mtshan dPal-bzaṅ-po of the Sa-skya-pa*. Leiden: Brill, 1983.

Huber, Toni. "A Tibetan Map of lHo-kha in the South-Eastern Himalayan Borderlands of Tibet." *Imago Mundi: The Journal of the International Society for the History of Cartography* 44 (1992), 9–23.

Humbach, Helmut. "New Coins of Fromo Kēsaro." *India and the Ancient World: His-*

tory, Trade and Culture before A.D. *650,* ed. Gilbert Pollet. Leuven: Departement Oriëntalistiek, 1987.

Kapstein, Matthew T. "The Indian Literary Identity in Tibet." *Literary Cultures in History: Perspectives from South Asia,* ed. Sheldon Pollock. Berkeley: University of California Press, 2003.

———. *The Tibetan Assimilation of Buddhism: Conversion, Contestation and Memory.* New York: Oxford University Press, 2000.

Klong-chen Chos-dbyings-stobs-ldan-rdo-rje. *Mdo rgyud rin po che'i mdzod.* 5 vols. Chengdu: Si-khron-mi-rigs-dpe-skrun-khang (Sichuan Nationalities' Publishing House), 2000.

Kratz, Dennis M. *The Romances of Alexander.* Garland Library of Medieval Literature, Series B, vol. 64. New York: Garland, 1991.

Kuznetsov, Broneslav I. "The Article Jerusalem in an Old Tibetan Map," *A. Korosi Csoma Sandor Intezet Kozlemenyei* (Budapest), nos. 3–4 (1974), 53–54.

Larner, John. *Marco Polo and the Discovery of the World.* New Haven, Conn.: Yale University Press, 1999.

Lokesh Chandra, ed. *Dpag-bsam-ljon-bzaṅ.* New Delhi: International Academy of Indian Culture, 1959.

Lopez, Donald S., Jr. *Buddhism and Science: A Guide for the Perplexed.* Chicago: University of Chicago Press, 2008.

———. *The Madman's Middle Way: Reflections on Reality of the Tibetan Monk Gendun Chopel.* Chicago: University of Chicago Press, 2006.

Mansel, Philip. *Constantinople: City of the World's Desire, 1453–1924.* London: Penguin, 1997.

Martin, Dan. "'Ol-mo-lung-ring, the Original Holy Place." *Sacred Spaces and Powerful Places,* ed. Toni Huber. Dharamsala: Library of Tibetan Works and Archives, 1999.

———. "Tibet at the Center: A Historical Study of Some Tibetan Geographical Conceptions Based on Two Types of Country Lists Found in Bon Histories." *Tibetan Studies,* vol. 1, ed. Per Kvaerne. Oslo: Institute for Comparative Research in Human Culture, 1994.

Moseley, C. W. R. D., trans. *The Travels of Sir John Mandeville.* Harmondsworth: Penguin, 1983.

Muir, Ramsay. *The Making of British India, 1756–1858.* Manchester: Manchester University Press, 1923.

Nakamura, Hirosi. "Old Chinese World Maps Preserved by the Koreans." *Imago Mundi: The Journal of the International Society for the History of Cartography* 4 (1947), 3–22.

Nebenzahl, Kenneth. *Mapping the Silk Road and Beyond: 2,000 Years of Exploring the East.* London: Phaidon, 2004.

Norwick, Braham. "The First *Tsha-tsha* Published in Europe." *Soundings in Tibetan Civilization,* ed. Barbara Nimri Aziz and Matthew Kapstein. Delhi: Manohar, 1985.

Orofino, Giacomella. "Apropos of Some Foreign Elements in the Kālacakratantra."

Tibetan Studies: Proceedings of the Seventh Seminar of the International Association for Tibetan Studies, vol. 2, ed. Ernst Steinkellner et al. Vienna: Austrian Academy of Sciences, 1997.

Ostrowski, Donald. *Muscovy and the Mongols: Cross-Cultural Influences on the Steppe Frontier, 1304–1589*. Cambridge: Cambridge University Press, 1998.

Padmakara Translation Committee. *The Words of My Perfect Teacher*. San Francisco: Harper Collins, 1994.

Petech, Luciano. Preface to *Dpag-bsam-ljon-bzaṅ*, ed. Lokesh Chandra. New Delhi: International Academy of Indian Culture, 1959.

Peters, F. E. *The Hajj: The Muslim Pilgrimage to Mecca and the Holy Places*. Princeton, N.J.: Princeton University Press, 1994.

Ricard, Matthieu, et al., trans. *The Life of Shabkar: The Autobiography of a Tibetan Yogin*. Albany: State University of New York Press, 1994.

Roerich, Georges. *Biography of Dharmasvāmin: A Tibetan Monk Pilgrim*. Patna: K. P. Jayaswal Research Institute, 1959.

Ruegg, David Seyfort. *Ordre spirituel et ordre temporel dans la pensée bouddhique de l'Inde et du Tibet*. Paris: De Boccard, 1995.

Schram, Louis J. *Monguors of the Kansu-Tibetan Frontier*. 3 vols. Philadelphia: American Philosophical Society, 1954–61.

Schwartzberg, Joseph E. "Maps of Greater Tibet." *Cartography in the Traditional East and Southeast Asian Societies*, ed. J. B. Harley and David Woodward. Chicago: University of Chicago Press, 1994.

Seyller, John. *Pearls of the Parrot of India: The Walters Art Museum Khamsa of Amir Khusraw of Delhi*. Seattle: University of Washington Press, 2002.

Snellgrove, David L. *The Nine Ways of Bon*. London: Oxford University Press, 1967.

Soucek, Svat. *A History of Inner Asia*. Cambridge: Cambridge University Press, 2000.

Stoneman, Richard. *The Greek Alexander Romance*. Harmondsworth: Penguin, 1991.

Sum-pa mkhan-po. *'Dzam gling spyi bshad*. *Collected Works of Sum-pa-mkhan-po*, vol. 2. New Delhi: International Academy of Indian Culture, 1975.

———. *'Dzam gling spyi bshad*. *'Dzam gling rgyas bshad dang spyi bshad gnyis*, ed. Nor-brang O-rgyan. Lhasa: Bod rang skyong ljongs spyi tshogs tshan rigs khang bod yig dpe rnying dpe sgrig khang, 1986.

Tucci, Giuseppe. *Travels of Tibetan Pilgrims in the Swat Valley*. Calcutta, 1940.

van Bekkum, Wout Jac. *A Hebrew Alexander Romance according to Ms London, Jews' College no. 145*. Orientalia Lovaniensia Analecta 47. Leuven: Peeters, 1992.

White, David. *Myths of the Dog-Man*. Chicago: University of Chicago Press, 1991.

Wujastyk, Dominik. "*Jambudvīpa*: Apples or Plums?" *Studies in the History of the Exact Sciences in Honour of David Pingree*, ed. Charles Burnett et al. Boston: Brill, 2004.

Wylie, Turrell V. "Dating the Tibetan Geography *'Dzam gling rgyas bshad* through Its Description of the Western Hemisphere." *Central Asiatic Journal* 4 (1958), 300–311.

———. *The Geography of Tibet according to the 'Dzam-gling rgyas-bshad*. SOR 25. Rome: Instituto italiano per il Medio ed Estremo Oriente, 1962.

———. *A Tibetan Religious Geography of Nepal*. SOR 42. Rome: Instituto italiano per il Medio ed Estremo Oriente, 1970.

Yang, Ho-chin. *The Annals of Kokonor*. Uralic and Altaic Series, vol. 106. Bloomington: Indiana University Press, 1969.

Zerubavel, Eviatar. *Terra Cognita: The Mental Discovery of America*. New Brunswick, N.J.: Rutgers University Press, 1992.

Contributors

Muzaffar Alam is a professor in the Department of South Asian Languages and Civilizations at the University of Chicago. His publications include *The Crisis of Empire in Mughal North India* (1986); *The Languages of Political Islam: India 1200–1800* (2004); and most recently, with Sanjay Subrahmanyam, *Indo-Persian Travels in the Age of Discoveries, 1400–1800* (2007).

Imre Bangha is a lecturer in Hindi at the University of Oxford and head of the Alexander Csoma de Koros Centre for Oriental Studies at the Hungarian University of Transylvania, Romania. He works with early Hindi text editing and is the author of *Saneh ko marag: Ananghan ka jivanvritt* (1999).

Aditya Behl taught Hindi and Urdu literature and sultanate and Mughal cultural history in the Department of South Asia Studies at the University of Pennsylvania. He translated fiction and poetry from Hindi, Urdu, and Panjabi into English, notably Shaikh Manjhan's *Madhumalati: An Indian Sufi Romance* (2000), and was the editor of *The Penguin New Writing in India* (1994). His monograph on the Hindavi Sufi romances, *Love's Subtle Magic: An Indian Islamic Literary Tradition, 1379–1545,* is forthcoming, as is a verse translation of Shaikh Qutban's Sufi romance *Mirigavati: The Magic Doe.*

Allison Busch is assistant professor of Indian literature at Columbia University. Her current research centers on courtly texts written in the classical Hindi dialect of Brajbhasha during the early modern period. She has published several essays on the poetry and intellectual life of writers of the seventeenth century. Her book on Mughal-period Hindi literature, *Poetry of Kings,* is forthcoming from Oxford University Press.

Sumit Guha is a professor of history at Rutgers University, New Brunswick. His research focuses on western India after 1200 CE.

Janet Gyatso has a Ph.D. from the University of California at Berkeley and taught at Amherst College before taking up her present position at Harvard University as the Divinity School's first Hershey Professor of Buddhist Studies. Her books include *Apparitions of the Self: The Secret Autobiographies of a Tibetan Visionary* (1998); *In the Mirror of Memory: Reflections on*

Mindfulness and Remembrance in Indian and Tibetan Buddhism (1992); and *Women of Tibet* (2005). She is currently working on a book on the intellectual history of Tibetan medicine in early modernity.

Matthew T. Kapstein is director of Tibetan Studies at the École Pratique des Hautes Études, Paris, and Numata Visiting Professor of Buddhist Studies at the University of Chicago. His recent publications include *The Tibetans* (2006) and *Buddhism between Tibet and China* (2009).

Françoise Mallison, Directeur d'Études Emerita at the École Pratique des Hautes Études, Section des Sciences historiques et philologiques, Paris, specializes in medieval religious culture of northern India and Gujarat. She has published translations of Narasimha Maheta and Swaminarayan lyrics, and essays on the Indian acculturation of Ismaili vernacular poetry. Her books include *Au point du jour: Les Prabhatiyam de Narasimha Maheta* (1987) and, as editor, *Littératures médiévales de l'Inde du nord* (1991) and *Constructions hagiographiques dans le monde indien: Entre mythe ẹt histoire* (2001).

Sheldon Pollock is the William B. Ransford Professor of Sanskrit and Indian Studies at Columbia University. His *The Language of the Gods in the World of Men: Sanskrit, Culture, and Power in Premodern India* (2006) has recently been issued in paperback. He is currently working on *Liberation Philology* (Harvard University Press), and *Reader on Rasa: A Historical Sourcebook in Indian Aesthetics*, the first in a new series of works to be published by Columbia University Press on classical Indian thought.

Velcheru Narayana Rao, currently a visiting professor at the University of Chicago, taught for more than three decades at the University of Wisconsin, Madison. He has published books on Telugu literature and Indian poetics and history.

Kurtis R. Schaeffer is a professor in the Department of Religious Studies at the University of Virginia, where he teaches and writes about the cultural and intellectual history of Tibet. His recent publications include *The Culture of the Book in Tibet* (2009).

Sunil Sharma is an assistant professor of Persianate and comparative literature at Boston University. He specializes in premodern Persian and Urdu literature.

David Shulman is the Renee Lang Professor of Humanistic Studies in the Department of Comparative Religion at the Hebrew University of Jerusalem. He is the author of several books, including *Textures of Time* (2003, with V. Narayana Rao and Sanjay Subrahmanyam) and *Spring, Heat, Rains* (2009).

Sanjay Subrahmanyam is a professor and the Doshi Chair of Indian History at UCLA. He has taught in Delhi and Paris and at Oxford. His recent publications include *Explorations in Connected History* (two volumes, 2005) and, with Muzaffar Alam, *Indo-Persian Travels in the Age of Discoveries, 1400–1800* (2007).

Mohamad Tavakoli-Targhi is a professor of history and Near and Middle Eastern civilizations at the University of Toronto and professor of historical studies at the University of Toronto, Mississauga, where he was chair of the Department of Historical Studies between 2004 and 2007. Since 2002 he has served as the editor of *Comparative Studies of South Asia, Africa and the Middle East*. He is the author of *Refashioning Iran: Orientalism, Occidentalism and Nationalist Historiography* (2001) and *Tajaddud-i Bumi* (Vernacular Modernity, 2003).

Index

Jews, 228–29; on Sikhs, 227; textual history of, 211, 235 n. 9; on Zoroaster, 219–20, 232; on Zoroastian texts, 219–20
Dādū Dayāl, 143
Dādū janma līlā, 143
Dakhani, 54, 55, 56
Dalai Lama, Fifth (Ngawang Lobsang Gyatso), 291, 293–301, 323–24
Dalpatrām, 173–76; textbooks quoted by, 177–78
Daṇḍanīti, 51
Daṇḍin (Dandin), 124, 297, 301
Dar Lotsawa Ngawang Püntsok Lhündrup, 294–96
Darmo Menrampa Lozang Drakpa, 295–97, 311–12, 319, 324
Daśarūpaka, 119
Delhi, 196, 240–41, 244, 246–47, 249, 251–52
Descartes, 3, 7, 11, 36–37, 262–65
Desi, the. *See* Gyatso, Sangyé
Dhananjaya, 119
dharmaśāstra (law and moral philosophy), 20, 25, 71, 82
dictionaries and lexicons, 24, 52, 55, 62, 177, 187–88, 212, 243, 244, 270, 292
Dinakara Bhaṭṭa, 22, 35
dissection, 311–12, 315, 328
dissimulation (*taqīya*), 214, 221–22
divination, 291–93, 297–98, 301
Divyasūricaritra, 77
Dnyānesvarī, 52
dohās (couplets), 145–46
Dohāvalī, 145

education: medical, 314–15, 324; monastic, 293, 314–15, 337; scribal, 186–88, 190, 194–95, 198, 200–201, 205–6. *See also* Bhuj Brajbhāṣā Pāṭhśālā
Eknath, 54–56
Enlightenment: European, 257–58, 263, 311; religious, 219, 318
ethnography, 10, 13, 205, 214, 226–29, 232, 241–43, 248–51
evidence, empirical. *See under* knowledge

Falak al-Sa'adah, 267–68
Farhang-i Rashidi, 270
food, 61, 76, 85, 95, 106, 217, 220, 230–31, 326, 345
Four Medical Tantras. See *Four Treatises*
Four Treatises (*Four Medical Tantras*; *rGyud bzhi*): author of, 300, 325–26, 328; on channels, 319–22; contents of, 299; commentaries on, 299, 301–2, 315, 317, 322–23, 326; codification and printing of, 297, 314, 324; importance of, 314–17, 323, 326
Francklin, W., 271

Gandhi, M. K., 69–70
Gelukpa school, 298–99, 301
genesis amnesia, 212, 258
geography and cartography, 12, 337–41
Ghālib, Asadullāh Khān, 249–50
Ghan-Ānand kabitt, 144
Gītāvalī, 145, 151, 152, 157
Gobineau, Arthur, 262
Goda Devi (Antal), 73, 77–79
Godararañja, 296
Gongmenpa, 316
grammar: Persian, 11, 188, 270, 272; Sanskrit, 25, 30–31, 295, 303; Tibetan, and in Tibet, 291, 302–3; vernacular, 24
Guidelines Clarifying Regulations, 297
Gujarati, 173–74, 177
Gyatso, Ngawang Lobsang. *See* Dalai Lama, Fifth
Gyatso, Pukpa Lhündrub, 298
Gyatso, Sangyé (the Desi): on forms of knowledge, 291–93, 297–99; Indian influence on, 302; major works by, 298–99; medical and historical works by, 299–302, 316, 322–24, 326; regency of, 291, 314; uniqueness of works, 297, 301

Haft Iqlīm, 243–44
hagiography, 143
Hamilton, Charles, 271
Hanumānbāhuk, 146–47, 152, 157–59, 161

Sheldon Pollock is the William B. Ransford Professor of Sanskrit and Indian Studies in the Department of Middle East and Asian Languages and Cultures at Columbia University. He is the author of *The Language of the Gods in the World of Men: Sanskrit, Culture, and Power in Premodern India* (2006) and the editor of a number of books, including *Literary Cultures in History: Reconstructions from South Asia* (2003) and, with Homi Bhabha, Carol Breckenridge, and Dipesh Chakrabarty, *Cosmopolitanism* (2002).

Library of Congress Cataloging-in-Publication Data
Forms of knowledge in early modern Asia : explorations in the intellectual history of India and Tibet, 1500-1800 / edited by Sheldon Pollock.
p. cm.
Includes bibliographical references and index.
ISBN 978-0-8223-4882-5 (cloth : alk. paper)
ISBN 978-0-8223-4904-4 (pbk. : alk. paper)
1. India—Civilization—16th century.
2. India—Civilization—17th century.
3. India—Civilization—18th century.
4. Tibet (China)—Civilization—17th century.
5. Tibet (China)—Civilization—18th century.
I. Pollock, Sheldon I.
DS423.F67 2011
954.02—dc22
2010035886